Optimal control theory is a technique being used increasingly by academic economists to study problems involving optimal decisions in a multiperiod framework. This textbook is designed to make the difficult subject of optimal control theory accessible to economists while at the same time maintaining rigor. Economic intuition is emphasized, and examples and problem sets covering a wide range of applications in economics are provided. Theorems are clearly stated, and their proofs carefully explained. The development of the text is gradual and fully integrated, beginning with simple formulations and progressing to advanced topics such as control parameters, jumps in state variables, and bounded state space. For greater economy, optimal control theory is introduced directly, without recourse to the calculus of variations. The connection with the latter and with dynamic programming is explained in a separate chapter.

A secondary purpose of the book is to draw a parallel between optimal control theory and static optimization. The first chapter provides an extensive treatment of constrained and unconstrained maximization, with emphasis on economic insight and applications. Starting from basic concepts, it derives and explains important results, including the envelope theorem and the method of comparative statics. This chapter may be used for a short course in static optimization.

The book is largely self-contained. No previous knowledge of differential equations is required.

Optimal control theory and static optimization in economics

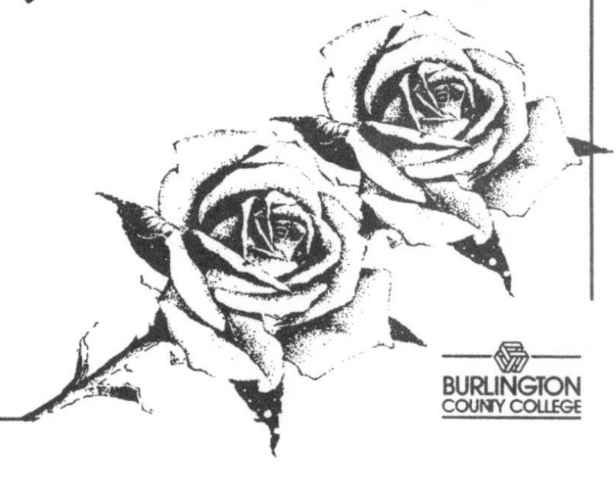

Optimal control theory and static optimization in economics

DANIEL LEONARD

NGO VAN LONG

CAMBRIDGE
UNIVERSITY PRESS

Published by the Press Syndicate of the University of Cambridge
The Pitt Building, Trumpington Street, Cambridge CB2 1RP
40 West 20th Street, New York, NY 10011-4211, USA
10 Stamford Road, Oakleigh, Melbourne 3166, Australia

First published 1992
Reprinted 1993

Printed in the United States of America

Library of Congress Cataloging-in-Publication Data

Léonard, Daniel.
Optimal control theory and static optimization in economics /
Daniel Léonard and Ngo Van Long.
p. cm.
Includes bibliographical references and index.
ISBN 0-521-33158-7 hardback. – ISBN 0-521-33746-1 paperback.
1. Mathematical optimization. 2. Control theory. 3. Statics and dynamics (Social sciences) I. Long, Ngo Van. II. Title.
HB143.7.L46 1992
330'.01'51–dc20 91–14126
CIP

A catalogue record for this book is available from the British Library.

ISBN 0-521-33158-7 hardback
ISBN 0-521-33746-1 paperback

Contents

Preface

As the range of problems tackled by economists expands, the curriculum of economics programs follows. Questions of choice in dynamic economic models are often an integral part of such programs. The most useful technique for dealing with these questions is optimal control theory. It was developed in the late 1950s as an outgrowth of the centuries-old calculus of variations, and it has been traditional to present an exposition of the latter as a preliminary to this more modern technique. Here we break with this tradition on the grounds that there is nothing to be learned from the calculus of variations that cannot be learned from optimal control theory, whereas the converse is not true. Our approach emphasizes the links between the methods of classical programming and those of optimal control theory. For this reason we begin with a thorough and lengthy exposition of static optimization techniques: unconstrained, equality-constrainted, and inequality-constrained problems (Chapter 1). After presenting some simple solution techniques for differential equations and their qualitative analysis through phase diagrams (Chapter 2), we proceed with a very short and informal chapter introducing various concepts related to optimization in dynamic models (Chapter 3). Chapter 4 describes the optimal control format for dynamic optimization problems and the core of its solution procedures, known as the maximum principle. We have attempted to make the reader's first encounter with a standard control problem as limpid as possible by relegating all complications to a later stage and emphasizing the links with the Lagrangean methods of static optimization. Chapter 5 diverges from the main line of argument to give a very brief account of the calculus of variations and the related method of dynamic programming. Chapter 6 deals with a much more general control problem, which involves several types of constraint. Chapter 7 extends the results by allowing for various boundary conditions at the beginning or the end of the planning horizon. Chapter 8 concentrates on a special class of models that might elicit discontinuities in the controls. Chapter 9 considers infinite-horizon problems, and Chapter 10 treats three separate topics.

The book is intended to be a very detailed exposition of static and dynamic optimization, beginning at an elementary level. (Some knowledge of calculus and matrix algebra is needed, but these are reviewed in the appendixes to Chapters 1 and 2.) The presentation gradually builds up to a degree of sophistication sufficient for readers to understand these topics as treated in most economic journal articles. Indeed, after a thorough study of the material presented here, including the exercises, readers should be able to use the techniques in their own research. Theorems and definitions are stated rigorously, but most proofs are chosen for their heuristic value.

The book is intended for university economists who feel a need to expand the array of techniques at their disposal without wishing to invest too much time in the study of more rigorous mathematical derivations. It can be used for self-instruction. Alternatively, the book can be used for a graduate course in economic optimization with more emphasis on the beginning or the end of the book, depending on students' backgrounds and the place of the course in the overall graduate program. It is essential that students attempt the exercises if they are to acquire a thorough grasp of the subject. The Bibliography includes a selected list of articles and monographs, including several volumes of collected papers that emphasize the use of optimal control theory in economics.

Many students and colleagues have contributed to this volume directly or indirectly; we are particularly grateful to Jeffrey Bernstein, Richard Cornes, Bruce Forster, Murray Kemp, T. H. Lou, the late Richard Manning, Frank Milne, John Pitchford, Hans-Werner Sinn, Mark Tippet, Sabine Toussaint, Stephen Turnovsky, and Neil Vousden. Our final thanks are reserved for Jan Anthony and Silvana Tomasiello, who produced many versions of the typescript, and for Mary Racine, who edited it.

CHAPTER 1

Static optimization

In this chapter we deal with problems involving the choice of values for a finite number of variables in order to maximize some objective. Sometimes the values the variables may take are unrestricted; at other times they are restricted by equality constraints and also by inequality constraints. In the course of the presentation an important class of functions will emerge; they are called *concave functions* and are closely associated with "nice" maximum problems. They will be encountered throughout this book. For this reason we weave the concept of concavity of functions through the exposition of maximization problems. This is done to suit our purposes, but concave functions have other important properties in their own right.

The notation we use is fairly standard. If in doubt, the reader should refer to the appendix to this chapter, which also contains a reminder of the basic notions of multivariate calculus and some matrix algebra needed to follow the exposition.

1.1 Unconstrained optimization, concave and convex functions

In what follows we assume all functions to have continuous second-order derivatives, unless otherwise stated. Strictly speaking, all domains of definitions should be open subsets of the multidimensional real space so that no boundary problems arise.

1.1.1 *Unconstrained maximization*

Consider the problem of finding a set of values $x_1, x_2, \ldots, x_n$ to maximize the function $f(x_1, \ldots, x_n)$. We often write this as

$$\underset{\mathbf{x}}{\text{Maximize}}\ f(\mathbf{x}), \tag{1.1}$$

where $\mathbf{x}$ is understood to be an n-dimensional vector. We refer to the problem of (1.1) as an unconstrained maximum because no restrictions are placed on $\mathbf{x}$.

Necessary conditions. Suppose we find a solution to this problem and denote the optimal vector by $\mathbf{x}^*$. Consider an arbitrarily small deviation from $\mathbf{x}^*$, say $d\mathbf{x}$. If we have a maximum at $\mathbf{x}^*$, then f must not increase for any $d\mathbf{x}$.

The change in f is approximated by

$$df = \sum_i f_{x_i}(\mathbf{x}^*)\, dx_i.$$

Clearly, $df \leq 0$ if we have a maximum at $\mathbf{x}^*$. Furthermore, suppose we found some $d\mathbf{x}$ vector such that $df < 0$; then by using the deviation $(-d\mathbf{x})$ we would obtain an increase in f. Therefore, it must be that for any $d\mathbf{x}$ vector, df is equal to zero. The only way this can be achieved for arbitrary deviations is to require each derivative $f_{x_i}(\mathbf{x}^*)$ to vanish. Formally,

$$f(\mathbf{x}) \text{ reaches a maximum at } \mathbf{x}^* \text{ implies } f_{x_i}(\mathbf{x}^*) = 0, \quad i = 1, \ldots, n. \tag{1.2}$$

This is called the *first-order condition*. Several remarks must now be made. First, the above reasoning, hence (1.2), also applies to minimization problems. Second, we have been lax in defining a maximum. We should have distinguished a global maximum from a local maximum. We say that $f(x)$ reaches a *global maximum* at $\mathbf{x}^*$ if $f(\mathbf{x}^*) \geq f(\mathbf{x})$ for all $\mathbf{x}$ on its domain of definition (assumed to be an open set). We say $f(\mathbf{x})$ reaches a *local maximum* at $\mathbf{x}^*$ if $f(\mathbf{x}^*) \geq f(\mathbf{x})$ for all $\mathbf{x}$ "close" to $\mathbf{x}^*$ (i.e., for all $\mathbf{x}$ within δ units of distance from $\mathbf{x}^*$, where δ is some positive number). The local maximum is a much weaker concept than the global one. However, because our argument relies on arbitrarily small deviations from $\mathbf{x}^*$, it applies to both cases. The first-order condition (1.2) follows from the existence of a maximum; hence, it is a necessary condition for a maximum, but it is not the only one, as we now show. As we noted previously, condition (1.2) is necessary for a local minimum as well. The following condition, called the *second-order necessary condition,* takes a different form for a maximum than for a minimum.

To establish it we must take a Taylor's expansion (with remainder) of the function f about the point $\mathbf{x}^*$:

$$\begin{aligned} f(\mathbf{x}^* + d\mathbf{x}) = f(\mathbf{x}^*) &+ \sum_{i=1}^{n} f_{x_i}(\mathbf{x}^*)(dx_i) \\ &+ \frac{1}{2} \sum_{i=1}^{n} \sum_{j=1}^{n} f_{x_i x_j}(\mathbf{x}^*)(dx_i)(dx_j) + \cdots + R, \end{aligned} \tag{1.3a}$$

or in vector notation (see the Appendix for details),

$$f(\mathbf{x}^* + d\mathbf{x}) = f(\mathbf{x}^*) + (d\mathbf{x})' \cdot f_{\mathbf{x}}(\mathbf{x}^*) + \tfrac{1}{2}(d\mathbf{x})' \cdot f_{\mathbf{x}\mathbf{x}'}(\mathbf{x}^*) \cdot (d\mathbf{x}) + \cdots + R, \tag{1.3b}$$

where $d\mathbf{x}$ is small enough (i.e., $\|d\mathbf{x}\| < \delta$) that higher-order terms vanish relative to second-order terms.

Suppose again that we have a (at least local) maximum, that is, $f(\mathbf{x}^*) \geq f(\mathbf{x}^*+d\mathbf{x})$, $\forall d\mathbf{x}$, $\|d\mathbf{x}\| < \delta$. Then $f_{\mathbf{x}}(\mathbf{x}^*)=0$, and neglecting terms higher than the second order we have

$$\begin{aligned} f(\mathbf{x}^*+d\mathbf{x})-f(\mathbf{x}^*) &= \tfrac{1}{2}(d\mathbf{x})'\cdot f_{\mathbf{x}\mathbf{x}'}(\mathbf{x}^*)\cdot d\mathbf{x} \\ &\leq 0, \quad \text{because } \mathbf{x}^* \text{ is a maximum.} \end{aligned}$$

Since $(d\mathbf{x})'\cdot f_{\mathbf{x}\mathbf{x}'}(\mathbf{x}^*)\cdot(d\mathbf{x})$ is negative or zero for all small deviation vectors $d\mathbf{x}$, the Hessian matrix of f evaluated at $\mathbf{x}^*$ must be negative-semidefinite. This is the *second-order necessary condition:*

$f(\mathbf{x})$ reaches a maximum at $\mathbf{x}^*$ implies $f_{\mathbf{x}\mathbf{x}'}(\mathbf{x}^*)$ is negative-semidefinite. (1.4)

Again, (1.4) applies to global as well as local maxima.

Sufficient conditions (for a local maximum). It is unfortunately not possible to state conditions that are both necessary and sufficient for a function to reach a maximum. We can, however, easily provide sufficient conditions:

If $f_{x_i}(\mathbf{x}^*)=0$, $i=1,\ldots,n$, and $f_{\mathbf{x}\mathbf{x}'}(\mathbf{x}^*)$ is negative-definite,
then $f(\mathbf{x})$ reaches a local maximum at $\mathbf{x}^*$. (1.5)

To prove this we shall consider again Taylor's expansion in (1.3) and let $d\mathbf{x}\to 0$, so that the second-degree term dominates those of higher order while the first-degree term vanishes; we obtain $f(\mathbf{x}^*+d\mathbf{x})<f(\mathbf{x}^*)$, thus establishing $\mathbf{x}^*$ as a local maximum.

1.1.2 *Global results and concave functions*

When we seek a maximum in an economic problem, it is most often a global one. Indeed, it is little comfort to know that we are doing the best we can but only if considering policies which differ minutely from the current one (local optimum). It is also clear that we will not be able to characterize a global maximum with conditions on the values of the function and its derivatives at the maximum itself; we will need to place restrictions on the overall shape of the function, restrictions that apply everywhere on the domain of definition, which we denote by X.

Consider the exact form of Taylor's expansion to the second degree: there exists a point $\mathbf{x}_t$ on the line segment between $\mathbf{x}$ and $\bar{\mathbf{x}}$ such that

$$f(\mathbf{x})=f(\bar{\mathbf{x}})+(\mathbf{x}-\bar{\mathbf{x}})'\cdot f_{\mathbf{x}}(\bar{\mathbf{x}})+\tfrac{1}{2}(\mathbf{x}-\bar{\mathbf{x}})'\cdot\mathbf{H}(\mathbf{x}_t)\cdot(\mathbf{x}-\bar{\mathbf{x}}), \tag{1.6}$$

where $\mathbf{H}(\mathbf{x}_t)$ denotes the Hessian matrix of f, evaluated at the point $\mathbf{x}_t$. If we were to restrict our attention to functions with a negative-semidefinite matrix everywhere on its domain of definition, then the last term of (1.6) would be guaranteed to be nonpositive for any $\mathbf{x}_t$ and the requirement that $\bar{\mathbf{x}}$ be a global maximum (i.e., $f(\mathbf{x})-f(\bar{\mathbf{x}})\leq 0\ \forall \mathbf{x}\in X$) would be equivalent to the first-order condition $f_{\mathbf{x}}(\bar{\mathbf{x}})=0$. We now formalize this argument.

Definition 1.1.1. A function with continuous second-order derivatives defined on a convex set X is concave if and only if its Hessian matrix is negative-semidefinite everywhere on its domain of definition X.

Theorem 1.1.1. Let $f(\mathbf{x})$ be a concave function; then it reaches a global maximum at $\bar{\mathbf{x}}$ if and only if $f_{\mathbf{x}}(\bar{\mathbf{x}})=0$.

Definition 1.1.1 applies only to functions with continuous second-order derivatives. It is useful to have a more general definition of concavity that does not require this assumption.

Definition 1.1.2. A function $f(\mathbf{x})$ with continuous first-order derivatives defined on a convex set X is concave if and only if

$$f(\mathbf{x}_2)-f(\mathbf{x}_1)\leq(\mathbf{x}_2-\mathbf{x}_1)'\cdot f_{\mathbf{x}}(\mathbf{x}_1),$$

for all $\mathbf{x}_1,\mathbf{x}_2$ on X.

Note that Definition 1.1.2 is less stringent than Definition 1.1.1 in terms of differentiability restrictions, since it requires continuity only for the first derivatives; this is the only difference between the two definitions. Indeed, if we assume that the function has continuous second-order derivatives, we can see that the two definitions are equivalent simply by writing down the exact form of Taylor's expansion. Given two arbitrary points $\mathbf{x}_1$ and $\mathbf{x}_2$, there exists a point $\mathbf{x}_t$ between them such that

$$f(\mathbf{x}_2)=f(\mathbf{x}_1)+(\mathbf{x}_2-\mathbf{x}_1)'\cdot f_{\mathbf{x}}(\mathbf{x}_1)+\tfrac{1}{2}(\mathbf{x}_2-\mathbf{x}_1)'\cdot\mathbf{H}(\mathbf{x}_t)\cdot(\mathbf{x}_2-\mathbf{x}_1),$$

$$f(\mathbf{x}_2)-f(\mathbf{x}_1)-(\mathbf{x}_2-\mathbf{x}_1)'\cdot f_{\mathbf{x}}(\mathbf{x}_1)=\tfrac{1}{2}(\mathbf{x}_2-\mathbf{x}_1)'\cdot\mathbf{H}(\mathbf{x}_t)\cdot(\mathbf{x}_2-\mathbf{x}_1)\leq 0.$$

The geometric interpretation is simply that a tangent plane to the graph of $f(\mathbf{x})$ must remain everywhere above the graph, the equation for the tangent plane at $\mathbf{x}_1$ being

$$y=f(\mathbf{x}_1)+(\mathbf{x}-\mathbf{x}_1)'\cdot f_{\mathbf{x}}(\mathbf{x}_1).$$

This is illustrated in Figure 1.1a for functions of one variable. Definition 1.1.2 does not cover functions that have "kinks" and as such are not differentiable everywhere. To admit this case, a more general definition is needed.

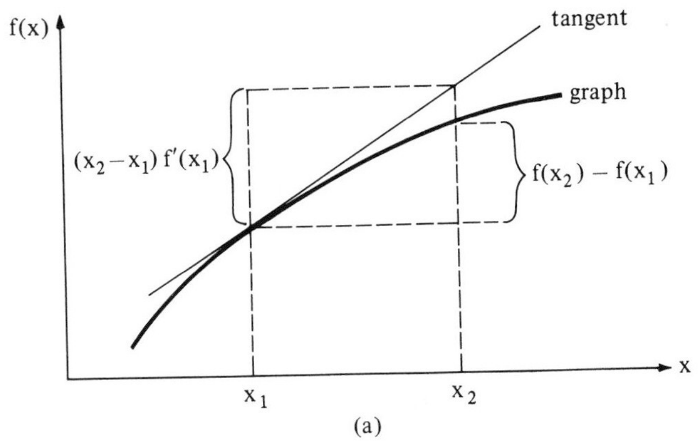

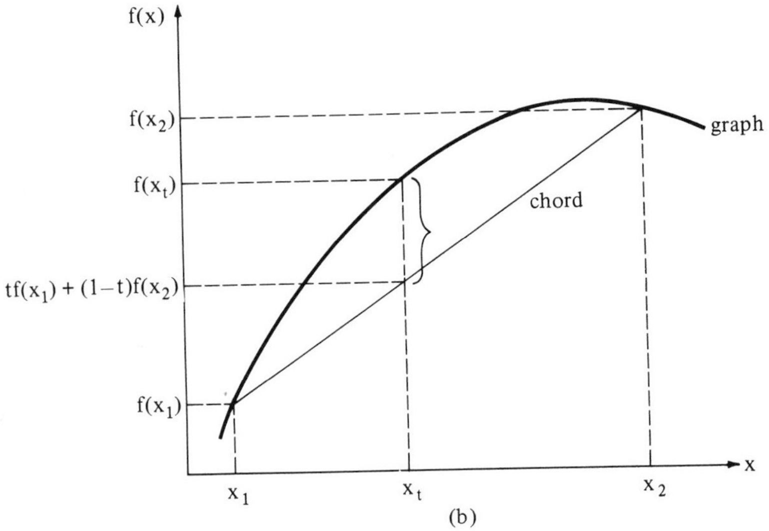

Figure 1.1

Definition 1.1.3. A function $f(\mathbf{x})$ defined on a convex set X is concave if and only if

$$f(\mathbf{x}_t) \geq tf(\mathbf{x}_1) + (1-t)f(\mathbf{x}_2),\ 0 \leq t \leq 1, \text{ all } \mathbf{x}_1, \mathbf{x}_2 \text{ in } X,$$
$$\text{where } \mathbf{x}_t = t\mathbf{x}_1 + (1-t)\mathbf{x}_2.$$

If a function satisfies Definition 1.1.2, it also satisfies Definition 1.1.3. To see this we state Definition 1.1.2 in two instances:

$$f(\mathbf{x}_2)-f(\mathbf{x}_t)\le(\mathbf{x}_2-\mathbf{x}_t)'\cdot f_{\mathbf{x}}(\mathbf{x}_t)$$

and

$$f(\mathbf{x}_1)-f(\mathbf{x}_t)\le(\mathbf{x}_1-\mathbf{x}_t)'\cdot f_{\mathbf{x}}(\mathbf{x}_t),$$

where

$$\mathbf{x}_t=t\mathbf{x}_1+(1-t)\mathbf{x}_2 \text{ for some } t, \quad 0\le t\le 1.$$

Since

$$\mathbf{x}_2-\mathbf{x}_t=t(\mathbf{x}_2-\mathbf{x}_1) \quad \text{and} \quad \mathbf{x}_1-\mathbf{x}_t=-(1-t)(\mathbf{x}_2-\mathbf{x}_1),$$

we have

$$f(\mathbf{x}_2)-f(\mathbf{x}_t)\le t(\mathbf{x}_2-\mathbf{x}_1)'\cdot f_{\mathbf{x}}(\mathbf{x}_t),$$
$$f(\mathbf{x}_1)-f(\mathbf{x}_t)\le-(1-t)(\mathbf{x}_2-\mathbf{x}_1)'\cdot f_{\mathbf{x}}(\mathbf{x}_t).$$

Multiplying the first inequality by $(1-t)$, the second by t, and adding yields (with $0\le t\le 1$)

$$tf(\mathbf{x}_1)+(1-t)f(\mathbf{x}_2)-f(\mathbf{x}_t)\le 0,$$

which was to be proved.

Note that no differentiability properties are required in Definition 1.1.3. The geometric interpretation of this definition is that a line (or chord) joining two points of the graph always lies below the graph, since the left-hand side of the inequality represents the value of f at a convex combination of $\mathbf{x}_1$ and $\mathbf{x}_2$ and the right-hand side is the same convex combination of the values of the function at $\mathbf{x}_1$ and $\mathbf{x}_2$ – hence the height of the point on the chord above $\mathbf{x}_t$. This is illustrated in Figure 1.1b for functions of one variable.

Concave functions have many notable properties; Theorem 1.1.2 lists some of the most useful ones.

Theorem 1.1.2

(i) Let $f(\mathbf{x})$ be a concave function and $k\ge 0$ a constant; then $kf(\mathbf{x})$ is a concave function.
(ii) Let $f(\mathbf{x})$ and $g(\mathbf{x})$ be concave functions; then $f(\mathbf{x})+g(\mathbf{x})$ is itself a concave function.
(iii) Let $f(\mathbf{x})$ be a concave function; then the upper contour set defined by $B(\bar{\mathbf{x}})\equiv\{\mathbf{x}\in R_n \mid f(\mathbf{x})\ge f(\bar{\mathbf{x}})\}$ is a convex set.

(The converse of (iii) is *not true*!)

The proofs of these results are straightforward; for instance, (iii) requires that we show that if $f(\mathbf{x}_1)\ge f(\bar{\mathbf{x}})$ and $f(\mathbf{x}_2)\ge f(\bar{\mathbf{x}})$, it follows that $f(\mathbf{x}_t)\ge f(\bar{\mathbf{x}})$; this is obvious from Definition 1.1.3.

Strictly concave functions: unique global maximum. While concave functions have the property that a solution of the first-order condition yields a global maximum, this does not ensure the uniqueness of that solution: a concave function may reach its global maximum at several points. For example, the following function is concave, but the first-order condition admits as a solution any point between 1 and 2; thus, the function reaches a global maximum at any x^* such that $1 \leq x^* \leq 2$.

$$f(x) = \begin{cases} x - 0.5x^2, & x < 1, \\ 0.5, & 1 \leq x \leq 2, \\ (x-1) - 0.5(x-1)^2, & 2 < x. \end{cases}$$

Other examples will be encountered in Section 1.1.5.

It is sometimes desirable to place more restrictions on the function so that if a maximum exists, it is the unique global maximum. We use this as a means of introducing a subclass of concave functions called *strictly concave functions.* Definitions 1.1.2 and 1.1.3 are adapted by simply requiring strict inequalities.

Definition 1.1.3′. A function $f(\mathbf{x})$ defined on a convex set X is strictly concave if and only if

$$f(\mathbf{x}_t) > t f(\mathbf{x}_1) + (1-t) f(\mathbf{x}_2), \quad 0 < t < 1,$$

for all $\mathbf{x}_1, \mathbf{x}_2$ in X, where $\mathbf{x}_1 \neq \mathbf{x}_2$ and $\mathbf{x}_t = t\mathbf{x}_1 + (1-t)\mathbf{x}_2$.

Definition 1.1.2′. A function $f(\mathbf{x})$ with continuous first-order derivatives defined on a convex set X is strictly concave if and only if

$$f(\mathbf{x}_2) - f(\mathbf{x}_1) < (\mathbf{x}_2 - \mathbf{x}_1)' \cdot f_{\mathbf{x}}(\mathbf{x}_1)$$

for all $\mathbf{x}_1$ and $\mathbf{x}_2$ in X, where $\mathbf{x}_1 \neq \mathbf{x}_2$.

It is obvious from Definition 1.1.2′ that $f_{\mathbf{x}}(\mathbf{x}_1) = 0$ is necessary and sufficient for $\mathbf{x}_1$ to be the unique global maximum of that function f.

We cannot claim that functions with continuous second-order derivatives are strictly concave if and only if their Hessian matrix is negative-definite, because some strictly concave functions have a Hessian matrix which becomes negative-semidefinite at some points. One instance is $f(x_1, x_2) = -(x_1)^4 - (x_2)^2$, which is negative-definite everywhere but at $x_1 = 0$, when it is negative-semidefinite. We must be content with the following theorem.

Theorem 1.1.3. A function that is defined on a convex set X and has a negative-definite Hessian matrix everywhere on X is strictly concave.

The reader is invited to prove this result using Definition 1.1.2′.

1.1.3 *Unconstrained minimization and convex functions*

Results for minimization problems are just mirror images of those for maximization problems and are obtained by replacing $f(\mathbf{x})$ by $-f(\mathbf{x})$. Thus, the *first-order necessary condition for a local minimum at* $\mathbf{x}^*$ is

$$f_i(\mathbf{x}^*) = 0, \quad i = 1, \ldots, n, \tag{1.7}$$

and the *second-order necessary condition* is

$$f_{\mathbf{x}\mathbf{x}'}(\mathbf{x}^*) \text{ is positive-semidefinite.} \tag{1.8}$$

The *sufficient conditions for a local minimum at* $\mathbf{x}^*$ are

$$f_{\mathbf{x}}(\mathbf{x}^*) = 0 \quad \text{and} \quad f_{\mathbf{x}\mathbf{x}'}(\mathbf{x}^*) \text{ is positive-definite.} \tag{1.9}$$

Similarly, we have to define convex functions in order to obtain global results on minimization. Corresponding to Definitions 1.1.1, 1.1.2, and 1.1.3 we now have the following (results on strictly convex functions are indicated in parentheses).

Definition 1.1.4. A function with continuous second-order derivatives defined on a convex set is (strictly) convex if and only if its Hessian matrix is positive-semidefinite (if its Hessian matrix is positive-definite).

Definition 1.1.5. A function $f(\mathbf{x})$ with continuous first-order derivatives defined on a convex set X is (strictly) convex if and only if

$$f(\mathbf{x}_2) - f(\mathbf{x}_1) \geq (\mathbf{x}_2 - \mathbf{x}_1)' \cdot f_{\mathbf{x}}(\mathbf{x}_1) \quad \text{for all } \mathbf{x}_1, \mathbf{x}_2 \text{ in } X,$$

$$(f(\mathbf{x}_2) - f(\mathbf{x}_1) > (\mathbf{x}_2 - \mathbf{x}_1)' \cdot f_{\mathbf{x}}(\mathbf{x}_1) \quad \text{for all } \mathbf{x}_1, \mathbf{x}_2 \text{ in } X, \text{ where } \mathbf{x}_1 \neq \mathbf{x}_2).$$

Definition 1.1.6. A function $f(\mathbf{x})$ defined on a convex set X is (strictly) convex if and only if

$$f(\mathbf{x}_t) \leq t f(\mathbf{x}_1) + (1-t) f(\mathbf{x}_2), \quad 0 \leq t \leq 1, \text{ all } \mathbf{x}_1, \mathbf{x}_2 \text{ in } X,$$

$$(f(\mathbf{x}_t) < t f(\mathbf{x}_1) + (1-t) f(\mathbf{x}_2), \quad 0 < t < 1, \text{ all } \mathbf{x}_1, \mathbf{x}_2 \text{ in } X, \ \mathbf{x}_1 \neq \mathbf{x}_2).$$

Theorem 1.1.4

(i) Let $f(\mathbf{x})$ be a convex function and $k \geq 0$ a constant; then $kf(\mathbf{x})$ is a convex function.
(ii) Let $f(\mathbf{x})$ and $g(\mathbf{x})$ be convex functions; then $f(\mathbf{x}) + g(\mathbf{x})$ is itself a convex function.
(iii) Let $f(\mathbf{x})$ be a convex function; then the lower contour set defined by $W(\bar{\mathbf{x}}) \equiv \{\mathbf{x} \in R_n \mid f(\mathbf{x}) \leq f(\bar{\mathbf{x}})\}$ is a convex set. (The converse of (iii) is *not* true!)
(iv) Let $f(\mathbf{x})$ be a (strictly) convex function; then $-f(\mathbf{x})$ is a (strictly) concave function.

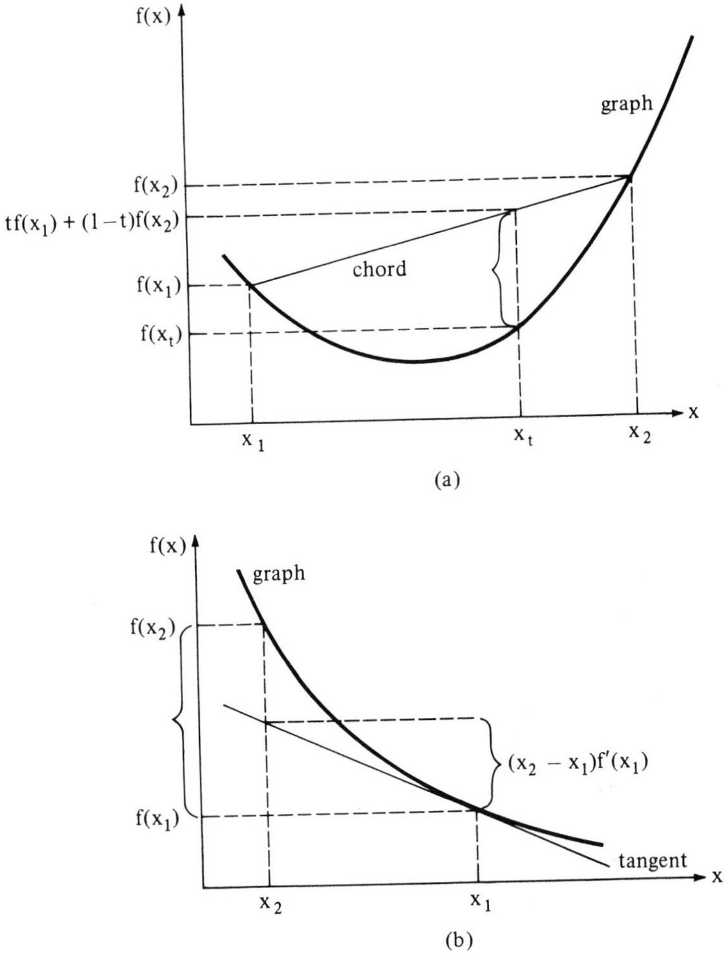

Figure 1.2

(v) A linear function is both convex and concave but not strictly either.

Definitions 1.1.5 and 1.1.6 are illustrated in Figure 1.2 for convex functions of one variable.

1.1.4 *Geometric representation*

Figures 1.3a and 1.3b represent the graphs of a concave and a convex function, respectively. It is important to realize that a concave function

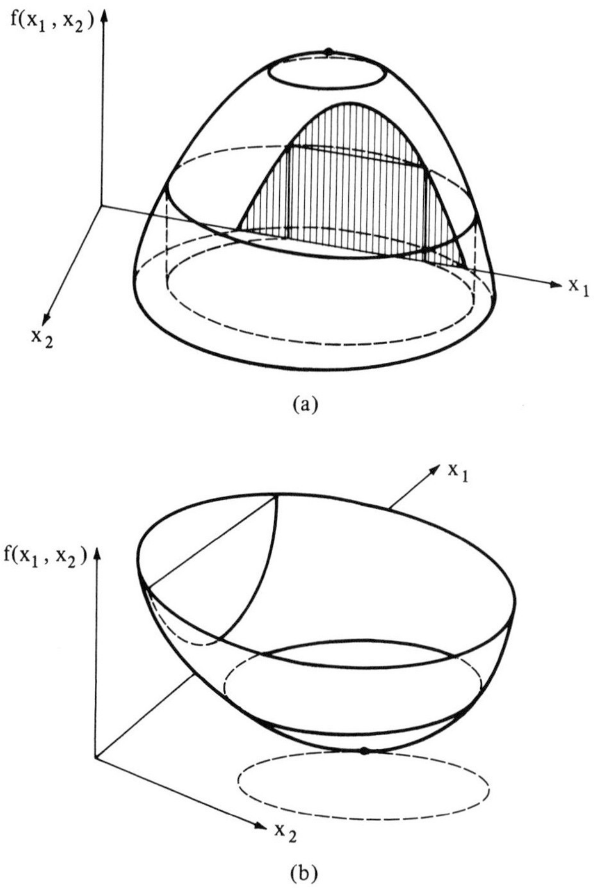

Figure 1.3

need not have a maximum, nor a convex function a minimum. If they do, then one is somewhat dome-shaped and the other bowl-shaped. It is then obvious that a rod connecting two points of the dome remains under it (Definition 1.1.3), while such a rod connecting two points of the bowl remains above its walls (Definition 1.1.6). It is clearly inconvenient to rely on three-dimensional diagrams; instead, we most often use level curves. We know that if a function is concave, its upper contour sets are convex sets. We use this information in Figure 1.4a to draw some level curves of a concave function, where the arrows indicate directions of increase of the function and one convex upper contour set is hatched. We can also verify that Definition 1.1.3 is satisfied: the function takes on the value c at points A and B; thus, it takes on a higher value at a point between

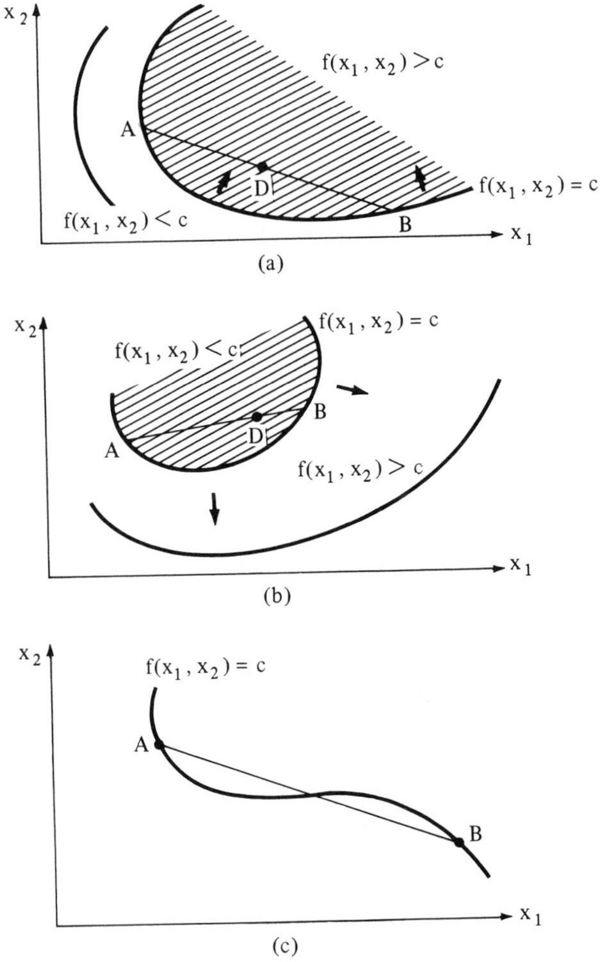

Figure 1.4

them, *D*, which is naturally within the convex upper contour set. A similar picture emerges for a convex function in Figure 1.4b, where a *lower* contour set is hatched and the arrows indicate directions of increase of the function. Finally note that a contour curve such as the one in Figure 1.4c cannot correspond to a concave (or a convex) function since it delineates no convex set on either side of it.

A word of warning is in order. Because concave functions have convex upper contour sets but some other functions do too, we cannot rely on this contour curve representation to *characterize concave functions* exactly. For many purposes, however, it will be adequate.

1.1.5 *Numerical examples and some useful functional forms*

It is useful to develop some "feel" for the concavity properties of functions so as to avoid always running back to the definitions. The knowledge of a few simple functions along with the composition rules already outlined and some more to follow is very helpful. We first list a few functions and the conditions for their concavity and/or convexity. The reader is invited to check these as exercises, using mainly Definitions 1.1.1 and 1.1.4.

$$f(\mathbf{x}) = \prod_{i=1}^{n} (x_i)^{\alpha_i} \text{ is concave for } x \geq 0$$
$$\text{if and only if } \alpha_i \geq 0,\ \forall i, \text{ and } \sum_{i=1}^{n} \alpha_i \leq 1. \tag{1.10}$$

$$f(\mathbf{x}) = (a_0 + a_1 x_1 + \cdots + a_n x_n)^{\alpha}, \text{ defined when } a_0 + a_1 x_1 + \cdots + a_n x_n > 0 \text{ is concave if and only if } 0 \leq \alpha \leq 1; \text{ it is convex if and only if } \alpha \geq 1 \text{ or } \alpha \leq 0. \tag{1.11}$$

$$f(\mathbf{x}) = \mathbf{x}' \cdot \mathbf{A} \cdot \mathbf{x} \text{ is concave if and only if } \mathbf{A} \text{ is negative-semidefinite; it is convex if and only if } \mathbf{A} \text{ is positive-semidefinite.} \tag{1.12}$$

$$f(\mathbf{x}) = \sum_{i=1}^{n} \alpha_i \ln(x_i + a_i) \text{ is concave whenever it is defined (i.e., } x_i + a_i > 0, \text{ all } i) \text{ if and only if } \alpha_i \geq 0,\ \forall i; \text{ it is similarly convex if and only if } \alpha_i \leq 0,\ \forall i. \tag{1.13}$$

Theorem 1.1.5. An increasing concave function of concave functions is concave.

Proof. Let

$$W(\mathbf{x}^1, \ldots, \mathbf{x}^n) \equiv V(U^1(\mathbf{x}^1), \ldots, U^n(\mathbf{x}^n)),$$

where $\mathbf{x}^i$ denotes a vector of arbitrary dimension, V is increasing and concave in all U^i jointly, and U^i is concave in $\mathbf{x}^i$, $\forall i$. We use the standard notation for convex combinations: $\mathbf{z}_t \equiv t\mathbf{z}_1 + (1-t)\mathbf{z}_2$, $0 \leq t \leq 1$,

$$\begin{aligned} W(\mathbf{x}_t^1, \ldots, \mathbf{x}_t^n) &= V(U^1(\mathbf{x}_t^1), \ldots, U^n(\mathbf{x}_t^n)) \\ &\geq V(tU^1(\mathbf{x}_1^1) + (1-t)U^1(\mathbf{x}_2^1), \ldots, tU^n(\mathbf{x}_1^n) + (1-t)U^n(\mathbf{x}_2^n)), \end{aligned}$$

because all U^i are concave and V is increasing,

$$\geq tV(U^1(\mathbf{x}_1^1), \ldots, U^n(\mathbf{x}_1^n)) + (1-t)V(U^1(\mathbf{x}_2^1), \ldots, U^n(\mathbf{x}_2^n)),$$

by the concavity of V,

$$= tW(\mathbf{x}_1^1, \ldots, \mathbf{x}_1^n) + (1-t)W(\mathbf{x}_2^1, \ldots, \mathbf{x}_2^n). \quad \square$$

Theorem 1.1.6. Let $f(\mathbf{x})$ be a function of n variables and let $\mathbf{z} = -\mathbf{x}$ and $h(\mathbf{z}) \equiv f(\mathbf{x})$; then if $f(\mathbf{x})$ is concave (convex), so is $h(\mathbf{z})$.

The proof is obvious using, for instance, Definition 1.1.3. As an example, $f(x) = 1 - e^{-x}$ is concave in x; hence, $h(z) = 1 - e^{z}$ is concave in z, where $z = -x$.

We now consider a few numerical examples that may or may not possess a global maximum.

Example 1.1.1. Let $f(\mathbf{x}) = \mathbf{x}' \cdot \mathbf{A} \cdot \mathbf{x} + \mathbf{a}' \cdot \mathbf{x}$, where

$$\mathbf{A} = \begin{bmatrix} -1 & 0.5 \\ 0.5 & -1 \end{bmatrix} \quad \text{and} \quad \mathbf{a} = \begin{bmatrix} -1 \\ 5 \end{bmatrix}.$$

It is concave since $\mathbf{A}$ is negative-definite and the linear term does not affect concavity. To find a maximum, set the first-order derivatives to zero and solve: $f_1 = -2x_1 + x_2 - 1 = 0$ and $f_2 = x_1 - 2x_2 + 5 = 0$ yield $x_1 = 1$, $x_2 = 3$, the point at which f reaches its global maximum.

As we mentioned earlier, a function may reach its global maximum at many points; that is, the solution may not be unique. This is illustrated in the following example of a concave but not strictly concave function.

Example 1.1.2. Let $f(\mathbf{x}) = (x_1)^{0.3}(x_2)^{0.7} - 0.3x_1 - 0.7x_2$. We know that this function is defined and concave for all $\mathbf{x}$ positive (e.g., use (1.10)):

$$f_1 = 0.3(x_1)^{-0.7}(x_2)^{0.7} - 0.3 = 0,$$
$$f_2 = 0.7(x_1)^{0.3}(x_2)^{-0.3} - 0.7 = 0.$$

These first-order conditions have many solutions; namely, any $\mathbf{x}$ satisfying $x_1 = x_2$ is a solution. The global maximum value of f is zero and the upper part of its graph is shaped like the inside of a tunnel.

Example 1.1.3: saddle point. In this example we emphasize the idea that a function may be concave in all its variables but not necessarily concave in those variables jointly. We also introduce the concept of a saddle point. The example involves the function $f(x_1, x_2) = -(x_1)^2 + ax_1x_2 - (x_2)^2$ for various values of a.

Case (a). Let $f(x_1, x_2) = -(x_1)^2 - (x_2)^2$; then

$$\mathbf{H} = \begin{bmatrix} -2 & 0 \\ 0 & -2 \end{bmatrix},$$

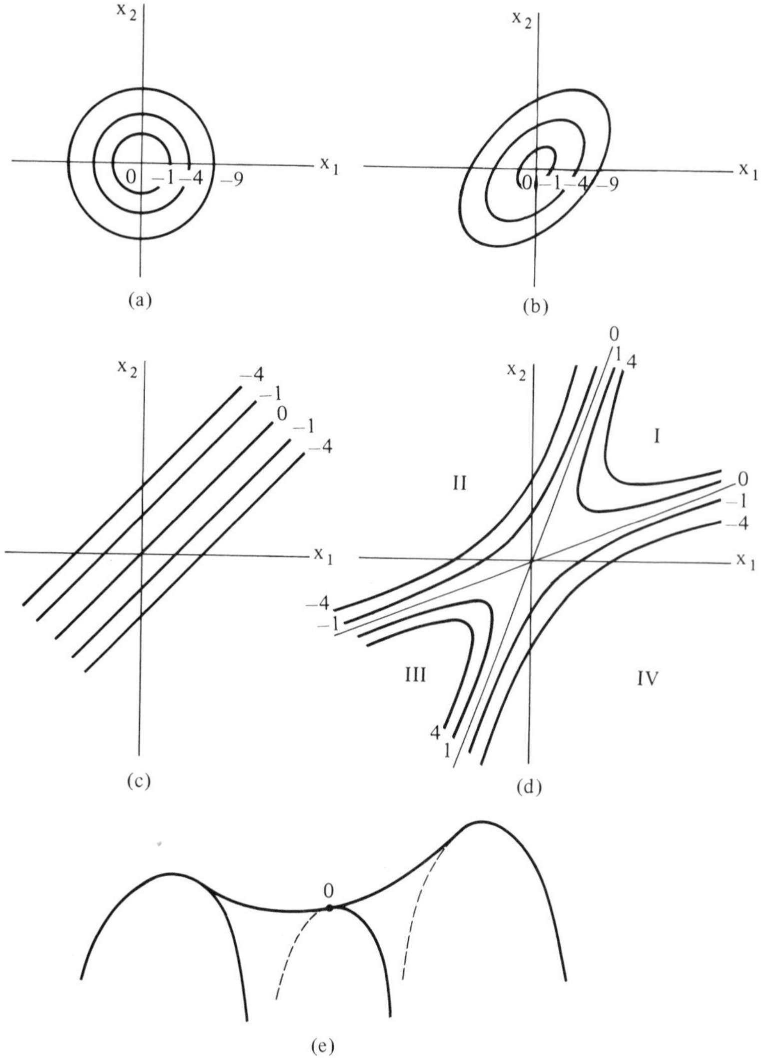

Figure 1.5

the function is concave, and $(0,0)$ is the global maximum. This is illustrated in Figure 1.5a. We proceed to "stretch" this function by introducing ever-increasing mixed terms.

Case (b). Let $f(x_1,x_2) = -(x_1)^2 + x_1x_2 - (x_2)^2$; then

$$\mathbf{H} = \begin{bmatrix} -2 & 1 \\ 1 & -2 \end{bmatrix},$$

the function is still concave, and $(0,0)$ is still a global maximum. The stretching is shown in Figure 1.5b.

Case (c). Let $f(x_1,x_2) = -(x_1)^2 + 2x_1x_2 - (x_2)^2$; then

$$\mathbf{H} = \begin{bmatrix} -2 & 2 \\ 2 & -2 \end{bmatrix},$$

the function is still concave, but the first-order conditions only imply $x_1 = x_2$; thus, there are many points at which the function reaches a global maximum. We again have a tunnel shape: the stretching has been carried out to an extent that we have a tubular shape with a horizontal top line. Note also that $|\mathbf{H}| = 0$. Further stretching will destroy concavity, as we now see.

Case (d). Let $f(x_1,x_2) = -(x_1)^2 + 3x_1x_2 - (x_2)^2$; then

$$\mathbf{H} = \begin{bmatrix} -2 & 3 \\ 3 & -2 \end{bmatrix};$$

the function is no longer concave in (x_1,x_2) because $|\mathbf{H}| = -5$, although it is still concave in x_1 and x_2 individually. The solution of the first-order conditions still is $(0,0)$, but we can no longer claim that it is a maximum. It is not a minimum either, but what we call a *saddle point*. The level curves are drawn in Figure 1.5d; the two straight lines corresponding to $f = 0$ delineate four regions, and when we move from region I to III the origin appears to be a minimum, but when we cross the origin while moving from region II to IV it appears as a maximum. This is the essential property of a saddle point configuration: it appears as a maximum in some directions and as a minimum in others. These directions need not be the axes as Figure 1.5d shows. Thus if we cross the origin following any one axis, it appears as a maximum with respect to that variable, which is as it should be since f is concave in x_1 and concave in x_2, separately. We have drawn a three-dimensional representation of the graph in Figure 1.5e. The additional mixed term has lifted the ends of the tunnel; it does look something like a saddle. A mountain pass is another, less common description.

1.1.6 *Some economic applications*

We are now able to tackle any economic problem in which the objective is to maximize some objective and where the entities to be chosen are many while their choice is unrestricted. One such problem is *profit maximization* by a competitive firm, to which we now turn.

Let $f(x_1,\dots,x_n)$ be the output obtainable from input levels $x_1,\dots,x_n$. If output price is p, the price of input i is w_i, and some fixed cost is k, the maximization of profit reduces to choosing $(x_1,\dots,x_n)$ to maximize

$$pf(x_1, \ldots, x_n) - \sum_{i=1}^{n} w_i x_i - k.$$

If we assume that f is concave, the global maximum will be the solution of the n equations

$$pf_i - w_i = 0, \quad i = 1, \ldots, n.$$

The first term is the rate of increase in output per unit of input i at the margin (called the marginal physical product of input i) multiplied by the output price; this is called the marginal value product of input i (MVP_i for short). The above condition equates it to the price of input i; thus, the price of the input is equal to the contribution to revenue made by the marginal unit. This seems sensible, yet fails to relate maximization of profit to the concavity of the profit expression. We now seek to clarify this relationship. In general, economic sense dictates that if $\text{MVP}_i > w_i$, we would gain by increasing the input level; conversely, $\text{MVP}_i < w_i$ would lead us to decrease input. Suppose now that f is concave and indeed that $f_{ii} < 0$ for all i; then the derivative of f with respect to x_i decreases when x_i increases; hence, if x_i were to rise above the level $\mathbf{x}^*$ indicated by the first-order condition, w_i would exceed MVP_i and we would bring x_i back down. Similar reasoning shows that if x_i strays below that level, we should bring it back up. If, in contrast, f_{ii} were positive at $\mathbf{x}^*$, that point could not be a maximum, for an increase in x_i from $\mathbf{x}^*$ would increase MVP_i above w_i and induce further increases in x_i. Indeed, with a strictly convex production function we could reach an arbitrarily large profit; in other words, the problem would have no solution. This possibility should always be kept in mind for any problem in analytical economics, since we work with unspecified functional forms and a precise solution is never derived. As a way of illustrating this point we consider profit maximization when the production function is homogeneous.

Homogeneous production functions and returns to scale. Suppose $q = f(\mathbf{x})$ is a production function that is homogeneous of degree h (see the Appendix for definitions). From a starting point of $\mathbf{x}$ units of input, suppose that we scale the operations up by a factor of $t > 1$, that is, employ $t\mathbf{x}$ units of input; we will obtain an output $f(t\mathbf{x}) = (t)^h f(\mathbf{x})$, and hence we will have scaled up output by a factor $(t)^h$. Depending on the value of h, this factor $(t)^h$ will be larger or smaller than t and output will increase more or less than the input vector. More precisely,

$h < 1 \rightarrow (t)^h < t$: $f(\mathbf{x})$ exhibits decreasing returns to scale.

$h = 1 \rightarrow (t)^h = t$: $f(\mathbf{x})$ exhibits constant returns to scale.

$h > 1 \rightarrow (t)^h > t$: $f(\mathbf{x})$ exhibits increasing returns to scale.

Theorem 1.1.7. Let $f(\mathbf{x})$ be homogeneous of degree h, positively valued, and concave; then $0 \leq h \leq 1$.

Proof. Since f_j is homogeneous of degree $(h-1)$, Euler's theorem yields

$$\sum_i x_i f_{ij} = (h-1) f_j.$$

Multiplying by x_j, summing, and applying Euler's theorem again yields

$$\sum_j \sum_i x_i x_j f_{ij} = (h-1) \sum_j x_j f_j = (h-1)hf.$$

The left-hand side is the quadratic form $\mathbf{x}'\mathbf{H}\mathbf{x}$, where $\mathbf{H}$ is the Hessian matrix of f. Concavity of f ensures that it is nonpositive; hence, $f > 0$ implies $h(h-1) \leq 0$. □

Note that our argument does not establish that if f is homogeneous of degree h and positively valued, then it is concave if and only if $h \leq 1$. This is because on the left-hand side of the preceding equation the values x_i and x_j are from the same vector at which f_{ij} is evaluated, a weaker requirement than Definition 1.1.1 of concavity. As a counterexample consider the function $f(x_1, x_2) = (x_1^2 + x_2^2)^{1/4}$, $x_1 > 0$, $x_2 > 0$. It is positively valued and homogeneous of degree $\frac{1}{2}$ but not concave since its upper contour sets are clearly not convex sets.

We are now ready to examine the implications of alternative assumptions regarding the degree of homogeneity of the production function on the profit of the firm. In what follows we assume that all $\mathbf{x}$ are positively valued and $f(\mathbf{x}) > 0$, unless otherwise indicated.

If the profit expression $\pi = pf(\mathbf{x}) - \mathbf{w}' \cdot \mathbf{x}$ has an unconstrained maximum, it will satisfy the necessary conditions

$$pf_i(\mathbf{x}) = w_i, \quad i = 1, \ldots, n. \tag{1.14}$$

If $\mathbf{x}^*$ solves equation (1.14), multiplying by x_i^*, summing, and applying Euler's theorem yields

$$p \sum_i x_i^* f_i(\mathbf{x}^*) = \sum_i w_i x_i^*,$$

$$hpf(\mathbf{x}^*) = \mathbf{w}' \cdot \mathbf{x}^*.$$

Substituting in the profit expression, we get

$$\pi = (1-h) pf(\mathbf{x}^*).$$

Therefore, at $\mathbf{x}^*$, profit will be positive, zero, or negative, depending on whether there exist decreasing, constant, or increasing returns to scale, respectively. In the case of increasing returns, first note that the objective function cannot be concave (if it were, h could not exceed 1, by Theorem

1.1.7) and Theorem 1.1.1 fails us. Furthermore, as in the proof of Theorem 1.1.7, we can show that

$$\mathbf{x}^{*\prime}\cdot\mathbf{H}(\mathbf{x}^*)\cdot\mathbf{x}^* = (h-1)hf(\mathbf{x}^*) > 0, \quad \text{since } h>1.$$

This demonstrates that $\mathbf{H}(\mathbf{x}^*)$ is not negative-semidefinite and violates the second-order necessary condition for a maximum. Let us remark that increasing returns to scale are often associated with unbounded profit and as such are not consistent with the hypothesis of a price-taking firm. The case of decreasing returns poses no special problems, since we can assume that f is concave, but the case of constant returns to scale is more difficult to handle, although the profit expression is concave under the additional assumption of concavity for f. The problem is with the first-order conditions (1.14):

$$pf_i(\mathbf{x}) = w_i, \quad i=1,\dots,n.$$

Recall that under constant returns, f_i is homogeneous of degree 0; therefore, if a vector $\mathbf{x}$ satisfies these conditions, so will any vector $t\cdot\mathbf{x}$, $t\geq 0$. The profit made with any of these vectors remains zero. The scale of operations is thus indeterminate and profit nil. This defect becomes a virtue when in some general equilibrium models such as those of international trade the focus is on the performance of each industry and the number and size of firms in each industry are not a matter of concern. There is, however, a further difficulty with the constant returns to scale assumption for an individual competitive firm. The problem is that for an arbitrary set of prices, equation (1.14) usually does not admit a solution, as we now demonstrate. Let $f(\mathbf{x})$ and $\mathbf{w}$ be fixed throughout, and suppose that at some price p^* (1.14) admits a vector $\mathbf{x}^*$ as a solution; then $\pi(\mathbf{x}^*)=0$ and $t\mathbf{x}^*$ is also a solution, $t\geq 0$; this is a global maximum, since we assumed π to be concave. Now consider another output price, say p; the profit expression can be written as

$$\begin{aligned}\pi &= pf(\mathbf{x}) - \mathbf{w}'\cdot\mathbf{x}\\ &= (p-p^*)f(\mathbf{x}) + [p^*f(\mathbf{x}) - \mathbf{w}'\cdot\mathbf{x}].\end{aligned} \tag{1.15}$$

We know that the second term in (1.15) has a global maximum of zero at $\mathbf{x}^*$ (and at $t\mathbf{x}^*$), but if $p>p^*$ we can make the first term infinitely large by increasing t; hence, there is no maximum, and the first-order necessary conditions (1.14) do not hold anywhere (if they did, a global maximum would exist by concavity of π). Note that an arbitrary $\mathbf{x}$ value may well make profit negative even in this case. Conversely, suppose that $p<p^*$; then the second term has a global maximum of zero at $t\mathbf{x}^*$, but the first term can be only negative or zero. Hence, the maximum is found at the

lower bound $\mathbf{x}=\mathbf{0}$ (i.e., $t=0$), but this is not an unconstrained maximum and again the necessary first-order conditions (1.14) fail to have a solution. Note that when $p<p^*$, any vector $\mathbf{x}>\mathbf{0}$ yields a negative profit.

In order to get a more intuitive grasp of these results, consider a firm with two inputs. The equations (1.14) are $pf_1(x_1,x_2)=w_1$ and $pf_2(x_1,x_2)=w_2$. However, if f is homogeneous of degree 1, then f_1 and f_2 are homogeneous of degree 0. Consequently, these derivatives are simply functions of a single argument x_2/x_1 (the only one that matters, since scale is irrelevant), and both determine a value for it; unless the prices are in a particular configuration, these values will differ and no solution exists. The exact relationship is that output price p be equal to the unit cost function $c(1,w)$; see the definition of cost functions in Section 1.2.3. Let us now briefly illustrate these results with a numerical example.

Example 1.1.4

$$f(x_1,x_2)=2(x_1)^{1/2}(x_2)^{1/2}, \quad w_1=1, \ w_2=2.$$

Equations (1.14) are

$$p(x_1)^{-1/2}(x_2)^{1/2}=1 \quad \text{and} \quad p(x_1)^{1/2}(x_2)^{-1/2}=2,$$

or

$$x_1/x_2=p^2 \quad \text{and} \quad x_1/x_2=4/p^2.$$

Therefore, (1.14) is satisfied if and only if $p=p^*=\sqrt{2}$; then the optimal input mix is $x_1=2x_2$, the scale is arbitrary, and profit is zero, a global maximum. If, however, $p<p^*$, say $p=1$, then $\pi=2(x_1)^{1/2}(x_2)^{1/2}-x_1-2x_2$ and letting $(x_1/x_2)^{1/2}=u$, $\pi=x_2[-u^2+2u-2]$. The bracketed expression is always negative and so is profit. Finally, if $p>p^*$, say $p=2$, then $\pi=x_2[-u^2+4u-2]$. This bracketed expression reaches a positive maximum of 2 when $u=2$, that is, $x_1=4x_2$, and by letting x_2 be large we can generate arbitrarily large profits. Finally, note that an arbitrary choice of u may generate a negative profit, for example, $u=4$, even with $p^*<p=2$.

To gain some geometric insight into the matter, try to visualize the graph of a linearly homogeneous function of two variables. Because of the property $f(tx_1,tx_2)=tf(x_1,x_2)$ we see that the graph is "ruled from the origin"; a half-line from the origin to any point of the graph lies on the graph in its entirety. Visualize now the graph of input costs $C=w_1x_1+w_2x_2$; it is a plane going through the origin. Let us now draw the graph of $pf(x_1,x_2)$ for low p values; it lies entirely below the cost plane: profit is everywhere negative. As p rises, the graph comes into contact with the plane, but it does so along an entire half-line from the origin. At this value of p, say p^*, profit is maximized anywhere along that half-line and is equal to zero. As p goes above p^*, the graph rises; it intersects the cost plane, and profit can be negative or positive depending on the choice of inputs. However, as the scale of operations is increased (moving away from the origin), the gap between the graph and the plane can be made arbitrarily large and so can profit: there is no maximum.

This concludes our brief survey of the economic applications of unconstrained optimization. Many economic problems involve constraints of some sort; this is taken up in the next section.

1.2 Optimization under equality constraints: the method of Lagrange

Economic agents typically face problems of choice subject to constraints. In many cases these take the form of equality constraints. Typical examples are the budget constraint of a utility-maximizing consumer or the resource constraints of a whole economy (land, labor, capital). This yields the classical equality-constrained problem:

Find $x_1^*, \ldots, x_n^*$ that maximize (alternatively minimize) $f(x_1, \ldots, x_n)$
subject to

$$\begin{aligned} &g^1(x_1, \ldots, x_n) = 0 \\ &\vdots \\ &g^m(x_1, \ldots, x_n) = 0, \quad m < n. \end{aligned} \tag{1.16}$$

We call f the *objective function* and g^j, $j = 1, \ldots, m$, the *constraints*. We require that there be fewer constraints than there are choice variables. Our task is to choose among all *feasible vectors* $\mathbf{x}$, that is, those satisfying the m-dimensional vector constraint $\mathbf{g}(\mathbf{x}) = 0$, the vector(s) that yield(s) the highest value for $f(\mathbf{x})$.

Although in a few simple cases it would be possible to use each constraint to eliminate one variable from the objective function and thereby obtain an unconstrained problem, we normally do not do this. Instead, we choose the seemingly more cumbersome method of introducing new variables, called multipliers, and solve for all variables at once. One minor reason for preferring this approach is that it preserves the structure and symmetry of the problem. The major reason is that these new variables, the multipliers, will be shown to provide important information on the sensitivity of the solution to parameter changes and on the operation of economic forces. Furthermore, this approach will be seen to be the prototype of methods used to solve more complicated problems such as nonlinear programming and optimal control.

1.2.1 *The method of Lagrange: Necessary conditions*

Consider problem (1.16). We introduce m new variables called Lagrange multipliers (one for each constraint) denoted by $\lambda_1, \dots, \lambda_m$ and form a new function called a *Lagrangean,*

$$\mathcal{L}(\lambda_1, \dots, \lambda_m, x_1, \dots, x_n) = f(x_1, \dots, x_n) + \sum_{j=1}^{m} \lambda_j \cdot g^j(x_1, \dots, x_n), \tag{1.17}$$

or more compactly,

$$\mathcal{L}(\boldsymbol{\lambda}, \mathbf{x}) = f(\mathbf{x}) + \boldsymbol{\lambda}' \cdot \mathbf{g}(\mathbf{x}). \tag{1.17$'$}$$

We can then state the main result.

Theorem 1.2.1. Let $\mathbf{x}^*$ be a solution to problem (1.16) and let the $m \times n$ matrix $\partial \mathbf{g}(\mathbf{x}^*)/\partial \mathbf{x}' = [\partial g^j(\mathbf{x}^*)/\partial x_i]$ have rank m (this is known as the *rank condition*). Then there must exist a unique set of values $\lambda_1, \dots, \lambda_m$ such that

$$\begin{aligned}
\frac{\partial \mathcal{L}}{\partial \lambda_1} &= g^1(\mathbf{x}^*) = 0, \\
&\vdots \\
\frac{\partial \mathcal{L}}{\partial \lambda_m} &= g^m(\mathbf{x}^*) = 0, \\
\frac{\partial \mathcal{L}}{\partial x_1} &= f_{x_1}(\mathbf{x}^*) + \sum_{j=1}^{m} \lambda_j g^j_{x_1}(\mathbf{x}^*) = 0, \\
&\vdots \\
\frac{\partial \mathcal{L}}{\partial x_n} &= f_{x_n}(\mathbf{x}^*) + \sum_{j=1}^{m} \lambda_j g^j_{x_n}(\mathbf{x}^*) = 0,
\end{aligned} \tag{1.18}$$

or more compactly (see the Appendix for matrix derivative notation),

$$\mathcal{L}_{\lambda} = \mathbf{g}(\mathbf{x}^*) = \mathbf{0} \quad \text{and} \quad \mathcal{L}_{\mathbf{x}} = f_{\mathbf{x}}(\mathbf{x}^*) + \mathbf{g}'_{\mathbf{x}}(\mathbf{x}^*) \cdot \boldsymbol{\lambda} = 0. \tag{1.18$'$}$$

(Note that $\mathbf{g}'_{\mathbf{x}}$ is the transpose of $\partial \mathbf{g}/\partial \mathbf{x}'$ and thus $n \times m$.)

Proof. We need to show that the column vector $f_{\mathbf{x}}(\mathbf{x}^*)$ of (1.18′) can be expressed as a linear combination of the columns of the $(n \times m)$ matrix $\mathbf{G}'(\mathbf{x}^*) \equiv \mathbf{g}'_{\mathbf{x}}(\mathbf{x}^*)$, the weights being identified as the multipliers $\boldsymbol{\lambda}$ of (1.18). We shall prove that the $n \times (m+1)$ matrix $[\mathbf{G}'(\mathbf{x}^*) \,\vdots\, f_{\mathbf{x}}(\mathbf{x}^*)]$ has at most rank m. Once we have done this our assumption that $\mathbf{G}(\mathbf{x}^*)$ has rank m (the rank condition) implies the desired result; that is, there exists a vector $\boldsymbol{\lambda}$ such that $f_{\mathbf{x}}(\mathbf{x}^*) + \mathbf{G}'(\mathbf{x}^*) \cdot \boldsymbol{\lambda} = 0$.

We shall use the implicit function theorem to prove that any feasible vector $\mathbf{x}^0$ such that $[\mathbf{G}'(\mathbf{x}^0) \vdots f_{\mathbf{x}}(\mathbf{x}^0)]$ has rank $m+1$ cannot yield a constrained maximum for (1.16). Suppose that the rank of the above matrix is $(m+1)$; then we can assume without loss of generality that the first $(m+1)$ rows of $[\mathbf{G}'(\mathbf{x}^0) \vdots f_{\mathbf{x}}(\mathbf{x}^0)]$ are linearly independent. By the implicit function theorem it follows that the set of $m+1$ equations

$$g^j(x_1, \dots, x_n) = 0, \quad j = 1, \dots, m,$$

$$f(x_1, \dots, x_n) - z = 0,$$

in $n+1$ variables $x_1, x_2, \dots, x_n$ and z, which by feasibility of $\mathbf{x}^0$ is known to have a solution $(x_1^0, x_2^0, \dots, x_n^0, z^0)$, where $z^0 = f(\mathbf{x}^0)$, also admits as a solution $(\bar{x}_1, \bar{x}_2, \dots, \bar{x}_n, \bar{z})$, for any arbitrarily chosen $(\bar{x}_{m+2}, \dots, \bar{x}_n, \bar{z})$ within a rectangular region around $(\mathbf{x}^0, z^0)$:

$$x_j^0 - \delta \leq x_j \leq x_j^0 + \delta, \quad j = m+2, \dots, n,$$

$$z^0 - \delta \leq z \leq z^0 + \delta,$$

where δ is some positive number. Note that $(\bar{x}_1, \bar{x}_2, \dots, \bar{x}_{m+1})$ depend on the choice of $(\bar{x}_{m+2}, \dots, \bar{x}_n, \bar{z})$, and in particular we can choose $\bar{z} = z^0 + \delta$. Then $f(\mathbf{x}^0) = z^0 < \bar{z} = f(\bar{\mathbf{x}})$ and $\mathbf{g}(\mathbf{x}^0) = \mathbf{0} = \mathbf{g}(\bar{\mathbf{x}})$. Hence, $\mathbf{x}^0$ does not yield a constrained maximum. Therefore, the rank of our $(m+1) \times n$ matrix is m. This completes the proof of the theorem. □

Remark (a). The assumption that $\mathbf{G}(\mathbf{x}^*)$ has rank m also guarantees that the vector $\boldsymbol{\lambda}$ that satisfies (1.18′) is unique. If the rank condition is relaxed, it is indeed possible (but not certain) that multipliers satisfying (1.18) exist, but they are not unique. To see why, suppose rank $\mathbf{G}(\mathbf{x}^*) = r < m$; our proof that at a maximum, rank $[\mathbf{G}'(\mathbf{x}^*) \vdots f_{\mathbf{x}}(\mathbf{x}^*)] < m+1$ still applies, but this matrix can now have rank r or $r+1$ (the only two possibilities). If its rank is r, then again $f_{\mathbf{x}}(\mathbf{x}^*)$ can be expressed as a linear combination of the columns of $\mathbf{G}'(\mathbf{x}^*)$, but the weights (the λ's) are no longer unique. If the rank is $r+1$, it is impossible to find a vector $\boldsymbol{\lambda}$ such that (1.18) is satisfied. To illustrate the latter case, we present the following example:

$$\text{Maximize } z = f(x_1, x_2) = x_2 - e^{x_1 - 1}$$

$$\text{subject to } (x_1^{1/2} - x_2^{1/2})^2 = 0.$$

In the positive quadrant the constraint can be represented by a 45° line, and for any z, the level curve of the objective function takes the form $x_2 = z + e^{x_1 - 1}$. Clearly, the constrained maximum occurs at $(x_1, x_2) = (1, 1)$, but at that point $f_1 = -1$, $f_2 = 1$, $g_1 = g_2 = 0$ and it is not possible to find any λ such that (1.18) is satisfied.

As an illustration of the case in which the λ's are not unique, consider any problem for which the constraints are tangent to one another at the constrained maximum:

$$\text{Maximize } \ln(x_1 x_2 x_3)$$
$$\text{subject to } x_1+x_2+x_3=3 \text{ and } 0.5(x_1^2+x_2^2+x_3^2)=1.5.$$

The first constraint, a plane, and the second constraint, a sphere centered at the origin, are tangent at the point $(1,1,1)$, which by symmetry is clearly the maximum. In addition to the constraints, the first-order conditions are

$$x_i^{-1}-\lambda_1-\lambda_2 x_i=0, \quad i=1,2,3.$$

At the maximum $x_i=1$, but $1=\lambda_1+\lambda_2$ is the only restriction placed on the multipliers; there are therefore many such λ values, while at $x^*=(1,1,1)$,

$$\mathbf{G}(\mathbf{x}^*)=\begin{bmatrix}-1 & -1 & -1\\ -1 & -1 & -1\end{bmatrix} \text{ has rank } 1<2.$$

Remark (b). We can provide a more intuitive derivation of conditions (1.18). There is a constrained maximum at $\mathbf{x}^*$, by assumption. Therefore, any small feasible change, that is, a small movement along the constraints, cannot improve the value of the objective function. This is the crux of the argument; the rest follows from taking first-order approximations and using linear algebra.

We represent a small movement by the differential notation $d\mathbf{x}$. If it is feasible, it does not change the value of the constraint; that is, it induces a zero small change in the vector $\mathbf{g}$: $d\mathbf{g}=0$. We stated earlier that such a change in $\mathbf{x}$ could not improve f; therefore, it must also induce $df=0$. In compact form,

$$(d\mathbf{g}=)\underset{m\times n}{\mathbf{G}(\mathbf{x}^*)}\cdot d\mathbf{x}=\mathbf{0} \Rightarrow df=f_{\mathbf{x}'}(\mathbf{x}^*)\cdot d\mathbf{x}=0. \tag{1.19}$$

The geometric meaning of the rank condition is that it rules out singularities (e.g., cusps, multiple points, isolated points) on the constraint surface, so that around $\mathbf{x}^*$ the curves differ but little from their tangents. $\mathbf{G}(\mathbf{x}^*)\cdot d\mathbf{x}=\mathbf{0}$ then puts an effective restriction on the change in $\mathbf{x}$ which takes into account all constraints and not one of them can be dispensed with. In other words, it makes precise our "small movement along the constraints."

Equation (1.19) more explicitly requires that all $(dx_1, \ldots, dx_n)$ that satisfy

$$
\begin{aligned}
(dg^1 =)\frac{\partial g^1(\mathbf{x}^*)}{\partial x_1}dx_1 + \cdots + \frac{\partial g^1(\mathbf{x}^*)}{\partial x_n}dx_n &= 0,\\
&\vdots\\
(dg^m =)\frac{\partial g^m(\mathbf{x}^*)}{\partial x_1}dx_1 + \cdots + \frac{\partial g^m(\mathbf{x}^*)}{\partial x_n}dx_n &= 0
\end{aligned}
\tag{1.20a}
$$

must also satisfy

$$
(df =)\frac{\partial f(\mathbf{x}^*)}{\partial x_1}dx_1 + \cdots + \frac{\partial f(\mathbf{x}^*)}{\partial x_n}dx_n = 0. \tag{1.20b}
$$

Thus, if $(dx_1, \ldots, dx_n)$ solves the system of equations (1.20a), it must also satisfy equation (1.20b). Therefore, the last equation adds nothing to what is already contained in (1.20a), and we must be able to duplicate it with an appropriate weighted sum of the equations of (1.20a). These weights are the multipliers. In the language of linear algebra, the vector of coefficient of (1.20b) and the vectors of coefficients of equations (1.20a) are linearly dependent. We have shown that the coefficients of the last equation must be a linear combination of those of the first m equations. Formally, there exist weights $\lambda_1, \ldots, \lambda_m$ such that

$$
\begin{aligned}
\lambda_1\frac{\partial g^1}{\partial x_1}(\mathbf{x}^*) + \cdots + \lambda_m\frac{\partial g^m}{\partial x_1}(\mathbf{x}^*) + \frac{\partial f(\mathbf{x}^*)}{\partial x_1} &= 0\\
&\vdots\\
\lambda_1\frac{\partial g^1}{\partial x_n}(\mathbf{x}^*) + \cdots + \lambda_m\frac{\partial g^m}{\partial x_n}(\mathbf{x}^*) + \frac{\partial f(\mathbf{x}^*)}{\partial x_n} &= 0.
\end{aligned}
\tag{1.21}
$$

These λ values are unique since we have assumed by the rank condition that $\mathbf{G}(\mathbf{x}^*)$, the coefficient matrix of (1.21), has full rank m. Equation (1.21) and the requirement that $\mathbf{x}^*$ actually be on the constraint $\mathbf{g}(\mathbf{x}^*) = \mathbf{0}$ are seen to be identical with the first-order conditions of (1.18).

Geometric interpretation. Consider the simplest case of the maximization of a function of two variables subject to one constraint. Find (x_1^*, x_2^*) to maximize $f(x_1, x_2)$ subject to $g(x_1, x_2) = 0$. This is illustrated in Figure 1.6. The thick curve is the constraint, while the thin curves are contour curves of the objective function; the arrow indicates the direction of increase of f. A point such as C is not feasible, while a point such as B is feasible but not optimal, since we can move along the constraint toward higher f values. At point A, however, any move along the constraint results in lower f values and A is the optimal solution: it is at the tangency of one of the level curves of f with constraint $g = 0$. Recall that the slope of a level curve is given by $dx_2/dx_1 = -f_1/f_2$ for f, and $dx_2/dx_1 = -g_1/g_2$ along $g = 0$; thus at a point of tangency such as A we have

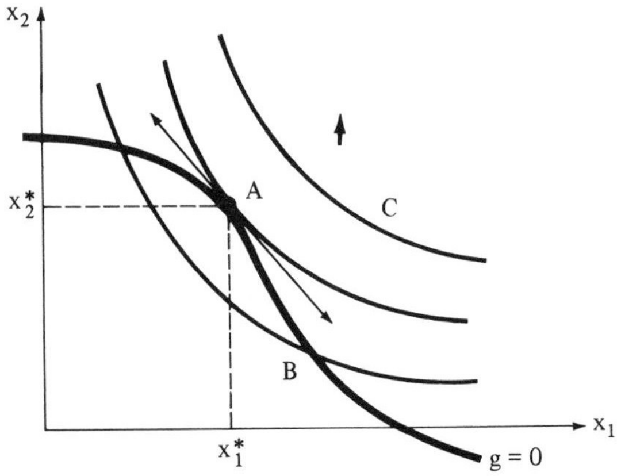

Figure 1.6

$$\frac{dx_2}{dx_1} = -\frac{f_1}{f_2} = -\frac{g_1}{g_2}$$

or

$$-\frac{f_1}{g_1} = -\frac{f_2}{g_2}.$$

If we define λ as the above ratio, then

$$-\frac{f_1}{g_1} = -\frac{f_2}{g_2} = \lambda$$

or

$$f_1 + \lambda g_1 = 0 \quad \text{and} \quad f_2 + \lambda g_2 = 0. \tag{1.22}$$

Equation (1.22) and the requirement that $g(x_1, x_2) = 0$ are the necessary conditions of (1.18) applied to this simple problem. From our derivation of this result, conditions (1.22) and more generally (1.18) are seen to reflect the tangency of a level surface of the objective function with the intersection of surfaces representing the constraints.

Example 1.2.1. Maximize $f(x_1, x_2) = \ln x_1 + \ln x_2$ subject to $2 - x_1^2 - x_2^2 = 0$. From the obvious symmetry of the problem the solution must be $x_1 = 1$, $x_2 = 1$. Let us then illustrate the necessity of (1.18) at that maximum. With one constraint there is only one multiplier and the Lagrangean is

$$\mathcal{L}(\lambda_1, x_1, x_2) = \ln x_1 + \ln x_2 + \lambda(2 - x_1^2 - x_2^2).$$

The necessary conditions are

$$\frac{\partial \mathcal{L}}{\partial \lambda} = 2 - x_1^2 - x_2^2 = 0,$$

$$\frac{\partial \mathcal{L}}{\partial x_1} = x_1^{-1} - 2\lambda x_1 = 0,$$

$$\frac{\partial \mathcal{L}}{\partial x_2} = x_2^{-1} - 2\lambda x_2 = 0.$$

To solve this, pass the terms involving λ to the right-hand side of the last two equations and divide through to get $x_2 x_1^{-1} = x_1 x_2^{-1}$; hence, $x_1^2 = x_2^2$ and $x_1 = x_2 = 1$, as predicted. Furthermore, substituting these values, we obtain $\lambda = 0.5$.

Sign indeterminacy of the multipliers. By the very nature of problem (1.16) the signs of the multipliers cannot be ascertained. To see this, suppose that we had written the first constraint as $-g^1(x) = 0$, quite an innocuous change. However, in the first-order conditions (1.18), whenever λ_1 appeared it now would be as $-\lambda_1 g^1_{x_i}$ instead of $\lambda_1 g^1_{x_i}$, $i = 1, \ldots, n$, nothing else having changed. Clearly, the solution to these equations would remain the same but for the sign of λ_1, which would be reversed. The reader is invited to rework Example 1.2.1 with the constraint written as $x_1^2 + x_2^2 - 2 = 0$ and the Lagrangean as $\mathcal{L}(\lambda, x_1, x_2) = \ln x_1 + \ln x_2 + \lambda[x_1^2 + x_2^2 - 2]$; the solution will be $(-0.5, 1, 1)$. This peculiarity is of little consequence; constraints must simply be written in a consistent way.

Remark. The first-order necessary conditions presented above apply equally well to a constrained minimum problem; to distinguish between the two, we must turn to second-order conditions.

1.2.2 *The method of Lagrange: second-order conditions*

Given a constrained local maximum $\mathbf{x}^*$ and associated multipliers $\boldsymbol{\lambda}$, there exists a set of second-order necessary conditions that must hold at that point. There also exists a set of slightly stricter conditions which if satisfied along with the first-order conditions at some $(\mathbf{x}^*, \boldsymbol{\lambda})$ ensures that this point is a local maximum. The situation is thus much the same as in the unconstrained case. These conditions involve the second-order derivatives of the Lagrangean $\mathcal{L}(\boldsymbol{\lambda}, \mathbf{x}) = f(\mathbf{x}) + \boldsymbol{\lambda}' \cdot \mathbf{g}(\mathbf{x})$. First we need to write down in detail the whole Hessian matrix of $\mathcal{L}$. It is important to keep precisely to the notation and the ordering of variables, since any change would usually alter the conditions. As before, let $\mathbf{x}$ be $n \times 1$ and $\boldsymbol{\lambda}$ be $m \times 1$. The Hessian of $\mathcal{L}$ will be $(m+n) \times (m+n)$, and since there are two sorts of variables we will often write this matrix down in partitioned form (check the Appendix for differentiation using matrices):

$$\mathbf{B}=\left[\begin{array}{c:c} \mathcal{L}_{\lambda\lambda'} & \mathcal{L}_{\lambda\mathbf{x}'} \\ \hdashline \mathcal{L}_{\mathbf{x}\lambda'} & \mathcal{L}_{\mathbf{x}\mathbf{x}'} \end{array}\right],$$

$$\mathbf{B}=\left[\begin{array}{c:c} \mathbf{0} & \mathbf{g}_{\mathbf{x}'} \\ \hdashline \mathbf{g}'_{\mathbf{x}} & f_{\mathbf{x}\mathbf{x}'}+\sum_{j=1}^{m}\lambda_j g^j_{\mathbf{x}\mathbf{x}'} \end{array}\right],$$

where $\mathbf{B}$ is the $(m+n)\times(m+n)$ Hessian matrix of $\mathcal{L}(\boldsymbol{\lambda},\mathbf{x})$ and the four submatrices $\mathcal{L}_{\lambda\lambda'}$, $\mathcal{L}_{\lambda\mathbf{x}'}$, $\mathcal{L}_{\mathbf{x}\lambda'}$, and $\mathcal{L}_{\mathbf{x}\mathbf{x}'}$ are of order $m\times m$, $m\times n$, $n\times m$, and $n\times n$, respectively. Matrix $\mathbf{B}$ can be written more precisely as

$$\mathbf{B}=\left[\begin{array}{ccc:ccc} 0 & \cdots & 0 & g^1_{x_1} & \cdots & g^1_{x_n} \\ \vdots & & & \vdots & & \\ 0 & \cdots & 0 & g^m_{x_1} & \cdots & g^m_{x_n} \\ \hdashline g^1_{x_1} & \cdots & g^m_{x_1} & f_{x_1x_1}+\sum_{j=1}^{m}\lambda_j g^j_{x_1x_1} & \cdots & f_{x_1x_n}+\sum_{j=1}^{m}\lambda_j g^j_{x_1x_n} \\ \vdots & & & \vdots & & \vdots \\ g^1_{x_n} & \cdots & g^m_{x_n} & f_{x_1x_n}+\sum_{j=1}^{m}\lambda_j g^j_{x_1x_n} & \cdots & f_{x_nx_n}+\sum_{j=1}^{m}\lambda_j g^j_{x_nx_n} \end{array}\right],$$

or

$$\mathbf{B}\equiv\left[\begin{array}{c:c} \mathbf{0} & \mathbf{G} \\ \hdashline \mathbf{G}' & \mathbf{L} \end{array}\right], \quad \text{where } G_{rs}=\left[\frac{\partial g^r}{\partial x_s}\right], \qquad (1.23)$$

$$L_{rs}=\left[\frac{\partial^2 f}{\partial x_r\partial x_s}+\sum_{j=1}^{m}\lambda_j\frac{\partial^2 g^j}{\partial x_r\partial x_s}\right].$$

Care must be taken to order the variables as shown, that is, $\lambda_1,\ldots,\lambda_m$, $x_1,\ldots,x_n$ since the following theorems are tailored to this format.

Theorem 1.2.2: necessity

(i) Let $\mathbf{x}^*$ be a local maximum for problem (1.16) and let $(\mathbf{x}^*, \boldsymbol{\lambda}^*)$ satisfy (1.18). Then the matrix $\mathbf{L}$ in (1.23) is negative-semidefinite for all vectors $\mathbf{z}$ satisfying $\mathbf{Gz}=\mathbf{0}$, where $\mathbf{L}$ and $\mathbf{G}$ are evaluated at $(\mathbf{x}^*, \boldsymbol{\lambda}^*)$.

(ii) If $\mathbf{x}^*$ is a local minimum for problem (1.16), modify (i) to positive-semidefiniteness.

Theorem 1.2.3: sufficiency

(i) Let $(\mathbf{x}^*, \boldsymbol{\lambda}^*)$ satisfy the first-order condition (1.18) and in addition let $\mathbf{L}$ in (1.23) be negative-definite for all vectors $\mathbf{z}\neq\mathbf{0}$ satisfying $\mathbf{Gz}=\mathbf{0}$, where $\mathbf{L}$ and $\mathbf{G}$ are evaluated at $(\mathbf{x}^*, \boldsymbol{\lambda}^*)$; then $\mathbf{x}^*$ is a local maximum for problem (1.16).

(ii) If we modify (i) to positive-definiteness we have sufficient conditions for a local minimum for problem (1.16).

One difficulty is that it is not straightforward to check the definiteness of a matrix under constraints. Fortunately there is a set of conditions equivalent to those of Theorem 1.2.3.

Theorem 1.2.4: sufficiency. Assume that the rank condition (rank $\mathbf{G} = m$) is satisfied.

(i) Let $(\mathbf{x}^*, \boldsymbol{\lambda}^*)$ satisfy the first-order condition (1.18) and in addition let the last $(n-m)$ leading principal minors of $\mathbf{B}$ alternate in sign beginning with that of $(-1)^{m+1}$, where $\mathbf{B}$ is evaluated at $(\mathbf{x}^*, \boldsymbol{\lambda}^*)$; then $\mathbf{x}^*$ is a local maximum for problem (1.16). (This sign sequence can also be characterized by requiring the last leading principal minor, i.e., $|\mathbf{B}|$, to have the sign of $(-1)^n$; alternatively, each leading principal minor of order $k \times k$ must have the sign of $(-1)^{(k-m)}$, $k = 2m+1, 2m+2, \ldots, m+n$.)

(ii) Let $(\mathbf{x}^*, \boldsymbol{\lambda}^*)$ satisfy the first-order condition (1.18) and in addition let the last $(n-m)$ leading principal minors of $\mathbf{B}$ be of the same sign as $(-1)^m$, where $\mathbf{B}$ is evaluated at $(\mathbf{x}^*, \boldsymbol{\lambda}^*)$; then $\mathbf{x}^*$ is a local minimum for problem (1.16).

The conditions of Theorem 1.2.4 are not necessary but can be made so with an additional restriction. The following result will be particularly useful in Section 1.3 (Takayama, 1985, p. 162).

Theorem 1.2.5. Suppose that $\mathbf{B}$, the Hessian matrix of the Lagrangean, is nonsingular. (We say that we have a *regular* maximum or minimum.) Then the conditions of Theorem 1.2.4 are also necessary for a maximum (resp. a minimum) for problem (1.16).

In order to clarify those rather complicated requirements, we shall look at some examples. First we represent diagrammatically which leading principal minors we are concerned with:

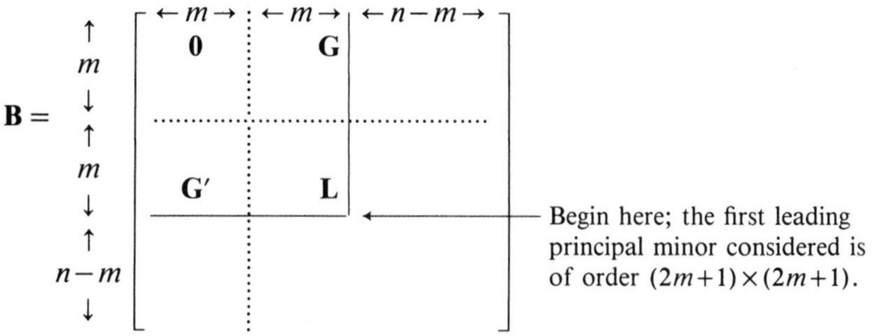

Examples 1.2.2

(i) Let $n=6$ and $m=3$; $\mathbf{B}$ is 9×9. The first leading principal minor to consider is 7×7; call it B_7. For a maximum it must have the sign of $(-1)^4>0$; $B_8<0$; B_9 is the last one and has the sign of $(-1)^6>0$. (Note that B_9 is $|\mathbf{B}|$.) In the case of a minimum all of these minors must have the sign of $(-1)^3<0$: $B_7<0$, $B_8<0$, $B_9<0$.

(ii) Let $n=3$ and $m=2$; $\mathbf{B}$ is 5×5. The first leading principal minor to consider is 5×5, thus simply the determinant of $\mathbf{B}$ which is required to be of the sign of $(-1)^3<0$ for a maximum, whether we look at it as the first or the last one. For a minimum it must have the sign of $(-1)^2>0$.

Now let us consider some numerical examples.

Example 1.2.3. Find the maximum and/or minimum of $f(x_1,x_2,x_3)=2x_1x_2x_3$ subject to $3-x_1^2-x_2^2-x_3^2=0$. Form the Lagrangean

$$\mathcal{L}(\lambda,x_1,x_2,x_3)=2x_1x_2x_3+\lambda[3-x_1^2-x_2^2-x_3^2].$$

The first-order conditions are

$$\frac{\partial\mathcal{L}}{\partial\lambda}=3-x_1^2-x_2^2-x_3^2=0,$$

$$\frac{\partial\mathcal{L}}{\partial x_1}=2x_2x_3-2\lambda x_1=0,$$

$$\frac{\partial\mathcal{L}}{\partial x_2}=2x_1x_3-2\lambda x_2=0,$$

$$\frac{\partial\mathcal{L}}{\partial x_3}=2x_1x_2-2\lambda x_3=0.$$

These conditions are easily solved and have, among others, the two solutions

$$\lambda=x_1=x_2=x_3=1 \quad\text{and}\quad \lambda=x_1=x_2=x_3=-1.$$

The Hessian matrix of $\mathcal{L}(\lambda,x_1,x_2,x_3)$ is

$$\mathbf{B}=\begin{bmatrix} 0 & -2x_1 & -2x_2 & -2x_3 \\ -2x_1 & -2\lambda & 2x_3 & 2x_2 \\ -2x_2 & 2x_3 & -2\lambda & 2x_1 \\ -2x_3 & 2x_2 & 2x_1 & -2\lambda \end{bmatrix}.$$

At the positive solution,

$$\mathbf{B}=\begin{bmatrix} 0 & -2 & -2 & -2 \\ -2 & -2 & 2 & 2 \\ -2 & 2 & -2 & 2 \\ -2 & 2 & 2 & -2 \end{bmatrix}$$

and

$$B_3=\begin{vmatrix} 0 & -2 & -2 \\ -2 & -2 & 2 \\ -2 & 2 & -2 \end{vmatrix}=32>0,$$

while $B_4=|\mathbf{B}|=-192<0$. Hence, the last $(3-1=2)$ two leading principal minors alternate in sign beginning with $(-1)^2>0$ (or alternatively ending with the sign of $(-1)^3<0$); the positive solution is a local constrained maximum.

At the negative solution,

$$\mathbf{B}=\begin{bmatrix} 0 & 2 & 2 & 2 \\ 2 & 2 & -2 & -2 \\ 2 & -2 & 2 & -2 \\ 2 & -2 & -2 & 2 \end{bmatrix}$$

and $B_3=-32$, $B_4=-192$. The last two leading principal minors all have the sign of $(-1)^2<0$; the negative solution is a local constrained minimum.

There are six other solutions to the first-order conditions for the vector (λ, x_1, x_2, x_3); they are $(1, 1, -1, -1)$, $(-1, -1, 1, 1)$, $(1, -1, 1, -1)$, $(1, -1, -1, 1)$, $(-1, 1, 1, -1)$, and $(-1, 1, -1, 1)$. The reader is invited to check whether any one of these is a constrained maximum or minimum for this problem. (*Hint:* All points at which f is positive are maxima; all those at which f is negative are minima.)

Example 1.2.4. Here we deal with two constraints. Find x_1, x_2, and x_3 that maximize $f(x_1, x_2, x_3)=4\ln x_1+2x_2+8x_3$ subject to $8-x_1-x_2-2x_3=0$ and $1-0.5x_1-x_3=0$. The Lagrangean is

$$\mathcal{L}(\lambda_1, \lambda_2, x_1, x_2, x_3)=4\ln x_1+2x_2+8x_3+\lambda_1[8-x_1-x_2-2x_3] \\ +\lambda_2[1-0.5x_1-x_3].$$

The first-order conditions are

$$\mathcal{L}_{\lambda_1}=8-x_1-x_2-2x_3=0,$$

$$\mathcal{L}_{\lambda_2}=1-0.5x_1-x_3=0,$$

$$\mathcal{L}_{x_1}=4x_1^{-1}-\lambda_1-0.5\lambda_2=0,$$

$$\mathcal{L}_{x_2} = 2 - \lambda_1 = 0,$$

$$\mathcal{L}_{x_3} = 8 - 2\lambda_1 - \lambda_2 = 0.$$

The fourth equation yields $\lambda_1 = 2$ and the fifth $\lambda_2 = 4$; from the third equation $x_1 = 1$, which substituted in the second equation yields $x_3 = 0.5$; finally, the first equation gives $x_2 = 6$. The solution is $(2, 4, 1, 6, 0.5)$. The Hessian matrix of the Lagrangean is as follows (remember to differentiate the first-order conditions with respect to $\lambda_1, \lambda_2, x_1, x_2, x_3$ in that order):

$$\mathbf{B} = \begin{bmatrix} 0 & 0 & -1 & -1 & -2 \\ 0 & 0 & -0.5 & 0 & -1 \\ -1 & -0.5 & -4x_1^{-2} & 0 & 0 \\ -1 & 0 & 0 & 0 & 0 \\ -2 & -1 & 0 & 0 & 0 \end{bmatrix}.$$

We need to look at the sign of the last $(3-2=1)$ leading principal minor; it must be $(-1)^3 < 0$. $|\mathbf{B}| = -4x_1^{-2}$, and when $\mathbf{B}$ is evaluated at the solution we have $|\mathbf{B}| = -4 < 0$, a local constrained maximum indeed.

1.2.3 *Some global results for equality-constrained problems*

All the results of the preceding subsection were valid for local optima only; in economics we are more often concerned with global optima. As one would expect, some forms of concavity restrictions will be useful in securing global results.

Theorem 1.2.6. Let $(\lambda^*, \mathbf{x}^*)$ be a solution to equation (1.18). If $\mathcal{L}(\lambda^*, \mathbf{x})$ is a concave (resp. convex) function of $\mathbf{x}$, then $\mathbf{x}^*$ is a global maximum (resp. minimum) for problem (1.16).

Proof. Concavity of $\mathcal{L}(\lambda^*, \mathbf{x})$ in $\mathbf{x}$ along with the last n equations of (1.18) implies that $\mathbf{x}^*$ is an unconstrained global maximum of $\mathcal{L}(\lambda^*, \mathbf{x})$; hence, $\mathcal{L}(\lambda^*, \mathbf{x}^*) \geq \mathcal{L}(\lambda^*, \mathbf{x})\ \forall \mathbf{x}$, or $f(\mathbf{x}^*) + \lambda^{*\prime} \cdot \mathbf{g}(\mathbf{x}^*) \geq f(\mathbf{x}) + \lambda^{*\prime} \cdot \mathbf{g}(\mathbf{x})\ \forall \mathbf{x}$. Clearly $\mathbf{x}^*$ satisfies the constraints and $\mathbf{g}(\mathbf{x}^*) = \mathbf{0}$. Therefore, $f(\mathbf{x}^*) \geq f(\mathbf{x}) + \lambda^{*\prime} \cdot \mathbf{g}(\mathbf{x})$ $\forall \mathbf{x}$ and finally $f(\mathbf{x}^*) \geq f(\mathbf{x})\ \forall \mathbf{x}$ for which $\mathbf{g}(\mathbf{x}) = \mathbf{0}$. □

An inconvenience here is that we have to solve the problem before we can ascertain the sort of optimum that is obtained. In some special cases we can do a little better.

Corollary 1.2.6′. Let $f(\mathbf{x})$ be concave (resp. convex) and all $g^j(\mathbf{x})$ be linear. Then a solution of (1.18) provides a global maximum (resp. minimum) for problem (1.16).

Corollary 1.2.6″. Let $f(\mathbf{x})$ and all $g^j(\mathbf{x})$ be concave (resp. convex) functions. Assume either that $f(\mathbf{x})$ is increasing in $\mathbf{x}$ and all $g^j(\mathbf{x})$ are decreasing in $\mathbf{x}$, or that $f(\mathbf{x})$ is decreasing while all $g^j(\mathbf{x})$ are increasing. If $(\lambda^*, \mathbf{x}^*)$ solves (1.18) and all the multipliers are of the same sign, $\mathbf{x}^*$ is a global maximum (resp. minimum) for problem (1.16). (If there is only one constraint, the requirement on the sign of the multipliers can be dispensed with.)

To prove the second corollary simply note that the restrictions on the monotonicity of f and $\mathbf{g}$ and the sameness of sign of the multipliers imply by (1.18) that the multipliers will be nonnegative.

Example 1.2.5: derivation of cost functions. Let $q = f(\mathbf{x})$ be the production function of a firm and $\mathbf{w}$ the vector of fixed input prices. In order to know the cost of production of q units of output the firm must solve the following problem, given $\mathbf{w}$ and q. Minimize $\mathbf{w}' \cdot \mathbf{x}$ subject to $q - f(\mathbf{x}) = 0$.

The objective function is linear, hence convex, and increasing for positive $\mathbf{w}$; the constraint is convex and decreasing if we make the usual assumptions about the production function; hence, Corollary 1.2.6″ applies. The first-order conditions are $q - f(\mathbf{x}) = 0$ and $\mathbf{w} - \lambda f_{\mathbf{x}}(\mathbf{x}) = 0$, the latter establishing the fact that λ^* will be positive and we can apply Theorem 1.2.6 to the Lagrangean $\mathcal{L}(\lambda^*, \mathbf{x}) = \mathbf{w}' \cdot \mathbf{x} + \lambda^*[q - f(\mathbf{x})]$, which is clearly convex in $\mathbf{x}$ under our assumptions. Solving will yield $\mathbf{x}^*(q, \mathbf{w})$, and the cost function will be $C(q, \mathbf{w}) = \mathbf{w}' \cdot \mathbf{x}^*(q, \mathbf{w})$. We now illustrate this with a numerical example.

Find the cost function associated with the production function $q = (x_1)^{1/4}(x_2)^{1/2}$. This is clearly increasing in (x_1, x_2) and also concave according to the result of (1.10). The Lagrangean for the minimum problem is

$$\mathcal{L}(\lambda, x_1, x_2) = w_1 x_1 + w_2 x_2 + \lambda[q - (x_1)^{1/4}(x_2)^{1/2}].$$

The first-order conditions are

$$\begin{aligned} q - (x_1)^{1/4}(x_2)^{1/2} &= 0, \\ w_1 - \tfrac{1}{4}\lambda(x_1)^{-3/4}(x_2)^{1/2} &= 0, \\ w_2 - \tfrac{1}{2}\lambda(x_1)^{1/4}(x_2)^{-1/2} &= 0. \end{aligned}$$

The last two yield $w_1/w_2 = x_2/2x_1$, and substitution into the first one gives $(x_1)^{3/4} = q(2)^{-1/2}(w_2/w_1)^{1/2}$; hence, $x_1 = (2)^{-2/3}(q)^{4/3}(w_2/w_1)^{2/3}$ and $x_2 = (2)^{1/3}(q)^{4/3}(w_1/w_2)^{1/3}$. Total cost is $w_1 x_1 + w_2 x_2$, and the cost function is

$$C(q, w_1, w_2) = ((2)^{-2/3} + (2)^{1/3})(q)^{4/3}(w_1)^{1/3}(w_2)^{2/3}$$
$$= 3(2)^{-2/3}(q)^{4/3}(w_1)^{1/3}(w_2)^{2/3},$$
$$C(q, w_1, w_2) \simeq 1.9(q)^{4/3}(w_1)^{1/3}(w_2)^{2/3}.$$

Note in passing that this is a convex function of q that will generate the usual increasing marginal cost feature. The value of the multiplier is $\lambda = (2)^{4/3}(q)^{1/3}(w_1)^{1/3}(w_2)^{2/3}$.

Example 1.2.6: derivation of demand functions. Let $U(\mathbf{x})$ be the utility function of an individual, $\mathbf{p}$ the vector of prices, and m the individual's income. This consumer seeks to maximize utility subject to the budget constraint, hence to choose $\mathbf{x}$ to maximize $U(\mathbf{x})$ subject to $\mathbf{p}' \cdot \mathbf{x} = m$.

If we assume that the utility function can be transformed into a concave function by a monotone-increasing transformation, we can take $U(\mathbf{x})$ to be concave, and since the constraint is linear in $\mathbf{x}$, Corollary 1.2.6′ applies. The first-order conditions are $m - \mathbf{p}' \cdot \mathbf{x} = 0$ and $U_{\mathbf{x}}(\mathbf{x}) - \lambda \mathbf{p} = 0$. Taking the ratio of two of the first-order conditions establishes that $U_i(\mathbf{x})/U_j(\mathbf{x}) = p_i/p_j$: the ratio of marginal utilities is equal to the corresponding price ratio. Solving will yield $\mathbf{x}^*(m, \mathbf{p})$, which are the quantities of goods the individual is willing to buy at price $\mathbf{p}$, with income m - hence the demand functions for this consumer. We now illustrate this with a numerical example.

Derive the demand functions associated with the utility function

$$U(\mathbf{x}) = \sum_{i=1}^{n} \beta_i \ln(x_i - \gamma_i),$$

assuming $\beta_i \geq 0$, $\sum_i \beta_i = 1$ (without loss), and $m > \mathbf{p}' \cdot \boldsymbol{\gamma} = \sum_{i=1}^{n} p_i \gamma_i$. The Lagrangean is

$$\mathcal{L}(\lambda, \mathbf{x}) = \sum_{i=1}^{n} \beta_i \ln(x_i - \gamma_i) + \lambda(m - \mathbf{p}' \cdot \mathbf{x}).$$

The first-order conditions are

$$m - \mathbf{p}' \cdot \mathbf{x} = 0$$

$$\beta_i = \lambda p_i (x_i - \gamma_i), \quad i = 1, \ldots, n.$$

Summing the last n equations yields

$$1 = \lambda(\mathbf{p}' \cdot \mathbf{x} - \mathbf{p}' \cdot \boldsymbol{\gamma}) = \lambda(m - \mathbf{p}' \cdot \boldsymbol{\gamma}),$$

and eliminating λ gives $\beta_i = [p_i(x_i - \gamma_i)]/(m - \mathbf{p}' \cdot \boldsymbol{\gamma})$ or finally $x_i(m, \mathbf{p}) = \gamma_i + \beta_i[(m - \mathbf{p}' \cdot \boldsymbol{\gamma})]/p_i$. These demand functions are known as the *linear expenditure system*.

1.2.4 *The method of Lagrange: economic applications*

We begin this section by providing an interpretation of the multipliers. Similar reasoning will be encountered throughout this book.

Economic interpretation of the multipliers. We first examine the simplest possible case of the efficient allocation of a resource between two industries. Let $q_i = F_i(x_i)$ and p_i be, respectively, the production function and output price in industry i, $i = 1, 2$; x_i is the amount of a resource used by industry i; the total amount of resource available to both industries is X. Assume that output prices are determined by the world market, and hence are fixed here. In order to allocate the resource efficiently, the central planner maximizes total revenue subject to the resource constraint:

$$\text{Maximize } R(x_1, x_2) = p_1 F_1(x_1) + p_2 F_2(x_2)$$
$$\text{subject to } x_1 + x_2 = X.$$

Under the standard assumption of concavity of production functions, the first-order conditions will be necessary and sufficient for a maximum. Those conditions are

$$\begin{aligned} X - x_1^* - x_2^* &= 0, \\ p_1 F_1'(x_1^*) - \lambda^* &= 0, \\ p_2 F_2'(x_2^*) - \lambda^* &= 0, \end{aligned} \tag{1.24}$$

where the primes denote derivatives, the asterisks denote optimality, and the Lagrangean is

$$\mathcal{L}(\lambda, x_1, x_2) = p_1 F_1(x_1) + p_2 F_2(x_2) + \lambda[X - x_1 - x_2].$$

The last two conditions look remarkably like the conditions for profit maximization with λ as the price of the resource input; indeed, the resemblance is not coincidental, as we now demonstrate.

Suppose that there is an exogenous small change in the amount of resource available: we now have $X + dX$ of it. How will the central planner allocate this extra amount, say dx_1 to industry 1 and dx_2 to the other? We are dealing with arbitrarily small changes, and thus linear approximations through total differentials are acceptable. The change in the maximum revenue is

$$dR^* = p_1 F_1'(x_1^*) dx_1 + p_2 F_2'(x_2^*) dx_2,$$
$$dR^* = \lambda^* [dx_1 + dx_2], \quad \text{by (1.24)},$$

and since the resource constraint must be satisfied, we have $dx_1 + dx_2 = dX$ and finally

$$dR^*/dX = \lambda^*. \tag{1.25}$$

This is a very important result. The total derivative of the maximum revenue with respect to the amount of resource available is equal to the optimal value of the multiplier attached to that resource. Speaking loosely, one extra unit of the resource would allow the planner to generate \$$\lambda^*$ more in revenue; therefore, λ^* is what it is worth to him, and he would be willing to pay \$$\lambda^*$ to get that extra unit. We see that λ^* emerges as the marginal worth of the hitherto unpriced resource; a common name for it is the *shadow price* or the *imputed value* of the resource. Thus, the method of Lagrange generates as a by-product of the solution a shadow pricing system that operates as a real one, as we shall soon demonstrate. First we state this result in a more general form.

Theorem 1.2.7. Let $(\lambda^*, \mathbf{x}^{1*}, \ldots, \mathbf{x}^{n*})$ be the solution to the problem of finding $(\mathbf{x}^1, \ldots, \mathbf{x}^n)$ to maximize $R(\mathbf{x}^1, \ldots, \mathbf{x}^n)$ subject to $\mathbf{x}^1 + \cdots + \mathbf{x}^n = \mathbf{X}$, where $\mathbf{X}$ is an $(m \times 1)$ vector of fixed amounts of resources, and $\mathbf{x}^i$ is the $(m \times 1)$ vector of inputs to the ith industry. Then

$$dR(\mathbf{x}^{1*}, \ldots, \mathbf{x}^{n*})/dX_j = \lambda_j^*, \quad j = 1, \ldots, m. \tag{1.26}$$

The proof is exactly the same as the one in the simplest case and is left as an exercise.

Let us continue with the single-resource, two-industry case. Consider a similar economy but a decentralized one in which each industry maximizes its own profit regardless of what the other one does and without any knowledge of the resource constraint. Let the prices be p_1, p_2 as before plus λ^* for the resource. The conditions for profit maximization are

$$p_1 F_1'(x_1) = \lambda^* \quad \text{and} \quad p_2 F_2'(x_2) = \lambda^*.$$

Clearly, the (x_1^*, x_2^*) solution of (1.24) applies here as well. Therefore, if we were to use the shadow price of the resource as the actual price, optimizing competitive behavior by the industries would lead to the resource being exactly exhausted without anyone paying any mind to it. This brings out the essential role of prices in decentralizing the decision-making process. Quite obviously, the result generalizes to many industries and many resources. This is a simple instance of the "invisible hand" phenomenon. To gain yet more insight into the process let us consider a numerical example.

Example 1.2.7. Let $F_1(x_1) = 12(x_1)^{1/3}$, $F_2(x_2) = (x_2)^{1/3}$, $p_1 = 1$, $p_2 = 3$, and $X = 72$. The central planner's problem is to maximize $R(x_1, x_2) = 12(x_1)^{1/3} + 3(x_2)^{1/3}$ subject to $x_1 + x_2 = 72$. The Lagrangean is

$$\mathcal{L}(\lambda, x_1, x_2) = 12(x_1)^{1/3} + 3(x_2)^{1/3} + \lambda[72 - x_1 - x_2],$$

and the first-order conditions are

$$72 - x_1 - x_2 = 0, \quad 4(x_1)^{-2/3} = \lambda, \quad \text{and} \quad (x_2)^{-2/3} = \lambda.$$

These yield $x_1 = 8x_2$ and finally $x_1^* = 64$, $x_2^* = 8$, and $\lambda^* = 0.25$.

To understand the role of the multiplier, suppose now that each industry maximizes profit while facing a price of w per unit for the resource. We have

(i) Maximize $\Pi_1 = 12(x_1)^{1/3} - wx_1$;
hence, $4(x_1)^{-2/3} = w$ or $x_1 = 8(w)^{-3/2}$.

(ii) Maximize $\Pi_2 = 3(x_2)^{1/3} - wx_2$;
hence, $(x_2)^{-2/3} = w$ or $x_2 = (w)^{-3/2}$.

The total demand for the resource is thus

$$x_1 + x_2 = 9(w)^{-3/2}.$$

If the resource is competitively priced (e.g., it is owned by many small independent agents), demand at equilibrium will simply equal supply: $9(w)^{-3/2} = 72$, or $w = 0.25$. Therefore, the competitive price of the resource is the same as the value of the multiplier that we had found earlier in the efficient resource allocation problem.

These results are special cases of a very general and extremely useful result, which we now state and prove. The proof follows the lines of the interpretation of λ but is more complicated; we shall use matrix notation, and the order of matrices must be carefully monitored but may be skipped on a first reading. (Again, see the Appendix for the notation.) Primes denote transposition.

Theorem 1.2.8: the envelope theorem. Let $(\lambda^*, \mathbf{x}^*)$ solve the following problem, for given $\mathbf{p}$. Find $\mathbf{x}$ that maximizes $u(\mathbf{x}; \mathbf{p})$ subject to $\mathbf{g}(\mathbf{x}; \mathbf{p}) = \mathbf{0}$, where λ, $\mathbf{x}$, and $\mathbf{p}$ are vectors. Then

$$\frac{du(\mathbf{x}^*; \mathbf{p})}{dp_s} = \frac{\partial \mathcal{L}(\lambda^*, \mathbf{x}^*; \mathbf{p})}{\partial p_s} \quad \text{for any } p_s \text{ element of } \mathbf{p},$$

where $du/dp_s = \partial u/\partial p_s + \partial u/\partial \mathbf{x}' \cdot d\mathbf{x}^*/dp_s$ denotes the total derivative of u with respect to p_s.

Note: This problem is the same as problem (1.16) with the addition of an $(S \times 1)$ vector of parameters $\mathbf{p}$.

Proof. Let $\boldsymbol{\lambda}$ and $\mathbf{x}$ be of order $m\times 1$ and $n\times 1$, respectively. The Lagrangean is $\mathcal{L}(\boldsymbol{\lambda},\mathbf{x};\mathbf{p})=u(\mathbf{x};\mathbf{p})+\mathbf{g}'(\mathbf{x};\mathbf{p})\cdot\boldsymbol{\lambda}$, and the first-order conditions are

$$\mathbf{g}(\mathbf{x}^*;\mathbf{p})=\mathbf{0}, \tag{1.27a}$$

$$u_{\mathbf{x}}(\mathbf{x}^*;\mathbf{p})+\mathbf{g}_{\mathbf{x}}'(\mathbf{x}^*;\mathbf{p})\cdot\boldsymbol{\lambda}^*=\mathbf{0}. \tag{1.27b}$$

Let the parameters change by $d\mathbf{p}$; the variables will change by $d\mathbf{x}$ and the objective function by

$$du(\mathbf{x}^*;\mathbf{p})=u_{\mathbf{x}'}(\mathbf{x}^*;\mathbf{p})\cdot d\mathbf{x}+u_{\mathbf{p}'}(\mathbf{x}^*;\mathbf{p})\cdot d\mathbf{p}. \tag{1.28}$$

For the constraint to still hold, the effects of changes in $\mathbf{p}$ and $\mathbf{x}$ must cancel out and we must have

$$\mathbf{g}_{\mathbf{x}'}(\mathbf{x}^*;\mathbf{p})\cdot d\mathbf{x}+\mathbf{g}_{\mathbf{p}'}(\mathbf{x}^*;\mathbf{p})\cdot d\mathbf{p}=\mathbf{0}\quad \text{(this is } m\times 1\text{)}. \tag{1.29}$$

Transposing (1.27b), we have

$$u_{\mathbf{x}'}(\mathbf{x}^*;\mathbf{p})=-\boldsymbol{\lambda}^{*\prime}\cdot\mathbf{g}_{\mathbf{x}'}(\mathbf{x}^*;\mathbf{p}),$$

and postmultiplying by $d\mathbf{x}$ we get

$$u_{\mathbf{x}'}(\mathbf{x}^*;\mathbf{p})\cdot d\mathbf{x}=-\boldsymbol{\lambda}^{*\prime}\cdot\mathbf{g}_{\mathbf{x}'}(\mathbf{x}^*;\mathbf{p})\cdot d\mathbf{x}=\boldsymbol{\lambda}^{*\prime}\cdot\mathbf{g}_{\mathbf{p}'}(\mathbf{x}^*;\mathbf{p})\cdot d\mathbf{p},\quad \text{by (1.29).}$$

Substituting the above result in (1.28) we obtain

$$du(\mathbf{x}^*;\mathbf{p})=[u_{\mathbf{p}'}(\mathbf{x}^*;\mathbf{p})+\boldsymbol{\lambda}^{*\prime}\cdot\mathbf{g}_{\mathbf{p}'}(\mathbf{x}^*;\mathbf{p})]\cdot d\mathbf{p},$$

$$du(\mathbf{x}^*;\mathbf{p})=\mathcal{L}_{\mathbf{p}'}(\boldsymbol{\lambda}^*,\mathbf{x}^*;\mathbf{p})\cdot d\mathbf{p},$$

or

$$du(\mathbf{x}^*;\mathbf{p})=\sum_{s=1}^{S}\frac{\partial\mathcal{L}(\boldsymbol{\lambda}^*,\mathbf{x}^*;\mathbf{p})}{\partial p_s}dp_s,$$

as claimed. □

This result establishes the importance of the method of Lagrange for equality-constrained problems, since this format enables us to gauge the sensitivity of the solution to exogenous parameter changes. It also applies, however, to unconstrained problems in which we simply have

$$\frac{du(\mathbf{x}^*;\mathbf{p})}{dp_s}=\frac{\partial u(\mathbf{x}^*;\mathbf{p})}{\partial p_s}. \tag{1.30}$$

This case makes it easier to see the common sense behind the envelope result since optimizing in an unconstrained problem required $u_{\mathbf{x}}(\mathbf{x}^*;\mathbf{p})=\mathbf{0}$; this term has been "optimized out" of the total differential (1.28). In the more complicated constrained case we need the multipliers to indicate how variables and constraints interact.

Maximum value functions: It is useful to think of the problem in Theorem 1.2.8 in the following way: We are to choose $\mathbf{x}$, given the vector of parameters $\mathbf{p}$; hence, our optimal choice depends on $\mathbf{p}$ and may formally be denoted by $\mathbf{x}^*(\mathbf{p})$. Clearly, in general, the optimal solution will not be unique and we cannot describe $\mathbf{x}^*(\mathbf{p})$ as a function. However, the maximum value of the objective function that we obtain must necessarily be unique, and we can define

$$V(\mathbf{p}) \equiv u(\mathbf{x}^*(\mathbf{p}); \mathbf{p}) \tag{1.31}$$

as a function of $\mathbf{p}$. This is the *maximum value function* for the problem of Theorem 1.2.8. *Minimum value functions* could be similarly defined. They are all sometimes called simply *value functions.* The envelope theorem can then be stated as $\partial V/\partial p_s = \partial \mathcal{L}/\partial p_s$. The study of maximum value functions has given rise to a branch of mathematical economics called *duality theory,* which is used extensively in microeconomic theory, international trade, and econometrics. A recent survey is that of Diewert (1982). It will sometimes be convenient in later chapters to use the concept of value functions. We shall see that it applies to a much broader class of problems than the one on which we defined it here.

We now illustrate the envelope theorem in a few instances.

Cost functions: Recall that in Section 1.2.3 we derived the cost function from the problem of minimizing $\mathbf{w}' \cdot \mathbf{x}$ subject to $q - f(\mathbf{x}) = 0$. The minimized objective function (a minimum value function) is $C(q, \mathbf{w}) = \mathbf{w}' \cdot \mathbf{x}^*(q, \mathbf{w})$, from which the derivatives with respect to q and $\mathbf{w}$ are not obvious. However, from the Lagrangean

$$\mathcal{L}(\lambda, \mathbf{x}; q, \mathbf{w}) = \mathbf{w}' \cdot \mathbf{x}^* + \lambda^*[q - f(\mathbf{x}^*)],$$

we see that

$$\frac{\partial C}{\partial q} = \frac{\partial \mathcal{L}}{\partial q} = \lambda^* \quad \text{and} \quad \frac{\partial C}{\partial w_s} = \frac{\partial \mathcal{L}}{\partial w_s} = x_s^*.$$

This establishes that λ^* is the marginal cost, while the first derivatives of C with respect to $\mathbf{w}$ are the input demand functions of the firm. The reader is invited to return to the numerical example of a cost function in Section 1.2.3 (Example 1.2.5) and verify the above results by differentiating the value function C with respect to q, w_1, and w_2.

Efficient allocation of resources: This is the problem of Theorem 1.2.7: find $(\mathbf{x}^1, \ldots, \mathbf{x}^n)$ to maximize $R(\mathbf{x}^1, \ldots, \mathbf{x}^n)$ subject to $\mathbf{x}^1 + \cdots + \mathbf{x}^n = \mathbf{X}$, where $\mathbf{x}^i$ and $\mathbf{X}$ are $(m \times 1)$ vectors. Forming the Lagrangean

$$\mathcal{L}(\lambda, \mathbf{x}^1, \ldots, \mathbf{x}^n; \mathbf{X}) = R(\mathbf{x}^1, \ldots, \mathbf{x}^n) + \lambda' \cdot [\mathbf{X} - \mathbf{x}^1 \ldots - \mathbf{x}^n],$$

we can apply the envelope theorem to obtain the result of Theorem 1.2.7 as a special case:

$$dR(\mathbf{x}^{1*}, \ldots, \mathbf{x}^{n*})/dX_j = \partial\mathcal{L}(\lambda^*, \mathbf{x}^{1*}, \ldots, \mathbf{x}^{n*}; \mathbf{X})/\partial X_j = \lambda_j^*,$$

or

$$\partial V(X_1, \ldots, X_m)/\partial X_j = \lambda_j^*,$$

where

$$V(X_1, \ldots, X_m) = R(\mathbf{x}^{1*}, \ldots, \mathbf{x}^{n*}).$$

Welfare economics. The problem of allocating resources to production activities and of determining the outputs of goods to various consumers in a way that is somehow desirable is a (some say "the") central problem in economics. We can use the method of Lagrange to formalize this problem and shed some light on the issues. First we must set down our notation carefully. There are I individuals, G goods, and R resources. The resources are used as inputs to produce the goods which are allocated to individuals so as to maximize welfare.

Resource r is available in fixed amount Y_r, $r = 1, \ldots, R$. $F_g(y_{1g}, \ldots, y_{rg}, \ldots, y_{Rg})$ is the production function of good g, $g = 1, \ldots, G$; the arguments are resource inputs. $U^i(x_1^i, \ldots, x_g^i, \ldots, x_G^i)$ is the utility function of individual i, $i = 1, \ldots, I$; the arguments are amounts of goods consumed. Total consumption of each good is required to equal the output of it: $\sum_{i=1}^I x_g^i = F_g$, $g = 1, \ldots, G$. The welfare function to be maximized has as arguments the utility levels of individuals: $W(U^1, \ldots, U^i, \ldots, U^I)$. We now set up the problem. In order to lighten the cumbersome notation, we write only the "general" argument of any function; for instance, the production function of good g becomes $F_g(\ldots y_{rg} \ldots)$, and the utility function of individual i becomes $U^i(\ldots x_g^i \ldots)$. In this fashion the subscripts r and g and the superscript i easily identify resources, goods, and individuals, respectively.

Find $(\ldots x_g^i \ldots)$ and $(\ldots y_{rg} \ldots)$ that maximize

$$W(\ldots U^i(\ldots x_g^i \ldots) \ldots)$$

subject to (1.32)

$$\sum_{i=1}^{I} x_g^i = F_g(\ldots y_{rg} \ldots), \quad g = 1, \ldots, G, \quad \text{and}$$

$$\sum_{g=1}^{G} y_{rg} = Y_r, \quad r = 1, \ldots, R.$$

The last constraint indicates that the total amount of each resource used in all industries equals the amount available. We shall use π_g as multipliers for the first set of constraints and λ_r for the second set. The Lagrangean is

$$\mathcal{L}(\boldsymbol{\pi}, \boldsymbol{\lambda}, \mathbf{x}, \mathbf{y}) = W(\cdots U^i(\ldots x_g^i \ldots) \ldots)$$
$$+ \sum_{g=1}^{G} \pi_g \left[F_g(\ldots y_{rg} \ldots) - \sum_{i=1}^{I} x_g^i \right] + \sum_{r=1}^{R} \lambda_r \left[Y_r - \sum_{g=1}^{G} y_{rg} \right].$$

Assuming that all functions are increasing and concave and all variables positively valued guarantees the optimality of the first-order conditions. We now derive these conditions for the four types of variables in the problem: π_g, λ_r, x_g^i, and y_{rg}:

$$F_g(\ldots y_{rg} \ldots) - \sum_{i=1}^{I} x_g^i = 0, \quad g = 1, \ldots, G, \tag{1.33}$$

$$Y_r - \sum_{g=1}^{G} y_{rg} = 0, \quad r = 1, \ldots, R, \tag{1.34}$$

$$W_i \times U_g^i - \pi_g = 0, \quad i = 1, \ldots, I, \; g = 1, \ldots, G, \tag{1.35}$$

$$\pi_g \times F_{rg} - \lambda_r = 0, \quad g = 1, \ldots, G, \; r = 1, \ldots, R, \tag{1.36}$$

where $W_i = \partial W / \partial U_i$, $U_g^i = \partial U^i / \partial x_g^i$, and $F_{rg} = \partial F_g / \partial y_{rg}$. There are $G + R + I \times G + G \times R$ conditions altogether. The multipliers λ_r and π_g are the shadow prices of resource r and good g, respectively. The meanings of (1.33) and (1.34) are clear. To interpret (1.35) we first write it twice: for the same individual i and two different goods g and g':

$$W_i U_g^i = \pi_g \quad \text{and} \quad W_i U_{g'}^i = \pi_{g'}.$$

Taking the ratio, we get

$$U_g^i / U_{g'}^i = \pi_g / \pi_{g'}.$$

The ratio of the marginal utilities of any two goods is equal to the ratio of their shadow prices. We now write (1.35) for the same good but two different individuals,

$$W_i U_g^i = \pi_g \quad \text{and} \quad W_{i'} U_g^{i'} = \pi_g,$$

which yields

$$W_i U_g^i = W_i' U_g^{i'}.$$

For any good, the marginal contribution of a unit of good g to welfare achieved through consumption by one individual is the same as that achieved through consumption by another individual. In other words, consumption is adjusted so that the marginal utilities of individuals are exactly balanced with their welfare weights.

We now turn to (1.36); again, writing it for the same good and two resources, we get

$$\pi_g F_{rg} = \lambda_r \quad \text{and} \quad \pi_g F_{r'g} = \lambda_{r'}.$$

The ratio is $F_{rg}/F_{r'g} = \lambda_r/\lambda_{r'}$; hence, the ratio of the marginal physical products of resources r and r' in any industry g is equal to the ratio of their shadow prices. Thus, the shadow prices play the same role here as do input prices for a profit-maximizing or cost-minimizing firm. Writing (1.36) for one resource but two different goods, we have

$$\pi_{g'} F_{rg'} = \lambda_r \quad \text{and} \quad \pi_g F_{rg} = \lambda_r,$$

which yields $\pi_{g'} F_{rg'} = \pi_g F_{rg}$. The marginal value product of any resource (using shadow prices) is the same in all industries. If it were smaller in one industry than in another, it would be beneficial to shift some of that resource from the latter to the former.

The above welfare maximum is also known as a Pareto optimum, and it is a classical result of welfare economics that, under some restrictions, it can be supported by a competitive equilibrium. Though we will not go through the proof of such a result, we will nonetheless illustrate it in order to show that the Lagrangean format itself suggests such a relationship. The notation we shall use for the prices of goods and factors is that of the shadow prices identified earlier, π_g and λ_r for all g and r. For simplicity we also assume that production functions exhibit constant returns to scale, so that all firms earn zero profit, the G industries act like so many competitive firms and maximize profit, and the I individuals each own a share of all factors of production, which they sell to obtain the income to be spent on goods so as to maximize their utility. We then have for each firm g, $g = 1, \ldots, G$: Choose $\{y_{rg}\}$ to maximize

$$\pi_g \cdot F_g(\ldots y_{rg} \ldots) - \sum_{r=1}^{R} \lambda_r y_{rg}.$$

This yields

$$\pi_g \cdot F_{rg} - \lambda_r = 0, \quad r = 1, \ldots, R, \;\; g = 1, \ldots, G. \tag{1.37}$$

The marginal value product of input r equals its price. Because firms earn zero profit consumers' incomes are just the value of their endowments of resources. Hence, for consumers we need only define their share of the resources (and not their ownership shares of industries). Suppose consumer i owns y_r^i of resource r, with (for feasibility)

$$\sum_{i=1}^{I} y_r^i = Y_r, \quad r = 1, \ldots, R. \tag{1.38}$$

Then the income of the representative consumer i is

$$\sum_{r=1}^{R} \lambda_r \cdot y_r^i \equiv \lambda' \cdot \mathbf{y}^i$$

and i's problem is to choose $\{x_g^i\}$ to maximize

$$U^i(\dots x_g^i \dots)$$

subject to

$$\sum_{g=1}^{G} \pi_g x_g^i = \lambda' \cdot \mathbf{y}^i.$$

The Lagrangean for this problem is

$$\mathcal{L}(\nu^i, x^i) = U^i(\dots x_g^i \dots) + \nu^i \left[\lambda' \cdot \mathbf{y}^i - \sum_{g=1}^{G} \pi_g x_g^i \right].$$

We require for each $i = 1, \dots, I$

$$\lambda' \cdot \mathbf{y}^i - \sum_{g=1}^{G} \pi_g x_g^i = 0, \tag{1.39}$$

$$U_g^i - \nu^i \pi_g = 0, \quad g = 1, \dots, G. \tag{1.40}$$

Thus, given preferences and technologies, once a distribution of resources $\{y_n^i\}$ has been selected, equations (1.37) to (1.40) characterize demand and supply for goods and resources. A competitive equilibrium $\{x_g^{i*}, y_{rg}^*\}$ is said to exist at prices $\{\pi_g^*, \lambda_r^*\}$ if these equations are consistent, that is, if demand equals supply for both goods and resources without anyone trying to achieve these equalities. Specifically $\{y_{rg}^*\}$ from (1.37) are required to satisfy

$$\sum_{g=1}^{G} y_{rg}^* = Y_r, \quad r = 1, \dots, R, \tag{1.41}$$

and $\{F_g^*\}$ from (1.37) and $\{x_g^{i*}\}$ from (1.39)–(1.40) must satisfy

$$\sum_{i=1}^{I} x_g^{i*} = F_g^*, \quad g = 1, \dots, G, \tag{1.42}$$

where

$$F_g^* = F_g(\dots y_{rg}^* \dots).$$

Clearly, the existence of such an equilibrium is a substantial question. It occupied economists for many decades in one form or other. The similarity between this problem and the previous welfare optimum gives us a clue. Consider the welfare optimum solution $\{x_g^i, y_{rg}, \pi_g, \lambda_r\}$ defined by (1.33)–(1.36) and try it as a competitive equilibrium. We can see that

(1.42), (1.41), and (1.37) are identical with (1.33), (1.34), and (1.36), respectively. There remain (1.39) and (1.40). Suppose that the distribution of resources $\{y_r^i\}$ is such that (1.39) is satisfied for $\{x_g^i, \pi_g, \lambda_r\}$ of the welfare optimum, so that each consumer can exactly afford the bundle allocated to her under the welfare optimum. Since this allocation satisfies (1.35), it will also satisfy (1.40) with $\nu^i = W_i^{-1}$. Therefore, the particular welfare optimum (dependent on the welfare function), which we have assumed exists, provides us with a particular competitive equilibrium (dependent on the distribution of income), while the method of Lagrange provides a means to calculate it, at least in principle. This last application reinforces our earlier statement that this method of solving equality-constrained optimization problems provides insight into the structure of the problem.

The next section, although concerned with both constrained and unconstrained problems, will provide more instances of the usefulness of our approach.

1.3 Comparative statics

In this section we attempt to answer some questions often asked of economists: given some system in equilibrium, how will the variables respond to an exogenous change in some parameter? The question we shall answer is actually slightly different from this. We shall try to ascertain how the variable would have differed from what it is in this equilibrium had the parameter been slightly different. In other words, we do not indicate how the system responds to a change, since this would entail some dynamic movement from one state to another, but we compare two hypothetical static equilibria; this is why the method is called "comparative statics." We will often seem to forget this distinction, however, and speak loosely of response to change, for brevity.

It will be seen that much reliance is placed on second-order conditions. Therefore, the reactions of equilibrium values to outside changes depend crucially on which optimization problem the equilibrium was the outcome of. It is important to understand that this provides us with a framework in which to test the particular optimization hypothesis indirectly.

There are no general results – just a general method of doing comparative statics exercises. Before we embark on a general exposition, let us look at a simple case in detail. Recall the derivation of cost functions in Section 1.2.3; it is reproduced here in the case of three inputs. Minimize $w_1x_1+w_2x_2+w_3x_3$ subject to $q-f(x_1,x_2,x_3)=0$,

$$\mathcal{L}(\lambda,x)=w_1x_1+w_2x_2+w_3x_3+\lambda[q-f(x_1,x_2,x_3)].$$

The first-order conditions are

$$\begin{aligned} q-f(x_1,x_2,x_3)&=0,\\ w_1-\lambda f_1(x_1,x_2,x_3)&=0,\\ w_2-\lambda f_2(x_1,x_2,x_3)&=0,\\ w_3-\lambda f_3(x_1,x_2,x_3)&=0. \end{aligned} \tag{1.43}$$

We have also seen that the optimal values of $\mathbf{x}$, when seen as functions of q and $\mathbf{w}$, are the demand functions for inputs by a firm producing q units of output. We now wish to examine how the demand for inputs reacts to price changes. Specifically we want to know the sign of dx_3/dw_3, say. In order to do this, we proceed much as we did when investigating the economic significance of multipliers. In reaction to a change in w_3, all variables change. We are dealing with minute changes, so that linear approximations using total differentials evaluated at the original solution are appropriate. For any first-order equation, say $E(\mathbf{v},p)=0$, involving some variables $\mathbf{v}$ and some changing parameter p, we have after the change $E(\mathbf{v},p)+dE=0$, or $E(\mathbf{v},p)+E_{\mathbf{v}}\cdot d\mathbf{v}+E_p\cdot dp=0$, but since $E(\mathbf{v},p)=0$ the procedure simply amounts to setting the total differential of the equation to zero. We now proceed to do this for each equation in (1.43), keeping in mind that all variables (λ,x_1,x_2,x_3) and the parameter w_3 are changing; the other parameters remain fixed and have a zero differential. All derivatives are evaluated at the original equilibrium, and we skip all arguments for brevity:

$$\begin{aligned} -f_1\cdot dx_1-f_2\cdot dx_2-f_3\cdot dx_3&=0,\\ -f_1\cdot d\lambda-\lambda f_{11}\cdot dx_1-\lambda f_{12}\cdot dx_2-\lambda f_{13}\cdot dx_3&=0,\\ -f_2\cdot d\lambda-\lambda f_{21}\cdot dx_1-\lambda f_{22}\cdot dx_2-\lambda f_{23}\cdot dx_3&=0,\\ -f_3\cdot d\lambda-\lambda f_{31}\cdot dx_1-\lambda f_{32}\cdot dx_2-\lambda f_{33}\cdot dx_3&=-dw_3. \end{aligned}$$

We can divide both sides by dw_3, let this differential tend to zero to obtain derivatives, and rewrite the system in matrix form to have

$$\begin{bmatrix} 0 & -f_1 & -f_2 & -f_3 \\ -f_1 & -\lambda f_{11} & -\lambda f_{12} & -\lambda f_{13} \\ -f_2 & -\lambda f_{21} & -\lambda f_{22} & -\lambda f_{23} \\ -f_3 & -\lambda f_{31} & -\lambda f_{32} & -\lambda f_{33} \end{bmatrix} \begin{bmatrix} d\lambda/dw_3 \\ dx_1/dw_3 \\ dx_2/dw_3 \\ dx_3/dw_3 \end{bmatrix} = \begin{bmatrix} 0 \\ 0 \\ 0 \\ -1 \end{bmatrix}. \tag{1.44}$$

Equation (1.44) is the comparative statics equation for our problem. The alert reader will have recognized the matrix of coefficients as the Hessian matrix of the Lagrangean, **B** – hence the matrix on which second-order conditions are based. The most convenient method of solution for dx_3/dw_3 is Cramer's rule; it yields

$$\frac{dx_3}{dw_3}=\frac{\begin{vmatrix} 0 & -f_1 & -f_2 & 0 \\ -f_1 & -\lambda f_{11} & -\lambda f_{12} & 0 \\ -f_2 & -\lambda f_{21} & -\lambda f_{22} & 0 \\ -f_3 & -\lambda f_{31} & -\lambda f_{32} & -1 \end{vmatrix}}{\begin{vmatrix} 0 & -f_1 & -f_2 & -f_3 \\ -f_1 & -\lambda f_{11} & -\lambda f_{12} & -\lambda f_{13} \\ -f_2 & -\lambda f_{21} & -\lambda f_{22} & -\lambda f_{23} \\ -f_3 & -\lambda f_{31} & -\lambda f_{32} & -\lambda f_{33} \end{vmatrix}}=-\frac{\begin{vmatrix} 0 & -f_1 & -f_2 \\ -f_1 & -\lambda f_{11} & -\lambda f_{12} \\ -f_2 & -\lambda f_{21} & -\lambda f_{22} \end{vmatrix}}{\begin{vmatrix} \text{(as before)} \end{vmatrix}}. \tag{1.45}$$

The reader will be pleased to know that we have no intention of expanding these determinants. Instead, we rely on the following crucial argument: The input demand function $x_3(q, w)$ was not an arbitrary equation. According to our model it was derived by a firm in the process of minimizing the total cost of producing output q. Therefore, the necessary conditions for a constrained minimum apply. Furthermore, we will assume that we have a regular minimum in the sense that the Hessian of $\mathcal{L}$ is nonsingular (without this assumption we could not solve the system (1.44)). Then according to Theorem 1.2.5 we can use the conditions of Theorem 1.2.4(ii) as necessary for a minimum: the leading principal minors of **B** are of the same sign (it does not matter which sign it is for our purpose here). Inspection of (1.45) reveals that the numerator is the third such minor, while the denominator is, of course, the fourth – hence, $dx_3/dw_3<0$. Therefore, the third input demand function is a decreasing function of its own price precisely because the demand function was derived through cost minimization. Had it been derived in some other way we would not have been able to use those second-order conditions, and we might not have been able to establish the result. In particular, in the rather incredible case in which a firm would have maximized cost, the sign of dx_3/dw_3 would be positive. Finally, note that we made no assumptions on f and that our comparative statics results were derived from necessary conditions (under the regularity assumption) that follow from the existence of a solution to the problem. We must not make too much of this last remark, however, because some conditions are implicitly met by assuming the existence of a sensible solution.

We now describe the method on a general problem like that of (1.16) reproduced here using vector notation: Find $\mathbf{x}(n\times 1)$ to maximize $f(\mathbf{x};\mathbf{p})$ subject to $\mathbf{g}(\mathbf{x};\mathbf{p})=\mathbf{0}$; $\mathbf{g}$ is $(m\times 1)$ and $\mathbf{p}$ is a $(r\times 1)$ parameter vector. The Lagrangean is $\mathcal{L}(\lambda,\mathbf{x};\mathbf{p})=f(\mathbf{x};\mathbf{p})+\lambda'\cdot\mathbf{g}(\mathbf{x};\mathbf{p})$; the first-order conditions are

$$\mathbf{g}(\mathbf{x};\mathbf{p})=\mathbf{0} \quad \text{and} \quad f_{\mathbf{x}}(\mathbf{x};\mathbf{p})+\mathbf{g}'_{\mathbf{x}}(\mathbf{x};\mathbf{p})\cdot\boldsymbol{\lambda}=\mathbf{0}. \tag{1.46}$$

If we assume the rank condition (as in (1.18)) and regularity (as in Theorem 1.2.5), the second-order necessary conditions are given in Theorem 1.2.4 and apply to the Hessian matrix of the Lagrangean, which is reproduced here with the notation of (1.23):

$$\mathbf{B}=\begin{bmatrix}\mathbf{0} & \mathbf{G}\\ \mathbf{G}' & \mathbf{L}\end{bmatrix} \text{ is } (n+m)\times(n+m).$$

When taking the total differential of (1.46) with respect to $\boldsymbol{\lambda}$, $\mathbf{x}$, and $\mathbf{p}$, we always get

$$[\mathbf{B}]\begin{bmatrix}d\boldsymbol{\lambda}\\ d\mathbf{x}\end{bmatrix}+\begin{bmatrix}\mathcal{L}_{\boldsymbol{\lambda}\mathbf{p}'}\\ \mathcal{L}_{\mathbf{x}\mathbf{p}'}\end{bmatrix}d\mathbf{p}=\mathbf{0}, \tag{1.47}$$

or

$$\begin{bmatrix}\mathbf{0} & \mathbf{G}\\ \mathbf{G}' & \mathbf{L}\end{bmatrix}\begin{bmatrix}d\boldsymbol{\lambda}\\ d\mathbf{x}\end{bmatrix}+\begin{bmatrix}\mathbf{g}_{\mathbf{p}'}\\ f_{\mathbf{x}\mathbf{p}'}+\sum_{j=1}^{m}\lambda_j g^j_{\mathbf{x}\mathbf{p}'}\end{bmatrix}[d\mathbf{p}]=0. \tag{1.47'}$$

Normally we simply write down the first part of (1.47) and derive the second part directly from the first-order conditions by differentiating them with respect to the parameters $\mathbf{p}$. Equation (1.47) is the basic equation of comparative statics. Because we have assumed nonsingularity of $\mathbf{B}$, there is always a unique solution to the problem. Sometimes the equation is solved at one go, but more often we are interested in isolating the influence of one parameter on selected variables. Then, as in our introductory example, Cramer's rule is the best method of solution.

We must warn the reader that although the equation can be solved, it does not necessarily imply that we can answer our original question. In many instances the sign of some dx_i/dp_j will be indeterminate. At times this indeterminacy can be resolved at the cost of some additional assumptions; in other cases it is not possible to find economically sensible assumptions that will resolve it. We now illustrate the technique with a few examples.

Convex cost functions. Let us return to the problem of deriving a cost function. Minimize $\mathbf{w}'\cdot\mathbf{x}$ subject to $q-f(\mathbf{x})=0$. We have $\mathcal{L}(\lambda,\mathbf{x};\mathbf{w},q)=\mathbf{w}'\cdot\mathbf{x}+\lambda[q-f(\mathbf{x})]$ and the first-order conditions

$$\begin{aligned} q-f(\mathbf{x})&=0,\\ \mathbf{w}-\lambda f_{\mathbf{x}}(\mathbf{x})&=\mathbf{0}. \end{aligned} \tag{1.48}$$

We want to investigate the effect of an increase in q, the target output, on the value of λ, previously identified as the marginal cost of output. Using the above method we have

$$\begin{bmatrix} 0 & -f_{\mathbf{x}'} \\ -f_{\mathbf{x}} & -\lambda f_{\mathbf{xx}'} \end{bmatrix} \begin{bmatrix} d\lambda \\ d\mathbf{x} \end{bmatrix} = \begin{bmatrix} -1 \\ \mathbf{0} \end{bmatrix} dq, \tag{1.49}$$

where $\mathbf{0}$ on the right-hand side is $n \times 1$.

The matrix of coefficients is the $(n+1)\times(n+1)$ Hessian of the Lagrangean, and the right-hand-side vector is the negative of the partial derivative of (1.48) with respect to q. Solving by Cramer's rule, we have

$$\Delta \frac{d\lambda}{dq} = \begin{vmatrix} -1 & -f_{\mathbf{x}'} \\ \mathbf{0} & -\lambda f_{\mathbf{xx}'} \end{vmatrix}, \tag{1.50}$$

where Δ is the determinant of the matrix in (1.49). From Theorems 1.2.4 and 1.2.5 Δ has the sign of $(-1)^1 < 0$. However, we do not in general know the sign of the other determinant in (1.50). We need further assumptions. Assume that $f(\mathbf{x})$ is both increasing and concave; then from (1.48), $\lambda > 0$, $f_{\mathbf{xx}'}$ is negative-semidefinite and $-\lambda f_{\mathbf{xx}'}$ is positive-semidefinite. This implies

$$\frac{d\lambda}{dq} = \Delta^{-1}(-1)|-\lambda f_{\mathbf{xx}'}| \geq 0.$$

Therefore, an increasing concave function gives rise to the usual nondecreasing marginal cost (recall that λ is identified as the marginal cost by the envelope theorem); this is the same as $d^2C/dq^2 \geq 0$, and we have a cost function that is convex in the output q.

Compensated price changes. Consider a consumer maximizing utility under a budget constraint, with all prices and income fixed. The resulting consumption bundles treated as functions of prices and income are the demand functions. These are often called *Marshallian demand functions,* to be distinguished from *Hicksian demand functions* (which are derived from minimizing the cost of achieving a given level of utility, much as cost functions were derived subject to achieving a given level of output). Suppose now that the price of the last good, p_n, rises. This will alter the relative prices and lead to some substitution among goods, but it will also induce a drop in the consumer's real income, since the consumer will no longer be able to afford the previous bundle of goods. Suppose that we want to isolate the substitution effect of the relative price changes. To achieve this we compensate the consumer for the drop in income: we add $dy = x_n\, dp_n$ to the consumer's income, where x_n is the quantity of good n previously purchased. This will, of course, modify our comparative statics equation. We seek to maximize $U(\mathbf{x})$ subject to $y - \mathbf{p}'\cdot\mathbf{x} = 0$; we have $\mathcal{L}(\lambda, \mathbf{x}) = U(\mathbf{x}) + \lambda[y - \mathbf{p}'\cdot\mathbf{x}]$. The necessary conditions are

$$y - \mathbf{p}'\cdot\mathbf{x} = 0,$$

$$U_{\mathbf{x}}(\mathbf{x}) - \lambda\mathbf{p} = \mathbf{0}.$$

The compensated price change will affect the budget equation for which the total differential would normally be $dy - \mathbf{p}' \cdot d\mathbf{x} - x_n \cdot dp_n = 0$, but here reduces to $-\mathbf{p}' \cdot d\mathbf{x} = 0$. Thus, the comparative statics equation is

$$\begin{bmatrix} 0 & \vdots & -\mathbf{p}' \\ \cdots & \cdot & \cdots \\ -\mathbf{p} & \vdots & U_{\mathbf{xx}'} \end{bmatrix} \begin{bmatrix} d\lambda \\ \cdots \\ d\mathbf{x} \end{bmatrix} = \begin{bmatrix} 0 \\ \cdots \\ 0 \\ \vdots \\ 0 \\ \lambda dp_n \end{bmatrix}. \tag{1.51}$$

Solving (1.51) for dx_n/dp_n by Cramer's rule involves substituting the right-hand-side column for the last column of the matrix and taking the ratio of that determinant over the determinant of the matrix of coefficients. This yields $dx_n/dp_n = \lambda \times$(the ratio of the last two leading principal minors of the Hessian of $\mathcal{L}$). If we assume U is increasing in $\mathbf{x}$, λ will be positive and the second-order conditions (Theorem 1.2.5) imply that the above ratio is negative; thus, $dx_n/dp_n < 0$. This is one of the few results of demand theory: a compensated increase in the price of a good decreases the demand for that good.

Although we have used the method of comparative statics in equality-constrained problems only, it can be applied equally well to unconstrained problems. The method remains the same – we totally differentiate the first-order conditions. To display the versatility of this technique we illustrate it with a less conventional example.

Education and trade. Consider a small country initially populated by uneducated workers with a low level of productivity. It can bring in foreign workers who are educated and have a high level of productivity, but this is costly in terms of foreign exchange. It can also use foreign workers to train local ones, thereby transforming the latter into high-productivity workers. For simplicity we assume that there are only two export industries in the country; all other activities are taken as fixed and suppressed. The two export goods are traded at fixed world prices. The country wants to maximize net foreign exchange earnings while maintaining full employment of educated and uneducated workers alike. Before we can characterize the optimal policy we must introduce some notation:

$\bar{l}$ the number of local, initially uneducated workers (fixed),
l the number of local workers that will be trained,
$T(l)$ the number of educated workers it takes to train l uneducated ones,

L	the total number of immigrant (educated) workers,
w	the foreign exchange cost of an immigrant worker,
l_1, l_2	the number of uneducated workers in industries 1 and 2, respectively,
L_1, L_2	the number of educated workers in industries 1 and 2, respectively.

The prices and the production functions are, in the obvious notation, p_1, $f_1(l_1)$, $F_1(L_1)$, p_2, $f_2(l_2)$, and $F_2(L_2)$.

Prices are expressed in terms of the foreign currency. We assume that all production functions are increasing and concave while $T(l)$ is increasing and convex. Note that there is an implicit assumption that educated and uneducated workers work separately as the output of good i is given by $f_i(l_i)+F_i(L_i)$. The problem is thus to maximize

$$E=p_1[f_1(l_1)+F_1(L_1)]+p_2[f_2(l_2)+F_2(L_2)]-wL,$$

where

$$l=\bar{l}-l_1-l_2 \tag{1.52a}$$

and

$$L+l=L_1+L_2+T(l) \tag{1.52b}$$

are the full-employment conditions for uneducated and educated workers, respectively. Our model is necessarily static and as such has the usual shortcomings. Thus, the l workers being trained can simultaneously begin work as educated workers, whereas the $T(l)$ teachers cannot take part in the production of goods. Substituting (1.52a) and (1.52b) we obtain the following: choose L_1, L_2, l_1, and l_2 to maximize

$$\begin{aligned} E=&\,p_1[f_1(l_1)+F_1(L_1)]+p_2[f_2(l_2)+F_2(L_2)] \\ &-w[L_1+L_2+l_1+l_2-\bar{l}]-wT(\bar{l}-l_1-l_2). \end{aligned}$$

It is easy to verify that this is a concave function of all variables and that the following conditions are optimal:

$$\begin{aligned} p_1F_1'-w&=0, \\ p_2F_2'-w&=0, \\ p_1f_1'-w+wT'&=0, \\ p_2f_2'-w+wT'&=0. \end{aligned} \tag{1.53}$$

Assuming that a solution to (1.53) exists, the last two equations imply $1-T'>0$, which means that it takes less than one full-time educated worker to train an additional local worker; the last condition is thus seen to equate the marginal value product of an uneducated worker in industry 2 with the net benefit of training that worker (the saving of not taking another foreign worker, net of teaching cost). We want to examine the effects of increases in the domestic labor supply $\bar{l}$ and the cost of foreign workers w on all variables, including L and l. We begin with $\bar{l}$. Taking the total differential of (1.53) yields the basic equation

$$\begin{bmatrix} p_1F_1'' & 0 & 0 & 0 \\ 0 & p_2F_2'' & 0 & 0 \\ 0 & 0 & p_1f_1''-wT'' & -wT'' \\ 0 & 0 & -wT'' & p_2f_2''-wT'' \end{bmatrix} \begin{bmatrix} \dfrac{dL_1}{d\bar{l}} \\ \dfrac{dL_2}{d\bar{l}} \\ \dfrac{dl_1}{d\bar{l}} \\ \dfrac{dl_2}{d\bar{l}} \end{bmatrix} = \begin{bmatrix} 0 \\ 0 \\ -wT'' \\ -wT'' \end{bmatrix}. \tag{1.54}$$

We shall depart from our usual method of solution by Cramer's rule, for three reasons: first we want to solve for all variables, second the matrix of coefficients is block diagonal and thus easy to invert, and third we can use the inverse again when investigating the effects of changes in w. The inverse of the matrix in (1.54) is now used to define the solution

$$\begin{bmatrix} \dfrac{dL_1}{d\bar{l}} \\ \dfrac{dL_2}{d\bar{l}} \\ \dfrac{dl_1}{d\bar{l}} \\ \dfrac{dl_2}{d\bar{l}} \end{bmatrix} = \begin{bmatrix} \dfrac{1}{p_1F_1''} & 0 & 0 & 0 \\ 0 & \dfrac{1}{p_2F_2''} & 0 & 0 \\ 0 & 0 & \dfrac{p_2f_2''-wT''}{\Delta} & \dfrac{wT''}{\Delta} \\ 0 & 0 & \dfrac{wT''}{\Delta} & \dfrac{p_1f_1''-wT''}{\Delta} \end{bmatrix} \times \begin{bmatrix} 0 \\ 0 \\ -wT'' \\ -wT'' \end{bmatrix}, \tag{1.55}$$

where

$$\Delta = p_1f_1''p_2f_2'' - wT''(p_1f_1''+p_2f_2'') > 0.$$

This immediately yields

$$\frac{dL_1}{d\bar{l}} = \frac{dL_2}{d\bar{l}} = 0, \quad \frac{dl_1}{d\bar{l}} = \frac{-wT''(p_2f_2'')}{\Delta} > 0, \text{ and } \frac{dl_2}{d\bar{l}} = \frac{-wT''(p_1f_1'')}{\Delta} > 0. \tag{1.56}$$

Furthermore,

$$\frac{dl}{d\bar{l}}=1-\frac{dl_1}{d\bar{l}}-\frac{dl_2}{d\bar{l}}$$

from (1.52a) and

$$\frac{dl}{d\bar{l}}=\Delta^{-1}[p_1f_1''p_2f_2'']>0. \tag{1.57}$$

Finally, from (1.52b)

$$\begin{aligned}\frac{dL}{d\bar{l}}&=(-1+T')\left(\frac{dl}{d\bar{l}}\right)\\&=(-1+T')\Delta^{-1}[p_1f_1''p_2f_2'']<0.\end{aligned} \tag{1.58}$$

Equations (1.56)–(1.58) indicate that an increase in the domestic (uneducated) work force will not affect the number of educated workers in industry; it will increase the number of uneducated workers in each industry and the number of local trainees, and this last effect will reduce the demand for foreigners to be used as teachers and hence will reduce immigration.

We now turn to the effects of an increase in the cost of educated immigrants. We directly state the solution as in (1.55):

$$\begin{bmatrix}\dfrac{dL_1}{dw}\\[2ex]\dfrac{dL_2}{dw}\\[2ex]\dfrac{dl_1}{dw}\\[2ex]\dfrac{dl_2}{dw}\end{bmatrix}=\begin{bmatrix}\dfrac{1}{p_1F_1''}&0&0&0\\[2ex]0&\dfrac{1}{p_2F_2''}&0&0\\[2ex]0&0&\dfrac{p_2f_2''-wT''}{\Delta}&\dfrac{wT''}{\Delta}\\[2ex]0&0&\dfrac{wT''}{\Delta}&\dfrac{p_1f_1''-wT''}{\Delta}\end{bmatrix}\times\begin{bmatrix}1\\[2ex]1\\[2ex]1-T'\\[2ex]1-T'\end{bmatrix},$$

with Δ as in (1.55). We have

$$\frac{dL_1}{dw}=\frac{1}{(p_1F_1'')}<0,\qquad\frac{dL_2}{dw}=\frac{1}{(p_2F_2'')}<0,$$

$$\frac{dl_1}{dw}=\frac{(1-T')p_2f_2''}{\Delta}<0,\qquad\frac{dl_2}{dw}=\frac{(1-T')p_1f_1''}{\Delta}<0.$$

Therefore,

$$\frac{dl}{dw}>0\quad\text{and}\quad\frac{dL}{dw}=\frac{dL_1}{dw}+\frac{dL_2}{dw}+(-1+T')\frac{dl}{dw}<0.$$

We discover than an increase in the cost of educated immigrants decreases their number; this decrease is spread among industries but is compensated by an increase in the number of local workers trained. This in turn will decrease the number of local uneducated workers in both industries.

1.4 Optimization under inequality constraints: nonlinear programming

It is easy to think of instances in which equality constraints are too limiting: why insist that every last bit of a resource be used up when it actually has a negative productivity? To overcome this objection we turn our attention to inequality-constrained problems; this has the advantage of allowing us formally to take into account the nonnegativity of most economic variables. The general format of the problem is to choose $\mathbf{x}$ to

$$\begin{aligned}&\text{Maximize } f(\mathbf{x})\\&\text{subject to } \mathbf{g}(\mathbf{x}) \geqq \mathbf{0}, \ \mathbf{x} \geqq \mathbf{0},\end{aligned} \tag{1.59}$$

where $\mathbf{x}$ is $n \times 1$ and $\mathbf{g}$ is $m \times 1$. Several preliminary remarks are in order:

(i) This is called a nonlinear programming problem; the theory was developed after that of linear programming (where f and $\mathbf{g}$ are linear) – hence the strange name.

(ii) The economic meaning of an inequality constraint on the amount of a resource available is that this particular resource is freely disposable; that is, there is no penalty for discarding it. This is not always reasonable.

(iii) We no longer have any restrictions placed on the number of constraints relative to the number of variables (both cases $m \geq n$ and $m \leq n$ are admitted). At the optimum some constraints will hold as equalities; they are said to be *binding* constraints. Other constraints will hold as strict inequalities; they are said to be nonbinding or *slack*. The set defined by $\mathbf{g}(\mathbf{x}) \geqq \mathbf{0}$, $\mathbf{x} \geqq \mathbf{0}$, is called the *feasible set* and a point in it is called a *feasible point*.

(iv) It is essential, when formulating a problem, that the direction of the inequalities be carefully thought out. In many cases it will be some version of "demand cannot exceed supply."

Finally, note that we have not ruled out the usefulness of equality constraints; for example, accounts must balance exactly.

We shall see that the method of solution is a modification of the Lagrangean technique, but the necessary first-order conditions are now a mixture of equalities and inequalities, known as the Kuhn–Tucker conditions from the names of the original authors. Before proceeding to an

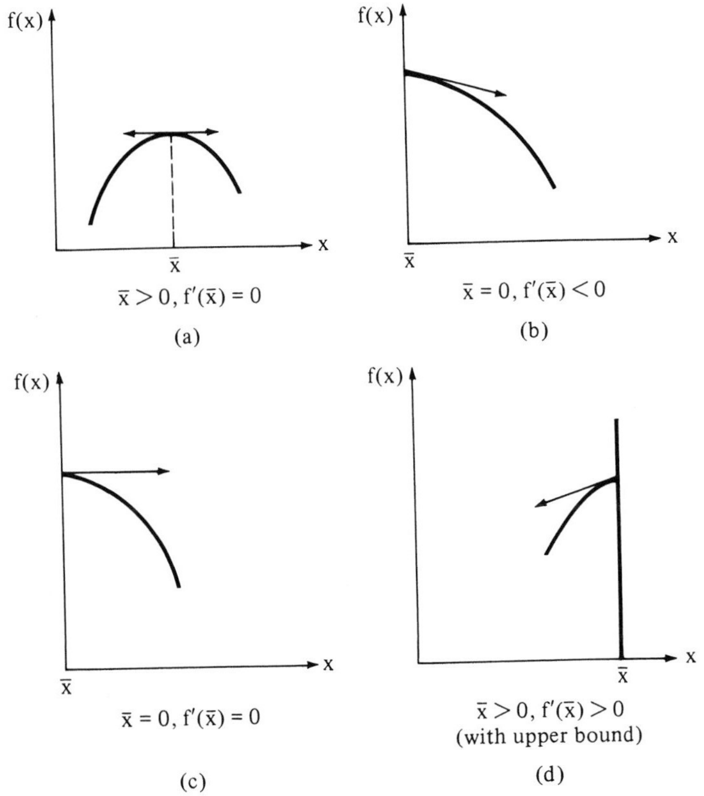

Figure 1.7

exposition of the theory, we illustrate the central idea of maximization within bounds.

Let $f(x)$ be a function of one variable to be maximized subject to $x \geq 0$. Whereas an outcome such as depicted in Figure 1.7a is possible, so is that of Figure 1.7b and even that of Figure 1.7c. The following condition covers all possible cases:

$$f'(\bar{x}) \leq 0 \text{ and } \bar{x} \geq 0; \text{ if } \bar{x} < 0 \text{ then } f'(\bar{x}) = 0, \text{ and if } f'(\bar{x}) < 0 \text{ then } \bar{x} = 0.$$

(Note that $f'(\bar{x}) = 0$ with $\bar{x} = 0$ is admitted.) This can be expressed more succinctly by

$$f'(\bar{x}) \leq 0, \qquad \bar{x} \geq 0, \qquad \bar{x}[f'(\bar{x})] = 0. \tag{1.60}$$

Note that the possibility of a negative rate of change in f at the optimum is associated with the existence of the lower bound of zero; if there

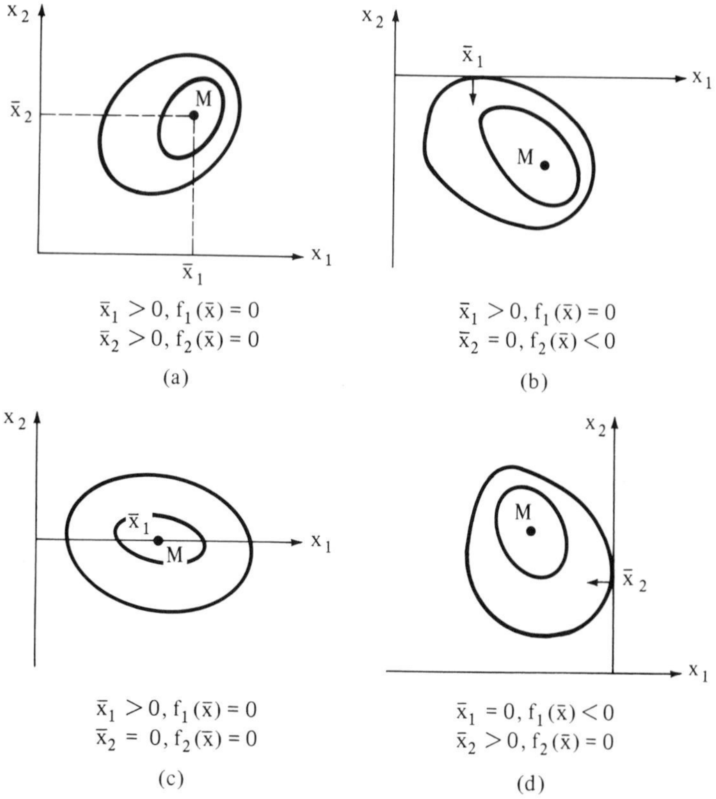

Figure 1.8

were an upper bound, the derivative might be positive as depicted in Figure 1.7d.

Condition (1.60) generalizes to functions of many variables. We can illustrate it for a function of two variables: to maximize $f(x_1, x_2)$ subject to $x_1 \geq 0$ and $x_2 \geq 0$ we must set

$$f_i(\bar{x}_1, \bar{x}_2) \leq 0, \quad \bar{x}_i \geq 0, \quad \bar{x}_i[f_i(\bar{x}_1, \bar{x}_2)] = 0, \quad i = 1, 2. \tag{1.61}$$

There are altogether nine different possibilities; we have represented four of them in Figure 1.8. The reader is invited to construct the rest. The constrained maximum characterized by (1.61) is called $\bar{\mathbf{x}}$, whereas the unconstrained maximum occurs at point M. These two coincide in cases (a) and (c). In case (b) the small arrow represents the direction of increase of the function at $\bar{\mathbf{x}}$: it points toward larger values of the function and is perpendicular to the tangent (the x_1 axis). This identifies $f_1(\bar{\mathbf{x}})$ as zero and $f_2(\bar{\mathbf{x}})$ as negative. Case (d) is similar to case (b).

We now proceed to state the Kuhn–Tucker theorem. The regularity condition required may take several forms and corresponds to the rank condition of Theorem 1.2.1. We shall discuss these at some length after the mechanics of the problem have been described.

Consider problem (1.59). We introduce m new variables similar to multipliers but usually called *dual variables,* one for each constraint; they are denoted by $\mu_1, \dots, \mu_m$. We form a new function similar to a Lagrangean:

$$\phi(\mu_1, \dots, \mu_m, x_1, \dots, x_n) = f(x_1, \dots, x_n) + \sum_{j=1}^{m} \mu_j g^j(x_1, \dots, x_n), \tag{1.62}$$

or more compactly,

$$\phi(\boldsymbol{\mu}, \mathbf{x}) = f(\mathbf{x}) + \boldsymbol{\mu}' \cdot \mathbf{g}(\mathbf{x}). \tag{1.62$'$}$$

Remark. We must at the outset emphasize one aspect of this format: it is essential that the problem be written exactly like this. In particular, the constraints require $\mathbf{g}(\mathbf{x}) \geqq \mathbf{0}$ and we *add* the products $\mu_j g^j(x)$ to the objective function to form the ϕ function. The first-order conditions are stated here for this format and would be affected by any change in it. Care must therefore be taken to state the problem and to define the ϕ function exactly in this way. Because of this need for precision we will confine our attention to maximum problems; minimum problems can be handled by reversing the sign of the objective function.

Theorem 1.4.1: Kuhn–Tucker. Let $\mathbf{x}^*$ be a solution to problem (1.59) and assume that a regularity condition applies (see Lemma 1.4.1). Then there must exist a set of values $\mu_1, \dots, \mu_m$ such that

$$\begin{aligned} &g^j(\mathbf{x}^*) \geq 0, \quad \mu_j \geq 0, \quad \text{and} \quad \mu_j g^j(\mathbf{x}^*) = 0, \quad j = 1, \dots, m, \\ &f_i(\mathbf{x}^*) + \sum_{j=1}^{m} \mu_j g_i^j(\mathbf{x}^*) \leq 0, \quad x_i^* \geq 0, \\ &x_i^* \cdot \left[f_i(\mathbf{x}^*) + \sum_{j=1}^{m} \mu_j g_i^j(\mathbf{x}^*) \right] = 0, \quad i = 1, \dots, n. \end{aligned} \tag{1.63}$$

These are known as the Kuhn–Tucker conditions. They can be written more compactly by using the ϕ function of (1.62):

$$\begin{aligned} &\frac{\partial \phi}{\partial \mu_j} \geq 0, \quad \mu_j \geq 0, \quad \text{and} \quad \mu_j \frac{\partial \phi}{\partial \mu_j} = 0, \quad j = 1, \dots, m, \\ &\frac{\partial \phi}{\partial x_i} \leq 0, \quad x_i^* \geq 0, \quad \text{and} \quad x_i^* \frac{\partial \phi}{\partial x_i} = 0, \quad i = 1, \dots, n, \end{aligned} \tag{1.63$'$}$$

or in vector notation,

$$\begin{aligned} &\phi_{\mu} \geqq \mathbf{0}, \quad \boldsymbol{\mu} \geqq \mathbf{0}, \quad \text{and} \quad \boldsymbol{\mu}' \cdot \phi_{\mu} = 0, \\ &\phi_{\mathbf{x}} \leqq \mathbf{0}, \quad \mathbf{x}^* \geqq \mathbf{0}, \quad \text{and} \quad \mathbf{x}^{*\prime} \cdot \phi_{\mathbf{x}} = 0. \end{aligned} \tag{1.63$''$}$$

The reason that only the inner product of μ and ϕ_μ is required to be zero (as opposed to each term as in (1.63′)) is that the sign restrictions on μ and ϕ_μ make each $\mu_j \cdot \phi_{\mu_j}$ term nonnegative. Hence, requiring each term to be zero is equivalent to requiring their sum to be zero. A similar argument applies to $\mathbf{x}^{*\prime} \cdot \phi_{\mathbf{x}} = 0$.

The inequalities and equalities that make up the Kuhn–Tucker conditions should be used as in the introductory example at the beginning of this section. Specifically, if μ_j is positive, then $g^j(\mathbf{x}^*) = 0$; by contrast, if $g^j(\mathbf{x}^*)$ is positive, then $\mu_j = 0$. Similarly, if $x_i^* > 0$, then $\phi_{x_i} = 0$, and if $\phi_{x_i} < 0$, then $x_i^* = 0$. There are therefore many possible outcomes, and these conditions provide no *a priori* indication as to which variables will be positive and which will be zero. This difficulty cannot be completely resolved, and the derivation of the solution may require some economic intuition into the problem (or plain guesswork). We now illustrate the use of these conditions with some examples, assuming for now that the Kuhn–Tucker conditions yield a constrained maximum.

Example 1.4.1. Find x_1 and x_2 that maximize $f(x_1, x_2) = \ln(x_1) + \ln(x_2 + 5)$ subject to

$$4 - x_1 - x_2 \geq 0, \quad x_1 \geq 0, \; x_2 \geq 0.$$

Form the ϕ function $\phi(\mu, x_1, x_2) = \ln(x_1) + \ln(x_2 + 5) + \mu[4 - x_1 - x_2]$, and the Kuhn–Tucker conditions are

$$\begin{aligned}
&\phi_\mu = 4 - x_1 - x_2 \geq 0, && \mu \geq 0, \; \mu[4 - x_1 - x_2] = 0, \\
&\phi_{x_1} = x_1^{-1} - \mu \leq 0, && x_1 \geq 0, \; x_1[x_1^{-1} - \mu] = 0, \\
&\phi_{x_2} = (x_2 + 5)^{-1} - \mu \leq 0, && x_2 \geq 0, \; x_2[(x_2 + 5)^{-1} - \mu] = 0.
\end{aligned}$$

We begin with a trial solution, with all variables positive; this results in three equations:

$$\begin{aligned}
&\mu > 0 \text{ implies } 4 - x_1 - x_2 = 0, \\
&x_1 > 0 \text{ implies } x_1^{-1} - \mu = 0, \\
&x_2 > 0 \text{ implies } (x_2 + 5)^{-1} - \mu = 0.
\end{aligned}$$

To solve, eliminate μ to get $x_1 = x_2 + 5$, and use the constraint to obtain $4 - x_2 - 5 - x_2 = 0$, or $x_2 = -0.5$. This is unacceptable, but it does give us a hint to set $x_2 = 0$ at the outset. Leaving all other variables positive, we still have three equations, counting $x_2 = 0$ as one of them. We obtain $x_1 = 4$, $x_2 = 0$, and $\mu = \frac{1}{4}$. We must check that $\phi_{x_2} \leq 0$: $\phi_{x_2} = 5^{-1} - \frac{1}{4} = -0.05$ indeed. The solution is illustrated in Figure 1.9a, where the feasible area is hatched.

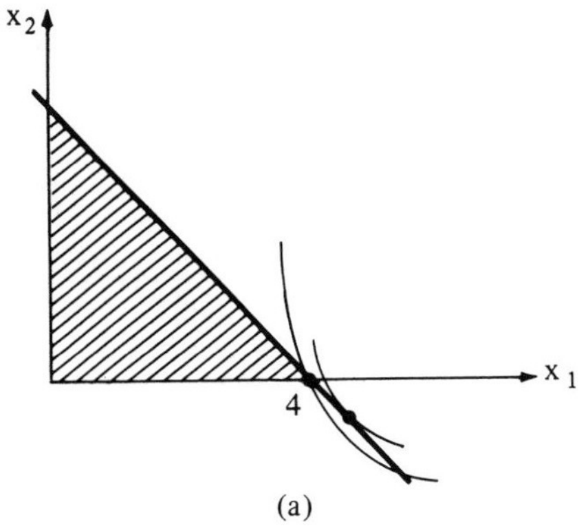

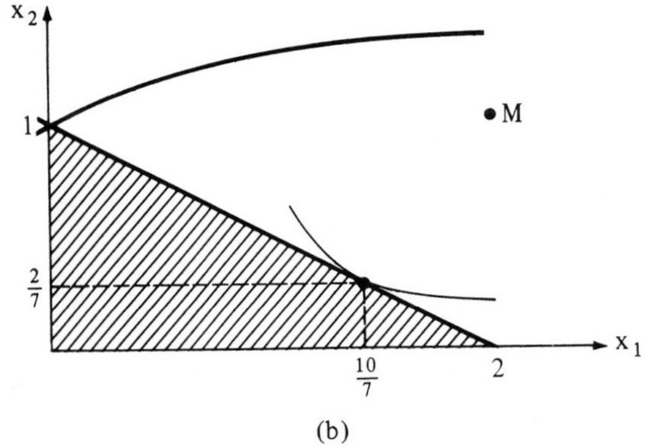

Figure 1.9

Example 1.4.2

Maximize $f(x_1, x_2) = 6x_1 - 2(x_1)^2 + 2x_1x_2 - 2(x_2)^2$

subject to $x_1 + 2x_2 \leq 2, \quad 1 + x_1 - (x_2)^2 \geq 0, \quad x_1 \geq 0, \ x_2 \geq 0.$

Form the ϕ function:

$$\phi(\mu_1, \mu_2, x_1, x_2) = 6x_1 - 2(x_1)^2 + 2x_1x_2 - 2(x_2)^2 + \mu_1[2 - x_1 - 2x_2] + \mu_2[1 + x_1 - (x_2)^2],$$

$$\phi_{\mu_1} = 2 - x_1 - 2x_2 \geq 0, \quad \mu_1 \geq 0, \ \mu_1[2 - x_1 - 2x_2] = 0,$$

$$\phi_{\mu_2} = 1 + x_1 - (x_2)^2 \geq 0, \quad \mu_2 \geq 0, \ \mu_2[1 + x_1 - (x_2)^2] = 0,$$

$$\phi_{x_1} = 6 - 4x_1 + 2x_2 - \mu_1 + \mu_2 \leq 0, \quad x_1 \geq 0, \ x_1\phi_{x_1} = 0,$$

$$\phi_{x_2} = 2x_1 - 4x_2 - 2\mu_1 - 2\mu_2 x_2 \leq 0, \ x_2 \geq 0, \ x_2\phi_{x_2} = 0.$$

If we suppose that all variables are positive, we obtain four equations: $\phi_{\mu_1}=0$, $\phi_{\mu_2}=0$, $\phi_{x_1}=0$, and $\phi_{x_2}=0$. Solving the first two yields $x_1=0$, $x_2=1$ (disregarding negative solutions); this is acceptable because it is possible that both $x_1=0$ and $\phi_{x_1}=0$. However, substituting these values in the last two equations yields $\mu_1=3$, but $\mu_2=-5$, which is unacceptable. Taking the hint, we set $\mu_2=0$, while all other variables are positive. This gives four equations: $\phi_{\mu_1}=0$, $\mu_2=0$, $\phi_{x_1}=0$, and $\phi_{x_2}=0$. Eliminating μ_1 from the last two and using $\mu_2=0$ yields $6-5x_1+4x_2=0$, which with $\phi_{\mu_1}=0$ is solved by $x_1=10/7$, $x_2=2/7$; substitution then yields $\mu_1=6/7$. We must check that $\phi_{\mu_2}\geq 0$: $\phi_{\mu_2}=10/7-(2/7)^2+1\simeq 2.35>0$. The solution is illustrated in Figure 1.9b, where M indicates the absolute maximum of the objective function.

1.4.1 *Regularity conditions (constraint qualifications)*

The necessity of (1.63) for a maximum to (1.59) was originally proved by Kuhn and Tucker under an assumption termed the *constraint qualification,* which was designed to avoid cusps in the feasible set. Later others (notably K. Arrow, L. Hurwicz, and H. Uzawa) refined these conditions. Here we list some of these results and illustrate others without claiming completeness. For a detailed survey the reader is referred to Takayama (1985) or Mangasarian (1969).

Lemma 1.4.1. The regularity condition of Theorem 1.4.1 is satisfied if any one of the following conditions is satisfied:

(i) $g^j(\mathbf{x})$ is linear, $j=1,\dots,m$.
(ii) $g^j(\mathbf{x})$ is concave and there exists $\bar{\mathbf{x}}>\mathbf{0}$ such that $g^j(\bar{\mathbf{x}})>0$, $j=1,\dots,m$.
(iii) The feasible set $(\mathbf{g}(\mathbf{x})\geqq\mathbf{0},\ \mathbf{x}\geqq\mathbf{0})$ is convex and has a nonempty interior, and $g^j_{\mathbf{x}}(\mathbf{x}^*)\neq 0$ if j is a binding constraint (i.e., $g^j(\mathbf{x}^*)=0$).
(iv) Renumber the constraints so that the first m' $(\leq m)$ are binding. The rank of the matrix $[\partial g^j(\mathbf{x}^*)/\partial\mathbf{x}]$, $j=1,\dots,m'$, is equal to m'.

We recognize in (iv) the rank condition of Theorem 1.2.1, which, of course, here applies only to the binding constraints. Condition (ii) is known as *Slater's condition.* These conditions are not equivalent to one another. For instance if $\mathbf{x}^*>0$, then (iv) guarantees the uniqueness of $\boldsymbol{\mu}$ in Theorem 1.4.1, but other conditions do not, as we now illustrate.

Example 1.4.3: nonuniqueness of dual variables. Maximize $\ln(x_1)+\ln(x_2)$ subject to $2x_1+x_2\leq 3$, $x_1+2x_2\leq 3$, and $x_1+x_2\leq 2$, with $x_1\geq 0$, $x_2\geq 0$.

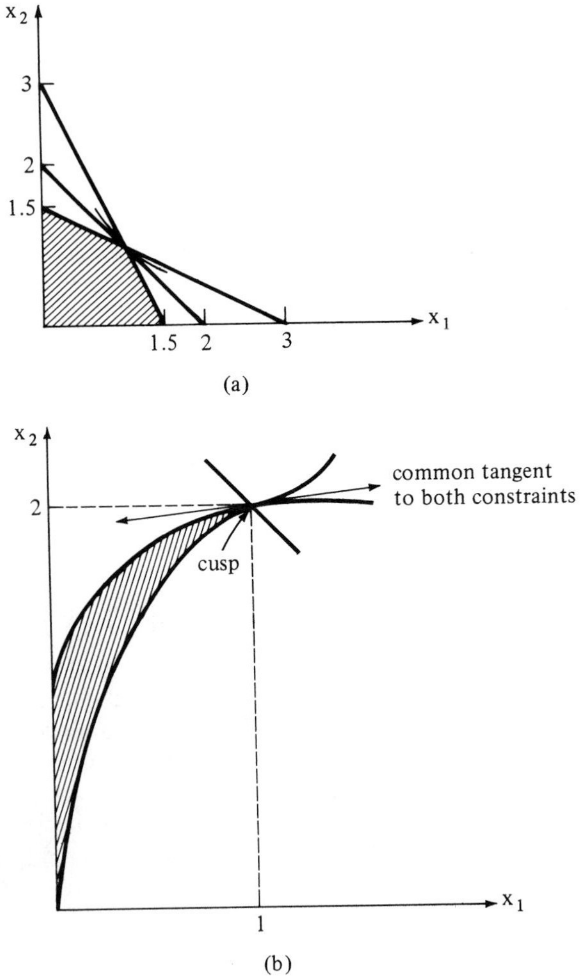

Figure 1.10

As shown in Figure 1.10a the point (1, 1) is optimal. Condition (i) in Lemma 1.4.1 applies, and there exist dual variables μ_1, μ_2, and μ_3 such that

$$\phi_{x_1} = x_1^{-1} - 2\mu_1 - \mu_2 - \mu_3 \leq 0, \quad x_1 \geq 0, \ x_1\phi_{x_1} = 0,$$
$$\phi_{x_2} = x_2^{-1} - \mu_1 - 2\mu_2 - \mu_3 \leq 0, \quad x_2 \geq 0, \ x_2\phi_{x_2} = 0,$$

where

$$\phi = \ln x_1 + \ln x_2 + \mu_1[3 - 2x_1 - x_2] + \mu_2[3 - x_1 - 2x_2] + \mu_3[2 - x_1 - x_2].$$

However, the μ's are not unique, since the rank of

$$\mathbf{g}_{\mathbf{x}'}=\begin{bmatrix}-2 & -1 & -1\\ -1 & -2 & -1\end{bmatrix}$$

is clearly less than 3. We can only give as a solution $\mu_1=\mu_2$ and $\mu_3=1-3\mu_2$ with the restrictions $\mu_1,\mu_2,\mu_3\geq 0$.

We will not state the Kuhn–Tucker constraint qualification because it is neither intuitively appealing nor easy to check; we will, however, give an example to illustrate why we want to eliminate cusps, because it is related to the above rank condition (see Figure 1.10b).

Example 1.4.4: cusp

Maximize x_1+x_2

subject to

$$1+(x_1)^{0.1}-x_2\geq 0,$$

$$-1.9(x_1)^3+5.7(x_1)^2-5.8x_1+x_2\geq 0,\quad x_1\geq,\ x_2\geq 0,$$

$$\phi=x_1+x_2+\mu_1(1+(x_1)^{0.1}-x_2)+\mu_2(-1.9(x_1)^3+5.7(x_1)^2-5.8x_1+x_2).$$

The feasible region is defined by the intersection of the region *above* the curve, $x_2=1.9x_1^3-5.7x_1^2+5.8x_1$, with the region *below* the curve, $x_2=1-(x_1)^{0.1}$; this feasible region is hatched in Figure 1.10b and exhibits a cusp at point $(1,2)$. Furthermore, this point dominates all other feasible points and is clearly optimal, since the objective function is increasing in all its arguments. We now show that conditions (1.63′) are not satisfied at this constrained maximum. Suppose that conditions (1.63′) hold; then, since $x_1>0$ and $x_2>0$, we would have $\phi_{x_1}=0$ and $\phi_{x_2}=0$. However, these equations are

$$1+0.1\mu_1(x_1)^{-0.9}+\mu_2(-5.7(x_1)^2+11.4x_1-5.8)=0,$$

$$1-\mu_1+\mu_2=0,$$

and when substituting $x_1=1,\ x_2=2$ we have

$$\begin{aligned}-0.1\mu_1+0.1\mu_2&=1,\\ \mu_1-\mu_2&=1,\end{aligned}\tag{1.64}$$

for which no solution exists. It is easy to verify that the rank condition is not satisfied here. This is because of the occurrence of the cusp, in which both constraints are tangent to one another.

Remark. Example 1.4.4 gives us an opportunity to bring out one technical detail which we have passed over. There would be no need for regularity conditions if (1.62) and Theorem 1.4.1 had been stated in the following form: Define $\phi^0(\mu_0, \boldsymbol{\mu}, \mathbf{x}) = \mu_0 f(\mathbf{x}) + \boldsymbol{\mu}' \cdot \mathbf{g}(\mathbf{x})$, and the theorem is modified to state that there exist $\mu_0, \mu_1, \ldots, \mu_m$, and so on. Clearly these μ's can be scaled up arbitrarily. In nearly all cases $\mu_0 \neq 0$; hence, it is convenient to set $\mu_0 = 1$, and the theorem appears as originally stated. However, in a few pathological cases such as in Example 1.4.4, μ_0 would be zero, as we now show, and the scaling is inappropriate. Let us redefine ϕ^0 as $(\mu_0 - 1)(x_1 + x_2) + \phi(\boldsymbol{\mu}, \mathbf{x})$. Then in order to get the new conditions, we need only attach a μ_0 factor to the right-hand side of (1.64) and we obtain

$$-0.1\mu_1 + 0.1\mu_2 = \mu_0,$$

$$\mu_1 - \mu_2 = \mu_0.$$

This system now admits the solutions $\mu_0 = 0$, $\mu_1 = \mu_2$. (There are many, because the rank condition is not satisfied.)

The same remark applies to Theorem 1.2.1, in which we could discard the rank condition if we had the Lagrangean $\mathfrak{L}^0(\lambda_0, \boldsymbol{\lambda}, \mathbf{x}) = \lambda_0 f(\mathbf{x}) + \boldsymbol{\lambda}' \cdot \mathbf{g}(\mathbf{x})$. We shall concentrate on regular problems, but the present remark should be kept in mind if a problem is encountered in which it is impossible to obtain a solution for the dual variables or multipliers.

We now turn our attention to a more restrictive class of problems for which we can obtain global results. This is not the only class for which we can obtain such results, but it is the simplest and most useful. For further results, the reader is referred to Mangasarian (1969) or Takayama (1985).

1.4.2 *Concave programming*

Definition 1.4.1. A *regular* concave programming problem satisfies the following:

(i) f and $\mathbf{g}$ are concave functions.
(ii) There exists $\bar{\mathbf{x}} \geqq \mathbf{0}$ such that $\mathbf{g}(\bar{\mathbf{x}}) \geqq \mathbf{0}$ and $g^j(\bar{\mathbf{x}}) > 0$ for all j constraints that are not linear. (This incorporates Slater's condition.)

Theorem 1.4.2. Assume that (1.59) is a regular concave programming problem. Then $\mathbf{x}^*$ is a solution to that problem if and only if there exists $\boldsymbol{\mu}$ such that $(\boldsymbol{\mu}, \mathbf{x}^*)$ satisfy the conditions of Theorem 1.4.1 (Kuhn–Tucker).

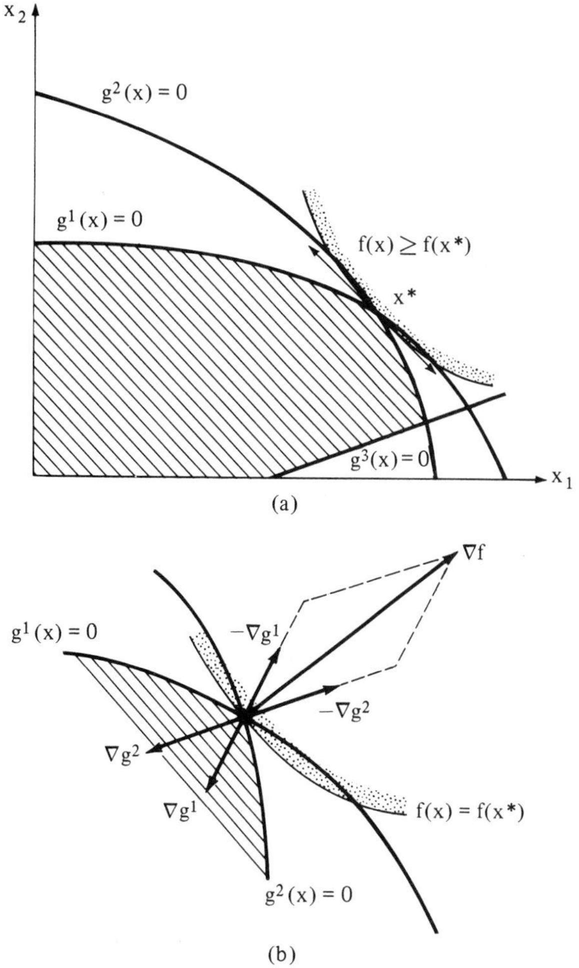

Figure 1.11

1.4.3 *Geometric interpretation*

In Figure 1.11a we have represented a concave programming problem with two variables and three constraints. The feasible set is hatched and the upper contour set of the objective function is lined with dots; the constrained maximum is at $\mathbf{x}^*$. In the case depicted the first two constraints are binding, while the third one is slack; thus, we expect $\mu_1 > 0$, $\mu_2 > 0$, and $\mu_3 = 0$. Some of the Kuhn–Tucker conditions hence require that at the optimum

$$f_{\mathbf{x}} + \mu_1 g^1_{\mathbf{x}} + \mu_2 g^2_{\mathbf{x}} + 0 g^3_{\mathbf{x}} = 0,$$

or in gradient notation,

$$\nabla f = -\mu_1 \nabla g^1_{\mathbf{x}} - \mu_2 \nabla g^2_{\mathbf{x}} - 0 \nabla g^3_{\mathbf{x}}.$$

This last relationship indicates that the gradient of the objective function is equal to a weighted sum of the negative of the gradients of the constraints where the weights are the dual variables: zero for slack constraints and positive for binding constraints. This is illustrated in Figure 1.11b, which is a blowup of 1.11a. It is interesting to compare Figure 1.11b with Figure 1.10, in which no values of the dual variables could be found in the ordinary way. In that case, the constraints are tangent at the optimum, their gradient vectors are colinear, and no weighted sum of them can duplicate the gradient of the objective function, which is on another line.

1.4.4 *Derivation of the Kuhn–Tucker conditions*

There are several proofs of the necessity of the Kuhn–Tucker conditions; these can be found, for instance, in either of the two references cited earlier, as well as in the original article. In order to emphasize similarities among optimization methods we derive them here using a modification of the method of Lagrange. This procedure will also highlight the links between the nonnegativity constraints on the x's and the inequality form of the Kuhn–Tucker conditions, as well as show the sign of the dual variables to be a consequence of the inequality form of the constraints. Consider the problem of finding $\mathbf{x}$ that maximizes $f(\mathbf{x})$ subject to

$$g^j(\mathbf{x}) \geq 0, \quad j = 1, \ldots, m,$$

$$x_i \geq 0, \quad i = 1, \ldots, n.$$

Suppose there is a maximum at $\mathbf{x}^*$ at which the following constraints are binding:

$$g^j(\mathbf{x}^*) = 0, \quad j = 1, \ldots, m' \leq m,$$

$$x_i^* = 0, \quad i = 1, \ldots, n' \leq n.$$

We can ignore the nonbinding constraints when characterizing the optimum (i.e., deriving the necessary conditions) and form a Lagrangean

$$\mathcal{L} = f(\mathbf{x}) + \sum_{j=1}^{m'} \mu^j g^j(\mathbf{x}) + \sum_{i=1}^{n'} \lambda_i x_i.$$

In the remainder of the derivation we shall omit asterisks and/or arguments of functions when convenient in order to simplify the notation. The first-order conditions are

$$\mathcal{L}_{\lambda_i} = x_i = 0, \qquad i = 1, \dots, n', \tag{1.65a}$$

$$\mathcal{L}_{\mu_j} = g^j(\mathbf{x}) = 0, \qquad j = 1, \dots, m', \tag{1.65b}$$

$$\mathcal{L}_{x_h} = f_{x_h} + \sum_{j=1}^{m'} \mu^j g^j_{x_h} + \lambda_h = 0, \quad h = 1, \dots, n. \tag{1.65c}$$

We now proceed to show that the multiplier associated with any binding restraint (either (1.65a) or (1.65b)) is positive or zero. Let us rewrite any restraint as

$$R^s(\mathbf{x}) - \alpha^s \geq 0, \quad s = 1, \dots, (n' + m'),$$

where α^s is initially zero but we aim to increase it slightly. We associate with such a restraint a multiplier π^s. (Naturally π^s can be a μ or a λ.) Consider a small increase in α^s from 0 to $\epsilon > 0$. This will shrink the feasible set, since points $\mathbf{x}$ that satisfy $\epsilon > R^s(\mathbf{x}) \geq 0$ are now excluded from it. This shrinking will result in a decrease (or no change) in the optimal value of the objective function, hence $df(\mathbf{x}^*)/d\alpha^s \leq 0$, all s. The Lagrangean in our modified notation is

$$\mathcal{L} = f(x) + \sum_{s=1}^{n'+m'} \pi^s \cdot [R^s(\mathbf{x}) - \alpha^s],$$

and by the envelope theorem

$$\frac{df(\mathbf{x}^*)}{d\alpha^s} = \frac{\partial \mathcal{L}}{\partial \pi^s} = -\pi^s \leq 0.$$

We have established that the multipliers associated with binding restraints are positive or zero. This is a consequence of the inequality form of the restraints, since this is what made it possible for some points previously included in the feasible set to become unfeasible when α^s increased.

We now revert to the original notation; we have established that

$$\begin{aligned} g^j(\mathbf{x}) = 0, \ \ \mu_j \geq 0, \quad j = 1, \dots, m', \\ x_i = 0, \ \ \lambda_i \geq 0, \quad i = 1, \dots, n'. \end{aligned} \tag{1.66}$$

We have ignored the slack restraints, but if we attach a zero-valued multiplier to them, their inclusion in the Lagrangean will not alter the first-order conditions. Let

$$\Psi = f(\mathbf{x}) + \sum_{j=1}^{m} \mu^j g^j(\mathbf{x}) + \sum_{i=1}^{n} \lambda_i \cdot x_i,$$

where to each

$$g^j(\mathbf{x})>0 \text{ assign } \mu_j=0, \quad j=m'+1,\dots,m \tag{1.67a}$$

and to each

$$x_i>0 \text{ assign } \lambda_i=0, \quad i=n'+1,\dots,n. \tag{1.67b}$$

Taking into account (1.66) and (1.67), the first-order condition (1.65b) becomes

$$\mu^j \geq 0, \quad g^j(\mathbf{x}) \geq 0, \quad \mu_j g^j(\mathbf{x})=0, \quad j=1,\dots,m. \tag{1.68}$$

The first-order condition (1.65c) is still $\mathcal{L}_{x_h}=0$ or $\Psi_{x_h}=0$ but now includes the slack restraints as well, with a zero multiplier:

$$f_{x_h}+\sum_{j=1}^{m} \mu^j g^j_{x_h}+\lambda_h=0, \quad h=1,\dots,n.$$

However, if $x_h>0$, then $\lambda_h=0$ and the condition is

$$f_{x_h}+\sum_{j=1}^{m} \mu^j g^j_{x_h}=0,$$

and if $x_h=0$, then $\lambda_h \geq 0$ from (1.66), and the condition is

$$f_{x_h}+\sum_{j=1}^{m} \mu^j g^j_{x_h}=-\lambda_h \leq 0.$$

Summing up, the first-order condition (1.65c) can be expressed as

$$f_{x_h}+\sum_{j=1}^{m} \mu^j g^j_{x_h} \leq 0, \quad x_h \geq 0, \quad x_h\left[f_{x_h}+\sum_{j=1}^{m} \mu^j g^j_{x_h}\right]=0, \quad h=1,\dots,n. \tag{1.69}$$

Note that (1.69) also incorporates (1.65a) and $\mathbf{x} \geqq \mathbf{0}$. Equations (1.68) and (1.69) are recognized as the Kuhn–Tucker conditions of Theorem 1.4.1. The inequality form of (1.69) has been shown to be a consequence of the nonnegativity restriction on $\mathbf{x}$, plus the asymmetric treatment of restraints like $x_i \geq 0$ and $g^j(\mathbf{x}) \geq 0$, the former not being included in the ϕ function. Hence, if a variable were to be unrestricted in sign, we would use the familiar equality form of the first-order condition for this variable. Similarly, if a constraint is an equality, one cannot ascertain the sign of the multiplier associated with it. This observation can be formalized.

Theorem 1.4.3: mixed problems. Consider the problem of maximizing $f(\mathbf{x},\mathbf{y})$ with respect to $\mathbf{x}$ and $\mathbf{y}$ subject to

$$\mathbf{g}(\mathbf{x},\mathbf{y}) \geqq \mathbf{0},$$
$$\mathbf{h}(\mathbf{x},\mathbf{y}) = \mathbf{0},$$
$$\mathbf{x} \geqq \mathbf{0}.$$

Let

$$\phi(\boldsymbol{\lambda},\boldsymbol{\mu},\mathbf{x},\mathbf{y}) = f(\mathbf{x},\mathbf{y}) + \boldsymbol{\mu}'\cdot\mathbf{g}(\mathbf{x},\mathbf{y}) + \boldsymbol{\lambda}'\cdot\mathbf{h}(\mathbf{x},\mathbf{y}),$$

where $\boldsymbol{\lambda}$ and $\boldsymbol{\mu}$ are vectors of appropriate orders. Assuming that the rank condition (on $\mathbf{h}$ and binding $\mathbf{g}$ constraints) is satisfied, the necessary conditions are

$$\begin{aligned}
&\phi_{\mathbf{x}} = f_{\mathbf{x}} + \boldsymbol{\mu}'\cdot\mathbf{g}_{\mathbf{x}} + \boldsymbol{\lambda}'\cdot\mathbf{h}_{\mathbf{x}} \leqq \mathbf{0}, \quad \mathbf{x} \geqq \mathbf{0}, \quad \mathbf{x}'\cdot\phi_{\mathbf{x}} = 0,\\
&\phi_{\mathbf{y}} = f_{\mathbf{y}} + \boldsymbol{\mu}'\cdot\mathbf{g}_{\mathbf{y}} + \boldsymbol{\lambda}'\cdot\mathbf{h}_{\mathbf{y}} = \mathbf{0},\\
&\phi_{\mu} = \mathbf{g}(\mathbf{x},\mathbf{y}) \geqq \mathbf{0}, \quad \boldsymbol{\mu} \geqq \mathbf{0}, \quad \boldsymbol{\mu}'\cdot\phi_{\mu} = 0,\\
&\phi_{\lambda} = \mathbf{h}(\mathbf{x},\mathbf{y}) = \mathbf{0}.
\end{aligned} \tag{1.70}$$

Unless $\mathbf{h}$ is made of linear functions, it is not possible to retain the concave programming format, since $h^j(\mathbf{x},\mathbf{y}) = 0$ can be duplicated by $h^j(\mathbf{x},\mathbf{y}) \geq 0$ and $-h^j(\mathbf{x},\mathbf{y}) \geq 0$, but both functions cannot be concave (unless they are linear). When the use of equality constraints is necessary, care should be taken that the solution is indeed optimal; for instance, Theorem 1.2.6 may be used if applicable.

Example 1.4.5. Find (x, y) to maximize $-(x)^2-(y)^2+20x-4y$ subject to

$$-x-(y)^2+9 \geq 0,$$
$$(x)^2+(y)^2-26=0 \quad \text{and} \quad x \geq 0,$$
$$\phi(\lambda,\mu,x,y) = -x^2-y^2+20x-4y+\mu[-x-y^2+9]+\lambda[x^2+y^2-26],$$
$$\phi_x = -2x+20-\mu+2\lambda x \leq 0, \quad x \geq 0, \quad x\phi_x = 0,$$
$$\phi_y = -2y-4-2\mu y+2\lambda y = 0,$$
$$\phi_\mu = -x-y^2+9 \geq 0, \quad \mu \geq 0, \quad \mu\phi_\mu = 0,$$
$$\phi_\lambda = x^2+y^2-26 = 0.$$

We shall "guess" that the solution involves $x > 0$ and $\mu = 0$ in order to save space. We then have the three equations

$$-2x+20+2\lambda x = 0,$$
$$-2y-4+2\lambda y = 0,$$
$$x^2+y^2-26 = 0.$$

The first two yield $x=-5y$ and, with the third, the two solutions $(x,y)=(5,-1)$ and $(-5,1)$. The latter is unacceptable, since $x<0$. Using $(5,-1)$ we obtain $\lambda=-1$. It remains to be checked that the solution satisfies the condition $\phi_\mu \geq 0$: $\phi_\mu=-5-1+9>0$ indeed. The rank condition requires that the vector of derivatives of the second constraint have rank 1, that is, not be nil, which is easily checked. The first constraint is slack and thus does not figure in this. As we previously noted we cannot cast this problem into a concave programming format. However, we can use Theorem 1.2.6, which requires $\phi(\lambda^*, \mu^*, x, y)$ to be a concave function of x and y. We get $\phi(-1,0,x,y)=-x^2-y^2+20x-4y-[x^2+y^2-26]$, which is easily seen to be concave and the solution $(-1, 0, 5, -1)$ is optimal.

Remark. In some problems it may be convenient to use the equality form of first-order conditions relating to the choice of $\mathbf{x}$. This is done by treating the nonnegativity restrictions on $\mathbf{x}$ formally as constraints, each with a nonnegative multiplier λ_h. We then have conditions like (1.65c).

1.5 Economic applications of nonlinear programming

One advantage of nonlinear programming over the method of Lagrange is that it formally takes into account some restrictions which had previously been ignored, although they were present in most cases: variables can now be restricted to nonnegative values, and resource constraints may indicate that there is no need to exhaust all of the resources available. These features increase the richness of the economic interpretation of many problems, as we now illustrate with a simple case.

1.5.1 *The pricing of resources and uneconomic activities*

Let there be n possible activities. The level of the ith activity is denoted by x_i and is constrained to be nonnegative, $i=1,\dots,n$. These activities use up resources; typically the vector of activities $\mathbf{x}$ uses up $h^j(\mathbf{x})$ of the jth resource, which is available in amount b_j. Resources are freely disposable, so that no cost is incurred if some of each is unused and discarded. The jth resource constraint is therefore $h^j(\mathbf{x}) \leq b_j$. Finally, the objective is to maximize net benefit $B(x_1,\dots,x_n)$. This can be formulated as nonlinear program, and we will assume B is concave, h^j is convex, all j, as well as some regularity condition. Find $x_1,\dots,x_n$ that maximizes $B(x_1,\dots,x_n)$ subject to

$$h^j(x_1,\dots,x_n) \leq b_j, \quad j=1,\dots,m,$$

$$x_i \geq 0, \quad i=1,\dots,n,$$

$$\phi(\mu, \mathbf{x}) = B(\mathbf{x}) + \sum_{j=1}^{m} \mu_j [b_j - h^j(\mathbf{x})].$$

The optimality conditions are

$$\phi_{x_i} = B_i - \sum_{j=1}^{m} \mu_j h_i^j \leq 0; \quad x_i \geq 0; \quad x_i \phi_{x_i} = 0, \quad i = 1, \dots, n, \tag{1.71a}$$

$$\phi_{\mu_j} = b_j - h^j(\mathbf{x}) \geq 0; \quad \mu_j \geq 0; \quad \mu_j \phi_{\mu_j} = 0, \quad j = 1, \dots, m, \tag{1.71b}$$

where

$$B_i \equiv \partial B / \partial x_i \quad \text{and} \quad h_i^j \equiv \partial h^j / \partial x_i.$$

First consider conditions (1.71b). The dual variable μ_j, like Lagrange multipliers previously, is the shadow price of resource j, but the interpretation is now richer. First of all, this shadow price is nonnegative, as we expect prices to be; if it happens to be positive, the resource will be exactly used up, since $\mu_j > 0$ implies $b_j = h^j(\mathbf{x})$; if, however, the resource is not all used up, then $h^j(\mathbf{x}) < b_j$ implies $\mu_j = 0$. A resource still available in a positive amount at the optimum has a zero price; in other words, if a resource is not scarce, it is free. This sensible pricing mechanism is the consequence of formally taking into account the assumption of free disposal of resources.

We now turn to conditions (1.71a). Since $h^j(\mathbf{x})$ is the amount of resource j used by the vector of activities $\mathbf{x}$, we interpret h_i^j as the marginal cost of activity i in terms of resource j; more loosely, it is the extra amount of resource j used by the marginal unit of activity i. When this is multiplied by the shadow price of the resource, μ_j, we have that cost in dollars (or whatever unit B is measured in). When these costs are summed for all resources, we have the overall marginal cost of activity i. The first part of (1.71a) then states that marginal benefit cannot exceed marginal cost for any activity. If marginal cost is greater than marginal benefit, the level of that activity is zero. Any activity carried out at a positive level has equal marginal benefit and marginal cost. The rationale behind this situation is that an excess of marginal cost over marginal benefit provides an incentive to reduce that level of activity until a balance has been reached; however, for some activities this balance is not possible and even when the level of activity has been reduced to zero, marginal cost still exceeds marginal benefit: this is an uneconomic activity, given existing resources and alternatives.

In order to gain yet more insight into the determination of shadow prices we now turn to a special case of the preceding problem in which there is a single linear constraint. We maximize $B(x_1, \dots, x_n)$ subject to

$$\sum_{i=1}^{n} x_i \leq b, \quad x \geqq 0,$$

$$\phi = B(\mathbf{x}) + \mu\left[b - \sum_{i=1}^{n} x_i\right],$$

$$\phi_\mu = b - \sum_{i=1}^{n} x_i \geq 0, \quad \mu \geq 0, \quad \mu\phi_\mu = 0,$$

$$\phi_{x_i} = B_i - \mu \leq 0, \quad x_i \geq 0, \quad x_i\phi_{x_i} = 0.$$

The second condition is the one that interests us most. It states that the shadow price is larger than or equal to the marginal benefit of each activity; furthermore, it is equal to the marginal benefit of every activity that is carried out at a positive level. Those activities that are not carried out usually have a marginal benefit that is less than the shadow price. Thus, if we think of the optimization as an iterative process, we can imagine scanning all possible activities and allocating some of the resource to those activities with a high marginal benefit just as an auctioneer would allocate shares of the flow of some resource (e.g., water, oil) to the highest bidders. When the flow has been completely allocated, its price *is* the highest marginal benefit obtainable from all bidders, the differential allocation of the resource itself smoothing out differences in the marginal benefits of the activities. This is what the shadow price is, the highest possible marginal return compatible with using up the resource. Any activity that cannot match this at any level of allocation simply misses out and is declared uneconomic.

We turn to a classic application of nonlinear programming. It has been selected because the inequality constraints are essential to the formulation of the problem.

1.5.2 *Peakload policy*

In many instances the supply of a product is limited by a capacity constraint. When the product is required in different amounts for different periods, the problem of choosing the right capacity as well as the schedule of supplies arises. Typical examples are the pricing of public utilities such as electricity, water, and telephone services. One can also think of the problem of choosing the appropriate size of a tent for a traveling circus or choosing the size of a power station for a planning horizon over which demand conditions will vary.

There are T periods; x_t is the supply in period t, X is the capacity, $R(x_1, \ldots, x_T)$ is the revenue from sales in all periods (note that this form allows the demand in one period to be affected by demand in other periods), $C(x_1, \ldots, x_T)$ is the variable operating cost, and $K(X)$ is the capital cost of a plant of capacity X. We assume concavity of R and convexity of C and K. The problem is to find $x_1, \ldots, x_T$ that maximize

$$R(x_1, \ldots, x_T) - C(x_1, \ldots, x_T) - K(X)$$

subject to

$$X - x_t \geq 0, \quad x_t \geq 0, \quad t = 1, \ldots, T.$$

From

$$\phi(\mu, \mathbf{x}, X) = R(\mathbf{x}) - C(\mathbf{x}) - K(X) + \sum_{t=1}^{T} \mu_t [X - x_t],$$

the necessary conditions are

$$\phi_{x_t} = \mathrm{MR}_t - \mathrm{MC}_t - \mu_t \leq 0, \quad x_t \geq 0, \quad x_t \phi_{x_t} = 0, \quad t = 1, \ldots, T,$$

$$\phi_X = -K' + \sum_{t=1}^{T} \mu_t \leq 0, \quad X \geq 0, \quad X\phi_X = 0,$$

$$\phi_{\mu_t} = X - x_t \geq 0, \quad \mu_t \geq 0, \quad \mu_t \phi_{\mu_t} = 0, \quad t = 1, \ldots, T,$$

where $\mathrm{MR}_t \equiv \partial R/\partial x_t$ and $\mathrm{MC}_t \equiv \partial C/\partial x_t$ are the marginal revenue and marginal operating cost of x_t, respectively. If supply is less than capacity in some period t, then $\mu_t = 0$, and if the product is supplied at all in that period, marginal cost equals marginal revenue. It is possible that for some periods marginal revenue is always below marginal cost, in which case the supply is zero in these periods. There will, however, be periods in which the capacity constraint binds unless nothing at all is produced. In the case where the capacity constraint never binds, all μ_t are zero, implying $-K' \leq 0$, which, assuming that K is an increasing function, must hold strictly, and this in turn implies $X = 0$ (taking into account that $(K')X$ must then be zero). There must therefore be some positive μ_t's if any production occurs ($X > 0$). In those periods $x_t = X > 0$ and the marginal revenue equals the marginal operating cost plus μ_t, which we interpret as a capacity surcharge. Furthermore, all the capacity surcharges add up to the marginal cost of capital needed to produce an extra unit of capacity K'. Note that we have assumed that the firm or public utility acts as a monopolist and maximizes profit, so that there is no implication of fairness in not using a capacity surcharge in offpeak periods; this is simply another instance of price discrimination. The monopolist makes a higher profit this way than if it charged a uniform price for all periods. Finally, note that we formulated the problem as if the monopolist chose the quantities, which is rather unreasonable. The actual problem is of peakload pricing, but if we let prices be the choice variables, we obtain the same overall results if we put reasonable restrictions on demand functions.

1.6 The special case of linear programming

The development of linear programming predates that of nonlinear programming, but the results are most easily obtained from concave programming theory. The great advantage of linear programming is that there exist very efficient solution algorithms suitable for very large problems; they are known as the simplex method or modifications of it. It is not our purpose here to present these algorithms but to make some analytical points. (For details of the simplex method see the classic Dorfman, Samuelson, and Solow, 1958, or Hadley, 1962.)

We define a linear programming problem that we call the *primal* as finding $\mathbf{x}$ to maximize $\mathbf{p}'\cdot\mathbf{x}$ subject to

$$\mathbf{Ax} \leqq \mathbf{c}, \quad \mathbf{x} \geqq \mathbf{0}; \quad \mathbf{A} \text{ is } m \times n. \tag{1.72}$$

The constraints and the objective function are linear. Therefore, using the concave programming results, we know that the Kuhn–Tucker conditions are necessary and sufficient for an optimum. We derive them in the usual way:

$$\phi(\mathbf{x}, \boldsymbol{\mu}) = \mathbf{p}'\mathbf{x} + \boldsymbol{\mu}'\cdot[\mathbf{c} - \mathbf{Ax}], \tag{1.73}$$

$$\phi_{\mu} = \mathbf{c} - \mathbf{Ax} \geqq \mathbf{0}, \quad \boldsymbol{\mu} \geqq \mathbf{0}, \quad \boldsymbol{\mu}'\cdot[\mathbf{c} - \mathbf{Ax}] = 0, \tag{1.74}$$

$$\phi_{\mathbf{x}} = \mathbf{p} - \mathbf{A}'\boldsymbol{\mu} \leqq \mathbf{0}, \quad \mathbf{x} \geqq \mathbf{0}, \quad \mathbf{x}'\cdot[\mathbf{p} - \mathbf{A}'\boldsymbol{\mu}] = 0. \tag{1.75}$$

Next consider another problem, called the *dual* and derived from the *primal* of (1.72) by interchanging the vector of coefficients $\mathbf{p}$ with the right-hand-side vector $\mathbf{c}$, transposing matrix $\mathbf{A}$, reversing the inequalities, and naming the variables for this problem $\boldsymbol{\mu}$. We then seek to find $\boldsymbol{\mu}$ to minimize $\mathbf{c}'\boldsymbol{\mu}$ subject to

$$\mathbf{A}'\boldsymbol{\mu} \geqq \mathbf{p}, \quad \boldsymbol{\mu} \geqq \mathbf{0}. \tag{1.76}$$

To obtain the optimality conditions for this problem, we change the sign of the objective function to have a maximum problem and obtain

$$\Psi(\boldsymbol{\mu}, \mathbf{x}) = -\mathbf{c}'\boldsymbol{\mu} + \mathbf{x}'\cdot[\mathbf{A}'\boldsymbol{\mu} - \mathbf{p}]. \tag{1.77}$$

The reason we have chosen to denote the multipliers by $\mathbf{x}$ will become obvious shortly. Bearing in mind that $\boldsymbol{\mu}$ is the vector of maximizing variables here and $\mathbf{x}$ the vector of multipliers, we obtain the following necessary and sufficient conditions:

$$\Psi_{\mathbf{x}} = \mathbf{A}'\boldsymbol{\mu} - \mathbf{p} \geqq \mathbf{0}, \quad \mathbf{x} \geqq \mathbf{0}, \quad \mathbf{x}'\cdot[\mathbf{A}'\boldsymbol{\mu} - \mathbf{p}] = 0, \tag{1.78}$$

$$\Psi_{\mu} = -\mathbf{c} + \mathbf{Ax} \leqq \mathbf{0}, \quad \boldsymbol{\mu} \geqq \mathbf{0}, \quad \boldsymbol{\mu}'\cdot[-\mathbf{c} + \mathbf{Ax}] = 0. \tag{1.79}$$

If we remember that quadratic forms such as $\mathbf{x}'\mathbf{A}'\boldsymbol{\mu}$ are scalar-valued, hence unaffected by transposition, it becomes clear that (1.78) and (1.75) are identical, as are (1.79) and (1.74). Solving (1.72) via the Kuhn–Tucker method uncovers a set of dual variables $\boldsymbol{\mu}$, which solves the dual problem (1.76). Conversely solving (1.76) uncovers the solution $\mathbf{x}$ to the primal

(1.72). Furthermore, the maximum value of the objective function of the primal (1.72a) is equal to the minimum value of the objective function of the dual (1.76a). Note that the name of dual variables adopted earlier for the multipliers of nonlinear programming problems comes from the dual of linear programming. It is also possible to define dual nonlinear programming problems, but this is not useful for our purposes here; for details one can consult Mangasarian (1969).

We now state and prove some important results.

Theorem 1.6.1. Let $\mathbf{x}^* \geqq \mathbf{0}$ and $\boldsymbol{\mu}^* \geqq \mathbf{0}$ be feasible vectors for the primal and the dual, respectively, and suppose that they yield the same value for their respective objective functions. Then $\mathbf{x}^*$ and $\boldsymbol{\mu}^*$ are optimal solutions to the primal and the dual, respectively.

Proof. By assumption we have

$$\mathbf{A}\mathbf{x}^* \leqq \mathbf{c} \quad \text{and} \quad \mathbf{A}'\boldsymbol{\mu}^* \geqq \mathbf{p}.$$

Multiply the first by $\boldsymbol{\mu}^{*\prime} \geqq \mathbf{0}$ and the second by $\mathbf{x}^{*\prime} \geqq \mathbf{0}$ to get

$$\boldsymbol{\mu}^{*\prime}\mathbf{c} \geq \boldsymbol{\mu}^{*\prime}\mathbf{A}\mathbf{x}^* \text{ and } \mathbf{x}^{*\prime}\mathbf{A}'\boldsymbol{\mu}^* \geq \mathbf{x}^{*\prime}\mathbf{p}, \quad \text{or} \quad \mathbf{c}'\boldsymbol{\mu}^* \geq \boldsymbol{\mu}^{*\prime}\mathbf{A}\mathbf{x}^* \geq \mathbf{p}'\mathbf{x}^*.$$

With the assumption that $\mathbf{c}'\boldsymbol{\mu}^* = \mathbf{p}'\mathbf{x}^*$, we have

$$\mathbf{c}'\boldsymbol{\mu}^* = \boldsymbol{\mu}^*\mathbf{A}\mathbf{x}^* \quad \text{and} \quad \boldsymbol{\mu}^{*\prime}\mathbf{A}\mathbf{x}^* = \mathbf{p}'\mathbf{x}^*,$$

and (1.74), (1.75), (1.78), and (1.79) are satisfied, which proves the result. □

Lemma 1.6.1. $\mathbf{x}^*$ and $\boldsymbol{\mu}^*$ are the solutions of the primal (1.72) and of the dual (1.76), respectively, if and only if they constitute a *saddle point* of the function ϕ of (1.73); more precisely,

$$\phi(\mathbf{x}, \boldsymbol{\mu}^*) \leq \phi(\mathbf{x}^*, \boldsymbol{\mu}^*) \leq \phi(\mathbf{x}^*, \boldsymbol{\mu}) \quad \text{for all } \mathbf{x} \geqq \mathbf{0},\ \boldsymbol{\mu} \geqq \mathbf{0}.$$

Proof. In Lemma 1.6.1, ϕ is being maximized with respect to $\mathbf{x} \geqq \mathbf{0}$, given $\boldsymbol{\mu}^*$, and at the same time is being minimized with respect to $\boldsymbol{\mu} \geqq \mathbf{0}$, given $\mathbf{x}^*$. Maximization under nonnegativity restrictions means

$$\phi_{\mathbf{x}}(\mathbf{x}^*, \boldsymbol{\mu}^*) \leqq \mathbf{0}, \quad \mathbf{x}^* \geqq \mathbf{0}, \quad \mathbf{x}^{*\prime} \cdot \phi_{\mathbf{x}}(\mathbf{x}^*, \boldsymbol{\mu}^*) = 0,$$

or

$$\mathbf{p} - \mathbf{A}'\boldsymbol{\mu}^* \leqq \mathbf{0}, \quad \mathbf{x}^* \geqq \mathbf{0}, \quad \mathbf{x}^{*\prime} \cdot [\mathbf{p} - \mathbf{A}'\boldsymbol{\mu}^*] = 0,$$

which is (1.75). Minimization under nonnegativity restrictions is much the same, but we require the derivatives to be positive or zero. It is easy to verify that this yields (1.74).

Conversely, if we have solutions $(\mathbf{x}^*, \boldsymbol{\mu}^*)$ to the primal and the dual, they satisfy (1.74) and (1.75) (or (1.78) and (1.79)), which implies that ϕ is minimized with respect to $\boldsymbol{\mu} \geqq \mathbf{0}$ and maximized with respect to $\mathbf{x} \geqq \mathbf{0}$; hence, $(\mathbf{x}^*, \boldsymbol{\mu}^*)$ is a saddle point of ϕ. □

We now turn to an economic interpretation of these linear programs. Let $\mathbf{p}$ be a price vector, $\mathbf{x}$ be the amount of goods produced, $\mathbf{c}$ be a vector of resources available, and $\mathbf{A}$ describe a linear technology. The primal (1.72) maximizes revenue subject to resource constraints. Conditions (1.74) have the usual interpretation for the pricing of scarce resources, where $\boldsymbol{\mu}$ is the shadow price vector of the resources. The first part of (1.75) states that the price of a good cannot exceed the resource cost of it; furthermore, if cost exceeds price, that good is not produced and for every good produced price equals cost. Note that this means that profit (per unit) is zero and that the first part of the condition requires profit to be nonpositive on any good. To investigate this last point further, consider the dual defined in (1.76). Clearly, it seeks a resource price vector that will minimize the total value of resources subject to the condition that profits on individual goods are nonpositive. The economic meaning of this problem may be surprising at first; however, we know that solving the primal is mathematically equivalent to solving the dual. Therefore, we must refine our interpretation of these two problems so that they are consistent with one another.

We wish to explain why the existence of an efficient allocation of resources is equivalent to the existence of a resource pricing system that rules out positive profits and minimizes aggregate resource cost. At the optimum the objective functions of the primal and dual are equal; hence, aggregate profit is zero. Furthermore, by (1.75) the profit made on each good is also zero, either because the unit profit is zero or because the good is not produced. This state of affairs resembles the outcome of competitive forces at work, and indeed such interpretation can be substantiated. Suppose the resources were held by agents with no market power and that the users of the resources and producers of the goods had no market power either. Then competition among resource owners would lead to price cutting, which would lower the value of resources as much as possible – meaning not lower than the point where resource users would make positive profits; if positive profits were allowed, the resource users would again bid up the resource prices by competing among themselves for the resources. Clearly, the goods for which unit profit is negative would not be produced, and the resources that are in excess supply would not command a positive price.

Thus, competitive pricing of resources in the way just described is equivalent to their efficient allocation. This became apparent through the connection between the dual and primal solutions.

The intricate dual relationship between allocation and pricing can be further highlighted by the use of the ϕ function and the saddle point of

Lemma 1.6.1. We know that $\mathbf{x}^*$ maximizes $\phi(\mathbf{x}, \boldsymbol{\mu}^*) = \mathbf{p}'\cdot\mathbf{x} + \boldsymbol{\mu}^{*\prime}\cdot\mathbf{c} - \boldsymbol{\mu}^{*\prime}\cdot\mathbf{A}\mathbf{x}$ over all $\mathbf{x} \geqq \mathbf{0}$ given $\boldsymbol{\mu}^*$. This is equivalent to maximizing aggregate profits, which are $\mathbf{p}'\cdot\mathbf{x} - \boldsymbol{\mu}^{*\prime}\mathbf{A}\mathbf{x}$ (the maximum is zero). Also, $\boldsymbol{\mu}^*$ minimizes $\phi(\mathbf{x}^*, \boldsymbol{\mu}) = \mathbf{p}'\cdot\mathbf{x}^* + \boldsymbol{\mu}'\cdot\mathbf{c} - \boldsymbol{\mu}'\cdot\mathbf{A}\mathbf{x}^*$ over all $\boldsymbol{\mu} \geqq \mathbf{0}$, given $\mathbf{x}^*$. This is equivalent to minimizing the aggregate value of unused resources $\boldsymbol{\mu}'\cdot\mathbf{c} - \boldsymbol{\mu}'\mathbf{A}\mathbf{x}^*$; the minimum is also zero, of course.

All of these results are valid in the special case of linear programming but are not generally valid in nonlinear programming. We can, however, duplicate any nonlinear programming solution with a linear program. Consider again the problem of choosing $\mathbf{x}$ to maximize $f(\mathbf{x})$ subject to $\mathbf{g}(\mathbf{x}) \geqq \mathbf{0}$, $\mathbf{x} \geqq \mathbf{0}$, where $\mathbf{x}$ is $n \times 1$ and $\mathbf{g}$ is $m \times 1$. We easily obtain the optimality conditions with $\phi(\boldsymbol{\mu}, \mathbf{x}) = f(\mathbf{x}) + \boldsymbol{\mu}'\cdot\mathbf{g}(\mathbf{x})$:

$$\phi_{\mathbf{x}}(\boldsymbol{\mu}^*, \mathbf{x}^*) = f_{\mathbf{x}}(\mathbf{x}^*) + \mathbf{g}'_{\mathbf{x}}(\mathbf{x}^*)\cdot\boldsymbol{\mu}^* \leqq \mathbf{0}, \quad \mathbf{x}^* \geqq \mathbf{0}, \quad \mathbf{x}^{*\prime}\cdot\phi_{\mathbf{x}} = 0, \tag{1.80}$$

$$\phi_{\mu}(\boldsymbol{\mu}^*, \mathbf{x}^*) = \mathbf{g}(\mathbf{x}^*) \geqq \mathbf{0}, \quad \boldsymbol{\mu}^* \geqq \mathbf{0}, \quad \boldsymbol{\mu}^{*\prime}\cdot\phi_{\mu} = 0. \tag{1.81}$$

Consider now the linear program to find $\mathbf{x}$ that maximizes

$$f_{\mathbf{x}'}(\mathbf{x}^*)\cdot\mathbf{x} \tag{1.82}$$

subject to

$$\mathbf{g}_{\mathbf{x}'}(\mathbf{x}^*)\cdot(\mathbf{x} - \mathbf{x}^*) \geqq \mathbf{0}, \quad \mathbf{x} \geqq \mathbf{0}, \tag{1.83}$$

where the $n \times 1$ vector $f_{\mathbf{x}}(\mathbf{x}^*)$ and the $m \times n$ matrix $\mathbf{g}_{\mathbf{x}'}(\mathbf{x}^*)$ are taken from (1.80) and (1.81). The optimality conditions are obtained from

$$\Psi(\boldsymbol{\mu}, \mathbf{x}) = f_{\mathbf{x}'}(\mathbf{x}^*)\cdot\mathbf{x} + \boldsymbol{\mu}'\cdot\mathbf{g}_{\mathbf{x}'}(\mathbf{x}^*)\cdot(\mathbf{x} - \mathbf{x}^*)$$

as

$$\Psi_{\mathbf{x}} = f_{\mathbf{x}}(\mathbf{x}^*) + \mathbf{g}'_{\mathbf{x}}(\mathbf{x}^*)\cdot\boldsymbol{\mu} \leqq \mathbf{0}, \quad \mathbf{x} \geqq \mathbf{0}, \quad \mathbf{x}'\cdot\Psi_{\mathbf{x}} = 0,$$

$$\Psi_{\mu} = \mathbf{g}_{\mathbf{x}'}(\mathbf{x}^*)\cdot(\mathbf{x} - \mathbf{x}^*) \geqq \mathbf{0}, \quad \boldsymbol{\mu} \geqq \mathbf{0}, \quad \boldsymbol{\mu}'\cdot\Psi_{\mu} = 0.$$

It is easy to verify that $\boldsymbol{\mu}^*$ and $\mathbf{x}^*$ of (1.80) and (1.81) satisfy these conditions. The geometric interpretation of (1.82) is that level curves of f are replaced by its tangent at $\mathbf{x}^*$ and (1.83) substitutes the tangents to the constraints at $\mathbf{x}^*$ for the constraints themselves, as illustrated in Figure 1.12. This result reinforces the point that information such as rigid prices or fixed coefficient technology, although globally incorrect, may be sufficient to support an equilibrium.

Appendix

Functions and sets

A *set* is a collection of objects. The sets encountered here are often subsets of the n-dimensional space of real numbers, denoted by R_n. If $\mathbf{x}$ is an

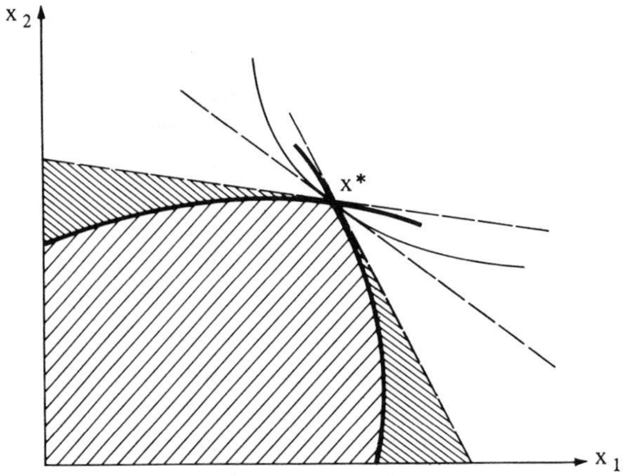

Figure 1.12

element of the set X, we write $\mathbf{x} \in X$, which can be read as "$\mathbf{x}$ belongs to X." The symbol $\forall \mathbf{x}$ means "for all elements $\mathbf{x}$." A set X in R_n is *bounded from above* if there exists a vector $\bar{\mathbf{b}}$ such that $\mathbf{x} \leq \bar{\mathbf{b}}$, $\forall \mathbf{x} \in X$. It is *bounded from below* if there exists a vector $\underline{\mathbf{b}}$ such that $\underline{\mathbf{b}} \leq \mathbf{x}$, $\forall \mathbf{x} \in X$. If a set is bounded from above and from below, it is said to be *bounded.* A set is *closed* if and only if the limits of all converging sequences contained in the set are in the set. The *complement* of a closed set is an *open* set. Examples of closed sets are closed intervals such as $-1 \leq x \leq 1$, and examples of open sets are open intervals such as $-1 < x < 1$. A set X is *convex* if and only if $x_1 \in X$ and $x_2 \in X$ imply $x_t \in X$, where $x_t = tx_1 + (1-t)x_2$ and $0 \leq t \leq 1$. We call x_t a convex combination of x_1 and x_2; it is a weighted average of x_1 and x_2 with nonnegative weights that sum to 1.

A real-valued *function* f defined on an n-dimensional subset of R_n, say D, is a rule that associates with each vector in D a single real value. We may write $f: D \rightarrow R$ or $f(\mathbf{x})$ or $f(x_1, \ldots, x_n)$. We may have multidimensional functions, say $f: R_n \rightarrow R_m$, but we still require that with each vector of R_n there be associated a single vector of R_m. A function f defined on a set D is said to be *continuous at* $\mathbf{x}^0$ if, for each sequence $\mathbf{x}^1, \mathbf{x}^2, \ldots$ in D converging to $\mathbf{x}^0$, we have $\lim_{n\rightarrow\infty} f(\mathbf{x}^n) = f(\lim_{n\rightarrow\infty} \mathbf{x}^n) = f(\mathbf{x}^0)$. A function is *continuous on* D if it is continuous at all points of D. A continuous function f defined on a set D that is closed and bounded has both a maximum and a minimum in that set; that is, there exists a point $\bar{\mathbf{x}} \in D$ such that $f(\bar{\mathbf{x}}) \geq f(\mathbf{x})$, $\forall \mathbf{x} \in D$, and there exists a point $\underline{\mathbf{x}} \in D$ such that $f(\mathbf{x}) \geq f(\underline{\mathbf{x}})$, $\forall \mathbf{x} \in D$. The *partial derivative* of $f(x_1, \ldots, x_n)$ with respect to x_i, evaluated at $\hat{\mathbf{x}}$, is defined to be

$$\lim_{\epsilon \to 0} \frac{f(\hat{x}_1, \ldots, \hat{x}_i + \epsilon, \ldots, \hat{x}_n) - f(\hat{x}_1, \ldots, \hat{x}_n)}{\epsilon}.$$

If the limit exists, the function is differentiable with respect to x_i at $\hat{\mathbf{x}}$. We say that $f(x_1, \ldots, x_n)$ is *differentiable on D* if it is differentiable with respect to $x_1, \ldots, x_n$ at all points of D. Partial derivatives are denoted by $\partial f/\partial x_i$ (read "del f del x_i") or f_{x_i}. The vector of partial derivatives is $\partial f/\partial \mathbf{x}$ or $f_{\mathbf{x}}$. These are first-order partial derivatives, and clearly second-order partial derivatives can be similarly defined. The second-order partial derivative of f with respect to x_i and x_j is denoted by $\partial^2 f/\partial x_i \partial x_j$ or $f_{x_i x_j}$. Whenever these derivatives are defined and continuous, it is always true that $\partial^2 f/\partial x_i \partial x_j = \partial^2 f/\partial x_j \partial x_i$: that is, the order of differentiation does not matter. Second-order partial derivatives are most conveniently arranged in a square matrix called the *Hessian matrix,* often denoted by $\mathbf{H}$. Because of the above property, Hessian matrices are always symmetric.

Matrix notation

We use the following conventions: lowercase letters denote column vectors; row vectors are identified by a prime as transposes of column vectors. For instance, $\mathbf{x}$ is a column vector $(n \times 1)$ and $\mathbf{x}'$ is a row vector $(1 \times n)$. When differentiating a scalar-valued function with respect to a vector of variables, we assume that *the derivatives are arranged exactly as the variables were.* For instance, if $f(x_1, \ldots, x_n)$ is a scalar-valued function of n variables, then $f_{\mathbf{x}}$ is the column vector of its first-order partial derivatives, while $f_{\mathbf{x}'}$ would be the same derivatives but arranged as a row vector. Similarly, if we take the derivatives of a vector-valued function, the same convention applies to each element of the function, and these vectors of derivatives are themselves arranged as the elements of the original function. To obtain the Hessian matrix of $f(\mathbf{x})$, we differentiate the first-order derivatives $f_{\mathbf{x}}$ with respect to $\mathbf{x}'$, $\mathbf{H} = f_{\mathbf{x}\mathbf{x}'}$, and the matrix is of the same order as $\mathbf{x}\mathbf{x}'$ $(n \times n)$. Suppose we have a function $\mathcal{L}(\boldsymbol{\lambda}, \mathbf{x})$, where $\boldsymbol{\lambda}$ is $m \times 1$ and $\mathbf{x}$ is $n \times 1$; then $\mathcal{L}_{\boldsymbol{\lambda}\mathbf{x}'}$ is an $m \times n$ matrix of derivatives such as $\partial^2 \mathcal{L}/\partial \lambda_j \partial x_i$, while $\mathcal{L}_{\mathbf{x}\boldsymbol{\lambda}'}$ is its $n \times m$ transpose. The *length* (or Euclidean norm) of a vector $\mathbf{x}$ is $\|\mathbf{x}\| = (\mathbf{x}' \cdot \mathbf{x})^{1/2}$.

Matrices are sometimes used to define functional forms; an important instance is a *quadratic form,* $f(\mathbf{x}) = \mathbf{x}'\mathbf{A}\mathbf{x} \equiv \sum_i \sum_j a_{ij} x_i x_j$, where $\mathbf{A}$ is assumed to be symmetric. We have $f_{\mathbf{x}} = 2\mathbf{A}\mathbf{x}$ and $f_{\mathbf{x}\mathbf{x}'} = 2\mathbf{A}$. Another instance is the *inner product* $f(\mathbf{x}) = \mathbf{a}'\mathbf{x} \equiv \sum_i a_i x_i$; this yields $f_{\mathbf{x}} = \mathbf{a}$. Some matrices give rise to quadratic forms that have an invariant sign. A matrix is *positive-definite* (resp. negative-definite) if and only if $\mathbf{x}'\mathbf{A}\mathbf{x} > 0$, $\forall \mathbf{x} \neq \mathbf{0}$ (resp. $\mathbf{x}'\mathbf{A}\mathbf{x} < 0$, $\forall \mathbf{x} \neq \mathbf{0}$). Similarly, a matrix is *positive-semidefinite* (resp. negative-semidefinite) if and only if $\mathbf{x}'\mathbf{A}\mathbf{x} \geq 0$, $\forall \mathbf{x}$ (resp. $\mathbf{x}'\mathbf{A}\mathbf{x} \leq 0$, $\forall \mathbf{x}$). These matrices can be characterized by the sign of their characteristic roots - all positive for a positive-definite matrix, all negative for a negative-definite matrix, positive or zero for positive-semidefiniteness, and negative or zero for negative-semidefiniteness. There is a more practical way, but we must first define some subdeterminants of square matrices. Consider an $n \times n$ matrix $\mathbf{B}$. The determinant of the matrix formed by deleting the last $n-r$ rows and the last $n-r$ columns of $\mathbf{B}$ is called the rth *leading principal minor;* it is the determinant of an $r \times r$ matrix and is denoted by B_r. There are n such minors in $\mathbf{B}$: $B_1, B_2, \dots, B_n$ and $B_n = |\mathbf{B}|$, while $B_1 = b_{11}$. The *principal minors* of $\mathbf{B}$ are obtained as the leading principal minors of any matrix obtained from $\mathbf{B}$ by a permutation of rows and columns. Alternatively, a principal minor of order r of $\mathbf{B}$ can be obtained by deleting any $n-r$ pairs of rows and columns from $\mathbf{B}$ and taking the determinant of the remainder. A matrix is positive-definite if and only if its leading principal minors are all positive; it is negative-definite if and only if its leading principal minors alternate in (strict) signs beginning with negative (or $(-1)^r B_r > 0 \;\; r = 1, \dots, n$). Positive-semidefiniteness requires all principal minors to be nonnegative, and negative-semidefiniteness requires all principal minors of order r to be such that $(-1)^r B_r \geq 0$, $r = 1, \dots, n$.

Taylor's expansion and total differentials

Matrix notation is useful for expressing Taylor's expansion for functions of several variables up to the second degree. Let f have continuous first- and second-order derivatives. Then

$$f(\mathbf{x}) = f(\mathbf{x}^*) + f_{\mathbf{x}'}(\mathbf{x}^*) \cdot (\mathbf{x} - \mathbf{x}^*) + \tfrac{1}{2}(\mathbf{x} - \mathbf{x}^*)' \cdot f_{\mathbf{x}\mathbf{x}'}(\mathbf{x}^*) \cdot (\mathbf{x} - \mathbf{x}^*) + R$$

with $R \to 0$ as $\mathbf{x} \to \mathbf{x}^*$; or in exact form,

$$f(\mathbf{x}) = f(\mathbf{x}^*) + f_{\mathbf{x}'}(\mathbf{x}^*) \cdot (\mathbf{x} - \mathbf{x}^*) + \tfrac{1}{2}(\mathbf{x} - \mathbf{x}^*)' \cdot f_{\mathbf{x}\mathbf{x}'}(\mathbf{x}_t) \cdot (\mathbf{x} - \mathbf{x}^*),$$

where

$$x_t = t\mathbf{x} + (1-t)\mathbf{x}^* \quad \text{for some } t \in [0, 1].$$

The total differential of a function $df = f_{\mathbf{x}'} \cdot d\mathbf{x} = \sum_{i=1}^{n} f_{x_i} \cdot dx_i$, which expresses the change in f induced by small changes in $\mathbf{x}$, can be viewed as a first-order Taylor's expansion with $\mathbf{x}$ close to $\mathbf{x}^*$ and $d\mathbf{x} = (\mathbf{x} - \mathbf{x}^*)$.

A useful device for representing functions of two variables is the concept of *level curves* (or *contour curves*). Let c be a value taken by the function $f(x_1, x_2)$. Then $f(x_1, x_2) = c$ implicitly defines a curve in the (x_1, x_2) plane along which f keeps the value c - it is called a level curve of

f. To each feasible value c corresponds one level curve. The level curves corresponding to two distinct values of a function cannot intersect one another. An expression for the slope of a level curve can be derived by noting that as x_1 and x_2 move along such a curve, the value of f does not change; hence, $df = f_1 dx_1 + f_2 dx_2 = 0$ and $dx_2/dx_1 = -f_1/f_2$. The total differential is also useful for obtaining an expression for total derivatives. Let the x_i variables in $f(x_1, \ldots, x_n)$ all depend on another variable, say t. Then we call df/dt the *total derivative* of f with respect to t. We can calculate it by "dividing" the total differential by dt:

$$\frac{df}{dt} = f_{\mathbf{x}'} \cdot \frac{d\mathbf{x}}{dt} = \sum_{i=1}^{n} f_{x_i}\left(\frac{dx_i}{dt}\right).$$

Homogeneous functions

A function $f(\mathbf{x})$ is said to be *homogeneous of degree h in* $\mathbf{x}$ if and only if $f(t\mathbf{x}) = (t)^h f(\mathbf{x})$, $\forall t \geq 0$, $\forall \mathbf{x}$. Euler's theorem for homogeneous functions states that if $f(\mathbf{x})$ is homogeneous of degree h in $\mathbf{x}$, then

$$\sum_{i=1}^{n} x_i \cdot f_{x_i}(\mathbf{x}) = hf(\mathbf{x}), \quad \forall \mathbf{x}.$$

It is also true that if $f(\mathbf{x})$ is homogeneous of degree h in $\mathbf{x}$, then $f_{x_j}(\mathbf{x})$ is homogeneous of degree $(h-1)$ in $\mathbf{x}$, $j = 1, \ldots, n$. One of the implications of these results is that a function that is homogeneous of degree 1 must have a singular Hessian matrix – hence a zero determinant. The slope of the level curves of homogeneous functions of two variables are given by

$$\frac{dx_2}{dx_1} = -\frac{f_1(x_1, x_2)}{f_2(x_1, x_2)} = -\frac{(x_1)^{h-1} f_1(1, x_2/x_1)}{(x_1)^{h-1} f_2(1, x_2/x_1)} = -\frac{f_1(1, x_2/x_1)}{f_2(1, x_2/x_1)}.$$

Therefore, the slope of such level curves depends only on the ratio x_2/x_1 and not on the value of the variables; hence, it is constant along a ray drawn from the origin across all level curves.

The implicit function theorem

Let $F(\mathbf{y})$ be an m-dimensional function defined on an open set and $\mathbf{y}$ be $n \times 1$ with $n > m$. Assume F to be continuously differentiable and the rank of the $m \times n$ matrix $F_{\mathbf{y}'}(\mathbf{y}^0)$ to be m at some point $\mathbf{y}^0$, where $F(\mathbf{y}^0) = 0$. Then for an arbitrarily chosen $(\bar{y}_1, \ldots, \bar{y}_m)$ within a rectangular region around $\mathbf{y}^0$ $(y_i^0 - \delta \leq \bar{y}_i \leq y_i^0 + \delta,\ \delta > 0)$, there exists a unique set of values $(\bar{y}_{m+1}, \ldots, \bar{y}_n)$ that depends on $(\bar{y}_1, \ldots, \bar{y}_m)$ so that $F(\bar{\mathbf{y}}) = 0$. In other words, we can express the last $n - m$ variables as a function of the first m variables in the neighborhood of $\mathbf{y}^0$.

Exercises

1. Indicate which of the following functions are concave or convex in (x_1, x_2). Which are strictly concave or convex? Find the maximum or minimum if it exists.

$$f(x_1,x_2)=(x_1)^2+(x_2)^2+x_1x_2+10x_1+10x_2,$$
$$f(x_1,x_2)=2(x_1)^2-(x_2)^2-4x_1+8x_2,$$
$$f(x_1,x_2)=2(x_1)^2-8x_1x_2+8(x_2)^2,$$
$$f(x_1,x_2)=6(x_1)^{1/3}(x_2)^{1/3}-x_1-x_2, \quad x_1>0,\ x_2>0,$$
$$f(x_1,x_2)=\ln(x_1-1)+\ln(x_2-2)-2x_1-3x_2,$$
$$f(x_1,x_2)=\ln(x_1+3)-2x_2,$$
$$f(x_1,x_2)=x_1(1-e^{-x_2/x_1}), \quad x_1>0,\ x_2>0.$$

2. There are two industries, hatching and laying, and two goods, eggs and chickens. The price of an egg is 1; the price of a chicken is p. The hatching industry produces y_1 chickens using x_1 eggs, which it buys from the laying industry. Its production function is $y_1=4(x_1)^{1/2}$. The laying industry produces x_2 eggs using y_2 chickens, which it buys from the hatching industry. Its production function is $x_2=(y_2)^{1/2}$. (We assume away any consideration of time lags, nonnegativity constraints, the indivisibility of live chickens or eggs, and claims to precedence by all parties.)

   Express in terms of p the profit-maximizing quantities of chickens and eggs produced or used in each industry. Assuming that all eggs produced in the laying industry are used as inputs by the hatching industry, what is the value of p? What, then, is the net output of chickens available for consumption?

3. Use the method of Lagrange to find all constrained extrema of the following functions:
   (i) $x_1^2+x_2^2+3x_3^2+3x_2x_3$ subject to $x_1+2x_2+3x_3=1$,
   (ii) $\ln x_1+\ln x_2$ subject to $(x_1+1)(x_2+1)=4$,
   (iii) $4x_1+2x_2-8$ subject to $(x_1-2)^2+(x_2-1)^2=80$,
   (iv) $(x_1-2)^2+3(x_2-1)^2+2x_3-2(x_1-4)(x_2+1)$, subject to $3x_1+x_2+5=0$ and $4x_1+2x_3-1=0$,
   (v) $2(x_1)^{1/2}(x_2)^{1/2}$ subject to $2(x_1)^{3/2}+16(x_2)^{3/2}=32$.

4. Find the vector $\mathbf{x}$ that maximizes $\sum_{i=1}^n \beta_i \ln(x_1-\gamma_i)$ subject to $\sum_{i=1}^n p_i x_i = y$. It is possible to interpret the objective function as a utility function and the constraint as a budget constraint: y is income, $\mathbf{p}$ is the price vector, and the positive parameters $\boldsymbol{\beta}$ and $\boldsymbol{\gamma}$ characterize the individual's tastes. The solution, which is to express $\mathbf{x}$ in terms of y, $\mathbf{p}$, $\boldsymbol{\beta}$, and $\boldsymbol{\gamma}$, yields the demand functions.

5. A consumer has the utility function $U=\ln C+\ln(24-N)$, where C is consumption and N is labor supply. Her budget constraint is $pC=\bar{M}+wN$, where p is the price of the consumption good, w the wage rate, and $\bar{M}$ the consumer's nonwage income.
   (a) Formulate the problem of utility maximization subject to the budget constraint, and derive the first-order conditions, using the Lagrange multiplier approach and ignoring the nonnegativity constraints.

(b) Find the demand function $C = C^*(p, w, \bar{M})$ and the labor supply function $N = N^*(p, w, \bar{M})$ (i.e., express C and N in terms of p, w, and $\bar{M}$). Show that N^* and C^* are homogeneous of degree zero in $(p, w, \bar{M})$.

(c) Let $U^* = \ln[C^*(p, w, \bar{M})] + \ln[24 - N^*(p, w, \bar{M})]$. Show that $\partial U^*/\partial \bar{M} > 0$ and $\partial U^*/\partial p < 0$. Show that U^* is concave in $\bar{M}$ and convex in p. What is the relationship between $\partial U^*/\partial \bar{M}$ and the Lagrange multiplier?

6. Two countries, 1 and 2, import a nonstorable resource from a third country in order to produce their domestic output. The production functions of the two countries are $q_1 = 4(x_1)^{1/2}$ and $q_2 = 2(x_2)^{1/2}$, where q_i is output and x_i input for country i, $i = 1, 2$. The amount of resource available is $X = 500$.

(a) Calculate the allocation of the resource that would yield the largest total output for the two countries under the constraint that all the resource is used up. Supposing that one unit of output is by definition worth \$1, how much would the importing countries be willing to pay for an extra unit of resource (the shadow price of the resource)? Supposing that this shadow price is the actual price paid and that it costs the third country \$0.1 to extract the resource, calculate the profit made by each of the three countries.

(b) Suppose now that the third country unilaterally chooses the resource price; denote it by p. Country 1 maximizes profit, given p; find its resource import as a function of p. Do the same for country 2. Use these results to express the third country's export revenue as a function of p. Recalling that it costs \$0.1 to extract the resource, find the value of p that maximizes the third country's profit. How much of the resource is unused? Would the first two countries be able to bribe the third country into returning to full use of the resource?

7. Let the production function be $q = x_1^{1/2} x_2^{1/2}$ and the input prices w_1 and w_2. There is no fixed cost. Derive the total cost function and check its concavity properties. Suppose now that the production function is $Q = f(x_1^{1/2} x_2^{1/2})$, where f is a strictly increasing function. Use the transformation $q = f^{-1}(Q) = x_1^{1/2} x_2^{1/2}$ and your earlier result to obtain the cost function.

8. In this exercise we must determine how many of two sorts of trees must be planted now with scarce labor and what are the optimal harvesting dates for the trees (harvesting is assumed not to require labor). The use of the land after harvesting is not considered here.

x	number of small trees,
$g(x)$	labor used to plant x small trees,
t	harvesting date of small trees,
X	number of large trees,
$G(X)$	labor used to plant X large trees,
T	harvesting date of large trees,
$f(t)$	revenue at time t from the sale of each small tree at time t,
$F(T)$	revenue at time T from the sale of each large tree at time T,
L	fixed labor supply, available now,
r	fixed exponential rate of interest (the present value of \$$a$ received at time θ is $ae^{-r\theta}$).

Both $g(x)$ and $G(X)$ are increasing and convex, while both $f(t)$ and $F(T)$ are increasing and concave.

(a) Derive the first-order conditions for maximizing the present value of total revenue from the sale of trees, subject to the labor constraint – x, X, t, and T are the choice variables. What is the economic meaning of these conditions?

(b) Assume that a regular maximum is obtained. Find the signs of dt/dL, dt/dr, dT/dL, and dT/dr. (*Hint:* Simplify the first-order conditions for t and T by canceling x, X, and the exponential terms before deriving comparative statics results.) Use these results to indicate what information is relevant to choosing harvesting dates and what information is relevant to choosing crop sizes. Finally, derive a numerical solution in the following case: $f(t)=2t^{1/2}$, $F(T)=8T^{1/2}$, $g(x)=2x$, $G(X)=X^2$, $L=20$, and $r=2\%$.

9. A firm produces one good in amount q for a price p. Its total cost of production is $C(q,k)$, where k is the amount of capital available and paid for. C is increasing and convex in all its arguments. Derive the profit-maximizing conditions in the short run (k is fixed) and the long run (k is chosen). Derive an expression for the firm's changes in output in response to a price change, dq/dp, in the short run and in the long run. Determine their signs and compare their absolute values. Interpret your results.

10. A manager is responsible for n machines. In any given period, the probability that exactly k machines will break down is $\pi_k>0$ $(\pi_0+\pi_1+\pi_2+\cdots\pi_n=1)$, where k is any whole number between 0 and n: $k=0,1,2,\ldots,n$. The π_k are "true" probabilities but are known only to the manager. The manager must report these probabilities to the control authority. In the absence of any reward or penalty, he might report some wrong probabilities $P_0, P_1, \ldots, P_n$ $(P_0+P_1+\cdots+P_n=1)$.

The central authority devises the following scheme: if k machines break down, and the reported probability is P_k, the authority will pay the manager $A+M\ln P_k$ dollars, where M and A are positive constants. The manager's expected reward is therefore

$$\text{ER}=\sum_{k=0}^{n}\pi_k(A+M\ln P_k). \tag{1}$$

The manager wants to choose $P_0,\ldots,P_n$ to maximize (1), subject to the constraints that

$$P_0+P_1+\cdots+P_n=1, \quad P_k\geq 0. \tag{2}$$

(a) Find the first-order conditions (ignore nonnegativity restrictions).

(b) Are the second-order conditions satisfied?

(c) Show that it is optimal for the manager to report the truth, i.e., $P_k=\pi_k$.

11. Consider an economy with two industries (each producing one good) and one resource. The resource can be used directly as an input in each industry, or it can be used indirectly to develop a technology applicable to both industries.

The level of technology is denoted by K; the amount of resource used directly as an input in industry i is denoted by x_i, and the industries' production functions and prices are $f^i(x_i, K)$ and p_i, $i=1,2$. In order to produce the level of technology K, it is necessary to use $H(K)$ units of the resource. We assume that f^1 and f^2 are strictly increasing and strictly concave in all their arguments and that $H(K)$ is strictly increasing and convex. The total amount of resource available is L, fixed.

(a) In the first instance suppose that the level of technology K is fixed. The amount of resource available for use as input is denoted by $X \equiv L - H(K)$.

(i) Formulate the central planner's problem of maximizing total revenue in both industries subject to the resource constraint (use X). Derive the first-order conditions and interpret them. Check that the second-order conditions are satisfied. Find the sign of the rate of change in x_2 if X changes exogenously; explain your finding. Find the expression for the rates of change in x_1 and in x_2 when K changes exogenously (X remains constant); interpret your results. Is it possible that the improvement in technology leaves the production of all goods unchanged?

(ii) Let the price of the resource input be denoted by w. Formulate the twin problems of maximizing profit separately in industries 1 and 2, taking all prices as given. Derive the optimality conditions. Can you use the solution in (i) to identify the input price that would generate a total input demand by industries just equal to X? Is this an instance of the decentralizing role of prices?

(b) Let

$$f^1(x_1, K) = 16\sqrt{2}(x_1)^{1/2}(K)^{1/4}, \quad p_1 = 1,$$

$$f^2(x_2, K) = 8\sqrt{2}(x_2)^{1/2}(K)^{1/4}, \quad p_2 = 2,$$

$$L = 160, \quad H(K) = 2K^2.$$

(i) Let $K=1$. Use the above data to solve the problem of (a)(i). (*Hint:* Make use of the symmetry of the example.)

(ii) Let $K=1$. Use the above data to derive total input demand as a function of w when firms act as in (a)(ii). Find the equilibrium value of w.

(c) In this section the level of technology K is to be optimally chosen. (Use L and $H(K)$ in the constraint, not X.)

(i) Formulate the central planner's problem as in (a)(i). Derive and interpret the first-order conditions from the central planner's point of view.

(ii) Use the data of (b) – but not $K=1$ – to solve the problem in (c)(i).

(iii) Let the price of the resource be denoted by w and the price of technology (per unit) by r. In an attempt to decentralize the allocation of resources, let each industry $i=1,2$ purchase x_i of the resource for use as input and K_i of the technology for the *common* use; thus, each industry can use the level (K_1+K_2) of technology, and the production function of industry i is $f^i(x_i, K_1+K_2)$, with $K_1+K_2=K$. In addition there is now a need for a "technology industry" with output K at price r and cost function $wH(K)$. Derive the profit-maximizing conditions for all three industries. Compare this solution with that of (c)(i) and attempt to explain any discrepancy. Does the "invisible hand" work its magic here?

(iv) Solve the problem of (c)(iii) using the data in (b) – but not $K=1$. (*Hint:* First eliminate r and w from the optimality conditions.) Compare your results with those obtained in (c)(ii).

12. A monopolist wishes to maximize profit, but misjudges demand conditions. We denote by q the quantity of output, $p(q)$ is the demand price, and $C(q)$ is the cost function. The monopolist thinks $\alpha p(q)$ is the demand function, where α is a positive parameter. We shall call $\alpha p(q)$ the expected price.
    (a) Derive the first-order and second-order conditions for a maximum expected profit, π^e (ignore nonnegativity restrictions). Suppose the second-order condition holds strictly and find the expression for $dq/d\alpha$ and its sign.
    (b) The actual profit π^a is defined using the actual price $p(q)$; how is it affected by α? Are your results sensible?
    (c) The difference $W=\pi^a-\pi^e$ is called the unexpected windfall profit. How does its sign depend on α? Find the sign of $dW/d\alpha$; when α is less than 1, is it possible that lowering it further increases windfall profit? Can you rationalize this?
13. Derive the Kuhn–Tucker conditions for the following nonlinear programming problems. Verify that they all are regular concave programming problems and derive the solution. (*Hint:* First illustrate graphically and attempt to guess which variables are positive and which are zero in the optimal solution.)

    (a) $$\underset{x_1\geq 0,\, x_2\geq 0}{\text{Maximize}}\ -8(x_1)^2-10(x_2)^2+12x_1x_2-50x_1+80x_2$$
    subject to $x_1+x_2\leq 1$ and $8(x_1)^2+(x_2)^2\leq 2.25$.

    (b) $$\underset{x_1\geq 0,\, x_2\geq 0}{\text{Maximize}}\ x_1+2x_2-(x_1)^2+3x_1x_2-3(x_2)^2$$
    subject to $2x_1+x_2\leq 2$ and $-x_1+x_2\leq -1$.

    (*Hint:* The optimal values of the dual variables may not be unique.)

    (c) $$\underset{x_1\geq 0,\, x_2\geq 0}{\text{Maximize}}\ 6x_1-2(x_1)^2+2x_1x_2-2(x_2)^2$$
    subject to $3x_1+4x_2\leq 6$ and $-x_1+4(x_2)^2\leq -\frac{1}{3}$.

    (d) $$\underset{x_1\geq 0,\, x_2\geq 0}{\text{Maximize}}\ 100+\ln x_1+\ln x_2$$
    subject to $98-(x_1)^2-(x_2)^2\geq 0$ and $418-(x_1)^2-6(x_2)^2\geq 0$.

14. Two people live together. They have separate incomes. They buy some good (X) for their own consumption (e.g., food or clothing). Some other good (Z) is enjoyed by both in the sense that each person enjoys the purchases of both people (e.g., heating or home improvements). Both people are selfish and maximize their own utility given their own budget constraint. Person i chooses $x_i \geq 0$ and $z_i \geq 0$ to maximize $U^i(x_i, z_1+z_2)$ subject to $px_i + \pi z_i \leq y_i$, $i=1,2$. Calculate the resulting equilibrium when $U^1 = \ln x_1 + \ln(z_1+z_2)$; $U^2 = \ln x_2 + 2\ln(z_1+z_2)$; $p=\pi=1$; $y_1=10$; $y_2=20$. Do they both free-ride (i.e., buy no good Z)? Does one? Can you suggest an improvement in their living arrangement?
15. Consider a profit-maximizing competitive firm that produces two goods with fixed amounts of land and capital; labor is also used as an input and is available at the going wage rate:

Amounts of goods sold	q_1, q_2 at prices $p_1 > 0$, $p_2 > 0$,
Labor used in the production of the goods	x_1, x_2 with wage rate $w > 0$,
Capital used in the production of the goods	k_1, k_2; total available $K > 0$,
Land used in the production of the goods	l_1, l_2; total available $L > 0$.

The production functions $F_1(x_1, k_1, l_1)$, $F_2(x_2, k_2, l_2)$ are concave.

(a) Set up the problem, derive the optimality conditions, and give an economic interpretation of all Lagrange multipliers and of the optimality conditions.

(b) Give a numerical solution to the problem when

$$p_1 = p_2 = 1, \quad w = 10, \quad K = 50, \quad L = 150,$$
$$q_1 = 110x_1 + 100k_1 + 100l_1 - x_1^2 - k_1^2 - l_1^2,$$
$$q_2 = 310x_2 + 300k_2 + 300l_2 - 5x_2^2 - 5k_2^2 - 5l_2^2.$$

What price would the firm pay for an extra unit of capital or land?

16. The object of this exercise is to calculate the total cost of producing a specified amount Q of a final good. To produce the final good, some labor input l_1 is needed, as is some input x of an intermediate good. The production function is denoted by $F(l_1, x)$ and is increasing and concave. To produce x units of the intermediate good, an input l_2 of labor is needed, as is some input q of the final good. The production function $f(l_2, q)$ is also increasing and concave. Note that the desired amount Q of the final good is the *net* output of that good; some of the gross output goes into producing x.

(a) Write down for each of the two goods a constraint whereby demand cannot exceed supply. Denote the wage rate by w and formulate the above minimum (labor) cost problem – no variable can be negative; Q is fixed. Derive the optimality conditions and interpret them.

(b) Give a numerical solution for the problem when $w=1$, $Q=1$, $F(l_1, x) = 2(l_1)^{1/2} + x - 4$, and $f(l_2, q) = l_2 + 4(q)^{1/2}$. What is total cost? How much would you pay for an extra unit of the intermediate good?

(c) Using the same production function derive the cost function for arbitrary values $w > 0$ and $Q \geq 0$. (You might have to distinguish between $Q \leq 2$ and $Q \geq 2$.)

17. A firm supplies a nonstorable product during several periods; demand differs from period to period. The inputs are labor and capital; labor can be adjusted freely from period to period but capital cannot. The firm chooses the capital and labor inputs to minimize total cost subject to meeting demand in all periods.

T	number of periods ($t=1,\dots,T$),
x_t	labor input in period t, $x_t \geq 0$, $t=1,\dots,T$,
k	capital input, $k \geq 0$,
r	price of capital, } given positive parameters
w	price of labor, } given positive parameters
q_t	demand in period t, $t=1,\dots,T$, } given positive parameters
$f(x_t,k)$	amount produced in period t, $t=1,\dots,T$.

The function f incorporates a capacity constraint in the sense that, given $k>0$, there exists a positive value of x_t, say $\bar{x}(k)$, that maximizes $f(x_t,k)$. Set up the problem, derive the Kuhn–Tucker conditions, and interpret them. Show that it is optimal to have excess capacity in every period (i.e., $x_t<\bar{x}(k)$ for all t), assuming the Kuhn–Tucker conditions are optimal. Solve the problem when $T=2$, $q_1=q_2=q$, $w=r=1$, and $f(x_t,k)=kx_t-0.5(x_t)^2$.

18. A consumer spends his income M on bread B, high-quality apples H, and low-quality apples L. His budget constraint is $P_H H+P_L L+P_B B \leq M$. Assume that $P_H=T+C_H>0$, $P_L=T+C_L>0$, and $P_B=C_B>0$, where C_H, C_L, and C_B are unit costs of production and T is the transport cost per apple. Suppose that his utility function is $U=U(H,L,B)$ $(H\geq 0,\ L\geq 0,\ B\geq 0)$, where U is strictly concave, with continuous and positive first-order partial derivatives, and, for all positive H and L, $U_H/U_L>\alpha>1$, where α is a constant.
   (a) Formulate the utility maximization problem using the Kuhn–Tucker method. Derive the first-order conditions. Show that all income is spent. Show that both H and L are consumed in positive quantities only if $(P_H/P_L)>\alpha$. If $P_H/P_L\leq\alpha$, which sort of apple is not consumed?
   (b) Let $K=H/L$. Suppose that for all $L>0$ and $H>0$, the ratio U_H/U_L depends on K alone, i.e., $U_H/U_L=\phi(K)$, and that $\phi'(K)<0$. Show that if $H>0$ and $L>0$, then $dK/dT>0$. What does this mean? (How does the transport cost affect the price differential?)
   (c) Verify that if $U(H,L,B)=V(H,L)W(B)$ and $V(H,L)$ is homogeneous, the assumption in (b) is satisfied.

19. A firm specializes in the processing of coconuts. The two main products are coprah and coconut oil. Further processing of the husks (using labor) yields floor matting and infant mattress filling. For simplicity, units are defined in such a way that one unit of raw coconuts can yield up to one unit *each* of coprah, oil, matting, and filling. The manager's problem is to decide how many units of coconuts to purchase and how much of each product to manufacture.

X	quantity of raw coconuts purchased,
q_i	quantity of product i manufactured, where $i=1$, 2, 3, and 4 refer to coprah, oil, matting, and filling, respectively,
$R_i(q_i)$	revenue from the sale of q_i units of product i,

c	unit cost of buying coconuts and processing them for coprah and oil, a positive constant,
$f(q_3, q_4)$	additional labor cost required to obtain q_3 units of matting and q_4 units of filling.

R_i is increasing and concave, $i = 1, \ldots, 4$, and f is increasing and convex.

(a) Set up the problem of maximizing total profit subject to the capacity constraints $q_i \leq X$, $i = 1, \ldots, 4$. Derive the optimality conditions.

(b) Analyze the optimality conditions. In particular, determine whether all coconuts purchased are always processed to yield all four products. How is the cost of coconuts allocated among the final products?

(c) Solve this problem numerically in the following case:

$$R_1(q_1) = 4(q_1)^{1/4}, \quad R_2(q_2) = q_2, \quad R_3(q_3) = 0.5q_3, \quad R_4(q_4) = 0.5q_4,$$
$$c = 2, \quad f(q_3, q_4) = 0.5(q_3)^2 + 0.5(q_4 + 1)^2.$$

CHAPTER 2

Ordinary differential equations

2.1 Introduction

The remainder of this book is devoted to the analysis of dynamic economic models. In these models time is an independent argument and the variables are all functions of time. The process of change in these variables over time has to be described. If we let the time argument be a real number, the description of these processes will necessarily involve the derivatives of some functions with respect to time.

Notation. Let t denote the real-valued time argument and let the value of some variable (which depends on t) be denoted by $x(t)$. Then the *total* derivative of x with respect to t is denoted by

$$\frac{dx(t)}{dt}, \quad \text{or } \dot{x}(t), \quad \text{or } \dot{x} \text{ for simplicity.} \tag{2.1}$$

In order to describe the process of change we must link the rate of change in x to the values of x and t; hence, we shall need equations of the type $f(x(t), \dot{x}(t), t) = 0$. This is an example of a differential equation. A solution to this equation is a function $x^*(t)$ such that $x^*(t)$ and $\dot{x}^*(t)$ satisfy the preceding expression identically. An ordinary differential equation links a single independent variable, a function (or functions) of that variable, and its derivatives. In this book the independent variable will always be time. (The term "ordinary" is used to indicate that there is only one independent variable. Otherwise, partial derivatives would appear and we would have a partial differential equation; these are not considered here.)

When the time variable is integer-valued we have equations of the type $f(x(t), x(t-1), t) = 0$, say. This is an example of a difference equation. We shall encounter some difference equations in Chapters 4 and 5, but these will be simple enough to be solved by recursion. The interested reader is referred to Goldberg (1958) for a thorough introduction.

Example 2.1.1. Consider a net investment stream that decreases in value over time according to the formula $I(t) = 100e^{-0.05t}$. Equating the value of net investment to the rate of change in capital stock at each date t, we get

$$\dot{K}(t)=\frac{dK}{dt}=100e^{-0.05t}, \tag{2.2}$$

where $K(t)$ is the value of capital stock at date t. This is a differential equation involving $\dot{K}(t)$ and t and can be solved by simple integration:

$$\int \frac{dK}{dt}\,dt=\int 100e^{-0.05t}\,dt,$$

$$K(t)=-\frac{100}{0.05}e^{-0.05t}+C, \tag{2.3}$$

where C is an arbitrary constant. The solution of differential equations always involves some integration procedure, and as a result arbitrary constants appear. Expression (2.3) is known as the general solution to equation (2.2) because of the presence of the undetermined constant C. In order to determine $K(t)$ exactly, we need to know its exact value at one particular date: say, at date $t=0$, we have $K=1{,}000$, or more compactly, the initial condition is $K(0)=1{,}000$. Substituting these values in (2.3) yields $1{,}000=-2{,}000(e^0)+C$ or $C=3{,}000$, thereby giving the particular solution

$$K(t)=3{,}000-2{,}000e^{-0.05t}. \tag{2.4}$$

Had we had another initial condition, we would have had another particular solution. Another feature of interest is that as time becomes large the solution in (2.4) converges to 3,000. In other examples the solution may become arbitrarily large and/or oscillate around some line.

We are now ready to state some formal results. More details can be obtained from many sources – for instance, Brauer and Nohel (1969), Coddington and Levinson (1955), or Pontryagin (1962).

2.2 Definitions and fundamental results

An ordinary differential equation of *order m* expresses a relation between a variable t, a function x of t, and the derivatives, up to the mth order of x with respect to t:

$$f(t,x,\dot{x},x^{(2)},\ldots,x^{(m)})=0, \tag{2.5}$$

where $x^{(i)}$ is the ith-order derivative of x with respect to t, $i=2,\ldots,m$. Thus, the *order* of a differential equation is given by the highest-order derivative entering the equation.

A differential equation of order m is said to be *normal* when it can be written as

$$x^{(m)} = g(x^{(m-1)}, \ldots, x^{(2)}, \dot{x}, x, t). \tag{2.6}$$

Only normal differential equations will be dealt with here.

If $\mathbf{x}^{(m)}, \mathbf{x}^{(m-1)}, \ldots, \dot{\mathbf{x}}$, and $\mathbf{x}$ are $(n \times 1)$ vectors, (2.6) represents an n-dimensional *system of differential equations of order m*.

If t does not appear as a distinct argument of g, equation (2.6) is said to be *autonomous*.

A *solution* to the differential equation (2.6) is a function $x^*(t)$ defined over some domain D that admits derivatives up to the mth order, also defined on D, such that this function and its derivatives satisfy (2.6) identically when t is in D.

A *general solution* to (2.6) will contain m arbitrary constants, and a *particular solution* to (2.6) will contain no arbitrary constants. If (2.6) is an n-dimensional system, we need $m \times n$ arbitrary constants for a general solution.

We call *initial conditions* for (2.6) the m conditions

$$x(t_0) = x_0,\ \dot{x}(t_0) = \dot{x}_0,\ x^{(2)}(t_0) = x_0^{(2)},\ \ldots,\ x^{(m-1)}(t_0) = x_0^{(m-1)}, \tag{2.7}$$

where $x_0, \dot{x}_0, x_0^{(2)}, \ldots, x_0^{(m-1)}$ are specified values. In the case of an n-dimensional system each of these conditions is itself $n \times 1$.

More generally, we call *boundary conditions* for equation (2.6) a set of m conditions that may or may not involve the derivatives of x; for instance,

$$x(t_1) = x_1, x(t_2) = x_2, \ldots, x(t_m) = x_m \tag{2.8}$$

could be used. In the case of an n-dimensional system each of these conditions would be $n \times 1$.

Theorem 2.2.1: existence and uniqueness. Consider the system of differential equations in (2.6) and assume that g has continuous partial derivatives on $X \times D$ with respect to all its arguments. Then for each set of initial conditions such as (2.7) that belongs to X, and if t_0 belongs to D, there exists a unique solution to system (2.6), valid for $t \in d$, where d is some subinterval of D.

A single equation of order m can always be reduced to a system of m first-order equations by the following very useful transformation:

$$y_1 = x, y_2 = \dot{x}, \ldots, y_{m-1} = x^{(m-2)}, y_m = x^{(m-1)}.$$

Equation (2.6) is equivalent to the system

$$\begin{aligned} \dot{y}_1 &= y_2, \\ \dot{y}_2 &= y_3, \\ &\vdots \\ \dot{y}_{m-1} &= y_m, \\ \dot{y}_m &= g(y_m, y_{m-1}, \ldots, y_2, y_1, t). \end{aligned} \tag{2.9}$$

Therefore, a differential equation of order m can be treated as a special case of a system of m first-order differential equations when it is convenient to do so. Hereafter, we turn our attention to systems of first-order differential equations.

An *equilibrium* of the autonomous first-order system of differential equations

$$\dot{\mathbf{x}} = \mathbf{F}(\mathbf{x}), \tag{2.10}$$

where $\mathbf{x}$, $\dot{\mathbf{x}}$, and $\mathbf{F}$ are n-dimensional vectors, is a point $\bar{\mathbf{x}}$ such that $\mathbf{F}(\bar{\mathbf{x}}) = \mathbf{0}$. (If the system were not autonomous the "equilibrium" would depend on t and would be much less amenable to analysis.)

An equilibrium point $\bar{\mathbf{x}}$ of (2.10) is said to be *stable* if any solution $\varphi(t)$ of (2.10) with initial condition $\mathbf{x}(t_0) = \mathbf{x}_0$, with $\mathbf{x}_0$ "close" to $\bar{\mathbf{x}}$, remains in the neighborhood of $\bar{\mathbf{x}}$ for all $t \geq t_0$.

An equilibrium point $\bar{\mathbf{x}}$ of (2.10) is said to be *asymptotically stable* if it is stable and if there exists a neighborhood of $\bar{\mathbf{x}}$ such that for any solution $\varphi(t)$ that begins in this neighborhood we have $\lim_{t \to +\infty} \varphi(t) = \bar{\mathbf{x}}$.

An equilibrium point is said to be *unstable* if it is not stable. Note that the concept of stability requires only that small perturbations of the equilibrium yield a solution that remains close to the equilibrium, whereas asymptotic stability requires that the solution eventually (in infinite time) return to the equilibrium.

An equilibrium point $\bar{\mathbf{x}}$ is said to be *globally asymptotically stable* if the neighborhood of $\bar{\mathbf{x}}$ in the definition of *asymptotic stability* is extended to the whole domain of definition of $\mathbf{F}$.

For the remainder of this chapter we have two objectives: (i) to present techniques that will enable readers to solve some simple differential equations such as the ones used in this volume (for this, readers should refresh their knowledge of the basic rules of integration; an appendix to this chapter is provided for that purpose); and (ii) to present an introduction to the qualitative theory of differential equations. We do not aim for a thorough coverage of these topics; for this readers are referred to the works cited at the end of the preceding section.

2.3 First-order differential equations

First-order differential equations (FODE) are single equations with x, $\dot{x}$, and t as the only arguments. We examine various simple types.

2.3.1 *Linear FODE with constant coefficients*

This type of equation has the form

$$\dot{x}+ax=b, \tag{2.11}$$

where a and b are constants. We first consider the case $a \neq 0$. To solve for $x(t)$ we multiply both sides by e^{at}, and the left-hand side becomes the derivative of $e^{at}x(t)$. Hence, integrating both sides yields

$$e^{at}x=\left(\frac{b}{a}\right)e^{at}+C,$$

where C is an arbitrary constant. Finally,

$$x(t)=\left(\frac{b}{a}\right)+Ce^{-at} \tag{2.12}$$

is the general solution. C can be determined with the help of an initial condition. In the above derivation we multiplied both sides by e^{at}; this term is called an *integrating factor.* From equation (2.11) the equilibrium is $\bar{x}=b/a$, and from the general solution (2.12) it is globally asymptotically stable if $a>0$; it is unstable if $a<0$. (If $a=0$, the general solution is obviously $x(t)=bt+C$ and is unbounded unless $b=0$.)

In the case where $b=0$ equation (2.11) reduces to

$$\dot{x}+ax=0. \tag{2.13}$$

This equation is said to be *homogeneous,* since if $x(t)$ is a solution, so is $k\cdot x(t)$, where k is any constant. The general solution to (2.13) is obtained by setting $b=0$ in (2.12): $x(t)=Ce^{-at}$. Note that the general solution to equation (2.11) (called a nonhomogeneous equation because of the presence of b) is the sum of the general solution to the homogeneous equation (2.13) plus the equilibrium solution of (2.11), which, because it involves no arbitrary constant, is seen to be a particular solution to (2.11). We shall see that this result also applies to higher-dimensional linear systems.

2.3.2 *Linear FODE with variable coefficients*

The coefficients of this equation depend on t; it takes the form

$$\dot{x}+a(t)x=b(t), \tag{2.14}$$

where $a(t)$ and $b(t)$ are known functions of t. The appropriate integrating factor is

$$I(t)=\exp\left[\int a(t)\,dt\right].$$

Multiplying both sides of (2.14) by $I(t)$ yields

$$I(t)\dot{x}+a(t)I(t)x=I(t)b(t), \tag{2.15}$$

and we see that the left-hand side of (2.15) is the derivative of $I(t)x$. Hence, integrating (2.15) yields

$$I(t)x=\int I(t)b(t)\,dt,$$

or

$$x(t)=[I(t)]^{-1}\int I(t)b(t)\,dt, \tag{2.16}$$

where an arbitrary constant is implicitly included in the integral and can be determined with an initial condition.

Example 2.3.1. Consider the equation

$$\dot{x}+2tx=5t.$$

Here $a(t)=2t$ and $b(t)=5t$; hence, $I(t)=e^{t^2+k}$, where k is arbitrary. Indeed, we can set $k=0$ without loss. By equation (2.16) the general solution is

$$\begin{aligned}x(t)&=e^{-t^2}\int 5te^{t^2}dt\\&=e^{-t^2}\left[\frac{5}{2}e^{t^2}+C\right]\\&=\frac{5}{2}+Ce^{-t^2}.\end{aligned}$$

Suppose that the initial condition is $x(t_0)=x_0$; then the particular solution can be determined by

$$x(t_0)=x_0=\tfrac{5}{2}+Ce^{-(t_0)^2},\quad C=(x_0-\tfrac{5}{2})e^{t_0^2},\quad\text{and}$$
$$x(t)=\tfrac{5}{2}+(x_0-\tfrac{5}{2})e^{-(t^2-t_0^2)}.$$

Note that if $x_0=\frac{5}{2}$, the solution reduces to $x(t)=\frac{5}{2}$ for all t. This identifies $\frac{5}{2}$ as the equilibrium solution: if the trajectory begins there, it stays there forever. Note also that it can be obtained from the original differential equation by setting $\dot{x}=0$. However, in most nonautonomous differential equations, of which linear FODE with variable coefficients are a special case, there is no fixed equilibrium point because of the presence of the t argument.

2.3.3 *Nonlinear FODE*

Nonlinear FODE cannot be put in the form of (2.14); in general they take the form $f(x,t)\dot{x}=g(x,t)$, where g is not linear in x. Nonlinear FODE are difficult to handle except for special cases, some of which are discussed below.

Separable equations. A differential equation is separable if it can be written as

$$f(x)\dot{x}=g(t). \tag{2.17}$$

Integrating both sides with respect to t yields

$$\int f(x)\frac{dx}{dt}\,dt=\int g(t)\,dt, \quad \text{or} \quad \int f(x)\,dx=\int g(t)\,dt. \tag{2.18}$$

This expression can be used to obtain the solution $x(t)$, at least in implicit form.

Example 2.3.2

$$3x^2\dot{x}=4t^3,$$

$$\int 3x^2\dot{x}\,dt=\int 3x^2\,dx=\int 4t^3dt, \quad \text{or} \quad x^3=t^4+C,$$

$$x(t)=[t^4+C]^{1/3}.$$

Bernoulli equations. An equation of the form

$$\dot{x}+u(t)x=w(t)x^n, \tag{2.19}$$

where n is any real number other than 0 or 1, is called a Bernoulli equation. It can be reduced to a FODE by a transformation of variables. Divide both sides by x^n to get

$$x^{-n}\dot{x}+u(t)x^{1-n}=w(t).$$

Define $z=x^{1-n}$ so that $\dot{z}=(1-n)x^{-n}\dot{x}$; then the preceding equation becomes

$$\left[\frac{1}{(1-n)}\right]\dot{z}+u(t)z=w(t),$$

which is a linear FODE in z. Once solved, $x(t)=[z(t)]^{1/(1-n)}$ is obtained.

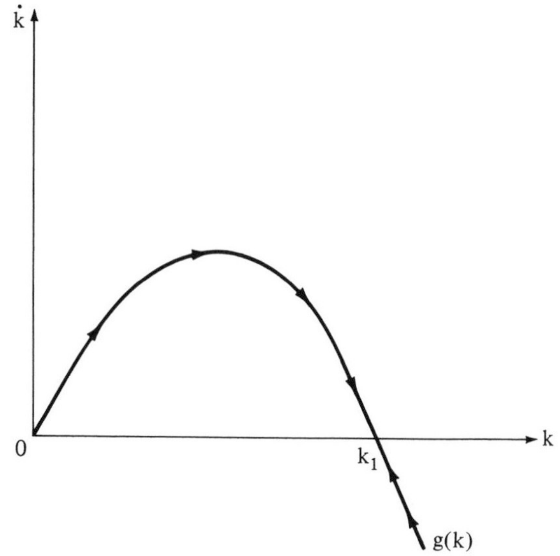

Figure 2.1

2.3.4 *Qualitative analysis on a phase diagram*

While many nonlinear FODE cannot be solved analytically, the qualitative properties of their solutions can sometimes be described by a graphic device. Suppose the equation is $\dot{x} = g(x)$. We can plot the graph of $\dot{x}$ against x; this graph is called the *phase line*. We illustrate the usefulness of this approach in the following example.

Example 2.3.3. An economy produces an output q from its capital stock k; the production function is $q = f(k)$, where f is an increasing and strictly concave function, with $f(0) = 0$, $\lim_{k\to\infty} f'(k) = 0$, and $\lim_{k\to 0} f'(k) = b > 0$. A constant fraction s of total output is saved and invested, so that gross investment is $sf(k)$. The capital stock depreciates at the rate $n > 0$ (a discussion of depreciation is deferred until Chapter 3); therefore, the net rate of capital accumulation is

$$\dot{k} = sf(k) - nk \equiv g(k). \tag{2.20}$$

According to the above assumptions we have $g(0) = 0$, $g'(0) = sb - n$, $\lim_{k\to\infty} g'(k) = -n$, and $g''(k) < 0$. If we assume that $sb - n > 0$, the graph of $g(k)$ cuts the horizontal axis twice at $k = 0$ and, say, $k = k_1 > 0$. This graph is drawn in Figure 2.1. The points k_1 and 0 are equilibrium points since $\dot{k} = 0$ at these points; this could also be seen from equation (2.20).

The phase diagram of Figure 2.1 can be used to examine the stability of these equilibria. On the phase line we have added some arrows. For all values of k within the open interval $(0, k_1)$, $g(k)$ is positive, indicating that $\dot{k} > 0$ – hence, k must be rising; this is why the arrows in this interval point toward large k values. For values of $k > k_1$, $g(k) < 0$, so that k decreases. The arrows point toward the left, accordingly. We can see that the equilibrium at k_1 is asymptotically stable, since for any starting point around k_1, $k(t)$ will approach k_1. The equilibrium at $k = 0$ is unstable, because there is no neighborhood of 0 from which $k(t)$ will tend to 0. (Indeed, if we rule out starting at $k = 0$, the equilibrium at k_1 is globally asymptotically stable.)

Had we assumed $sb < n$, the function $g(k)$ would have been negative for all $k > 0$ and the unique equilibrium at 0 would have been globally asymptotically stable.

2.4 Systems of linear FODE with constant coefficients

Consider the n-dimensional system of differential equations

$$\dot{\mathbf{x}} = \mathbf{A}\mathbf{x} + \mathbf{b} \tag{2.21}$$

and the associated homogeneous system

$$\dot{\mathbf{x}} = \mathbf{A}\mathbf{x}. \tag{2.22}$$

The general solution of (2.22) is relatively easy to obtain, and a particular solution of (2.21) is its equilibrium point given by $\bar{\mathbf{x}} = -\mathbf{A}^{-1}\mathbf{b}$ (defined if $\mathbf{A}$ is nonsingular). For this reason the following lemma is important.

Lemma 2.4.1. If $\mathbf{x}_1(t)$ is a solution to (2.21) and $\mathbf{x}_2(t)$ is a solution to (2.22), then $\mathbf{x}_1(t) + \mathbf{x}_2(t)$ is also a solution to (2.21).

The proof is obvious and follows from the linearity of the two systems. Suppose now that we have the general solution to (2.22), $\mathbf{x}^*(t)$ say, and a particular solution to (2.21) (e.g., the equilibrium $\bar{\mathbf{x}}$). Then $\mathbf{x}^*(t) + \bar{\mathbf{x}}$ is a solution to (2.21) by Lemma 2.4.1, and since it contains n arbitrary constants it is the general solution to (2.21). Thus, the behavior of the solution to (2.21) around its equilibrium point $\mathbf{x} = \bar{\mathbf{x}}$ is identical to that of the solution to (2.22) around its equilibrium point $\mathbf{x} = \mathbf{0}$. For this reason we can restrict our attention to homogeneous systems when examining the qualitative behavior of solutions. In what follows we restrict our attention to two-dimensional systems as a way of illustrating the types of trajectories that may emerge. Most results are given without proof; for a more detailed treatment the reader is referred to Brauer and Nohel (1969, sec. 2.8), for instance.

2.4.1 *Algebraic solutions*

We consider the homogeneous system

$$\begin{bmatrix} \dot{x}_1 \\ \dot{x}_2 \end{bmatrix} = \begin{bmatrix} a_{11} & a_{12} \\ a_{21} & a_{22} \end{bmatrix} \begin{bmatrix} x_1 \\ x_2 \end{bmatrix}, \tag{2.23}$$

or more compactly,

$$\dot{\mathbf{x}} = \mathbf{A}\mathbf{x}. \tag{2.23$'$}$$

As in the single-equation case the solution will involve exponential functions of the type $e^{\lambda t}$. To see how these emerge, suppose a particular solution is $\mathbf{x} = \mathbf{a}e^{\lambda t}$, where $\mathbf{a}$ is a vector of constants, not all zero. Deriving $\dot{\mathbf{x}}$ and substituting in (2.23′), we obtain $\dot{\mathbf{x}} = \lambda\mathbf{a}e^{\lambda t} = \mathbf{A}\mathbf{a}e^{\lambda t}$. Hence, $\lambda\mathbf{a} = \mathbf{A}\mathbf{a}$ or $[\mathbf{A} - \lambda\mathbf{I}]\mathbf{a} = \mathbf{0}$. For this system to have solutions other than $\mathbf{a} = \mathbf{0}$, the matrix $\mathbf{A} - \lambda\mathbf{I}$ must be singular; that is,

$$|\mathbf{A} - \lambda\mathbf{I}| = \mathbf{0},$$

or in the case of system (2.23),

$$\lambda^2 - \operatorname{tr}\mathbf{A}\lambda + |\mathbf{A}| = 0, \tag{2.24}$$

where

$$\operatorname{tr}\mathbf{A} = a_{11} + a_{22} \quad \text{and} \quad |\mathbf{A}| = a_{11}a_{22} - a_{12}a_{21}.$$

This equation is called the characteristic equation of $\mathbf{A}$. Its roots λ_1 and λ_2 will serve to construct the general solution of (2.23) and will appear in terms such as $e^{\lambda_1 t}$ and $e^{\lambda_2 t}$. Note that $\operatorname{tr}\mathbf{A} = \lambda_1 + \lambda_2$, while $|\mathbf{A}| = \lambda_1\lambda_2$. Since equation (2.24) is quadratic, it is possible that its roots are conjugate complex numbers such as $\alpha + i\beta$ and $\alpha - i\beta$, where i is the imaginary number defined by $i^2 = -1$ and α and β are real. In this case we shall transform the expressions to obtain real-valued solutions; these will involve trigonometric functions of βt, and hence will generate oscillatory trajectories. For simplicity we assume hereafter that *neither root is zero;* this guarantees that $\mathbf{A}$ is nonsingular and that the origin is the unique equilibrium. For our purposes we can simplify the problem further. It can be shown that for any 2×2 real matrix $\mathbf{A}$, there exists a real matrix $\mathbf{T}$ such that $\mathbf{T}^{-1}\mathbf{A}\mathbf{T} = \mathbf{B}$, where $\mathbf{B}$ can take only the forms

$$\begin{aligned} &\text{(a) } \mathbf{B} = \begin{bmatrix} \lambda_1 & 0 \\ 0 & \lambda_2 \end{bmatrix}, \qquad \text{(b) } \mathbf{B} = \begin{bmatrix} \lambda & 0 \\ 0 & \lambda \end{bmatrix}, \\ &\text{(c) } \mathbf{B} = \begin{bmatrix} \lambda & 0 \\ 1 & \lambda \end{bmatrix}, \qquad \text{(d) } \mathbf{B} = \begin{bmatrix} \alpha & \beta \\ -\beta & \alpha \end{bmatrix}, \end{aligned} \tag{2.25}$$

where λ_1 and λ_2 are distinct real roots, λ is a double root, and $\alpha \pm i\beta$ are conjugate complex roots of both **A** and **B**. Then we can define by a linear transformation a new variable **y** such that $\mathbf{x} = \mathbf{T}\mathbf{y}$. Accordingly, $\dot{\mathbf{x}} = \mathbf{T}\dot{\mathbf{y}}$ and (2.23′) reduces to $\mathbf{T}\dot{\mathbf{y}} = \mathbf{A}\mathbf{T}\mathbf{y}$, or $\dot{\mathbf{y}} = \mathbf{T}^{-1}\mathbf{A}\mathbf{T}\mathbf{y}$, and finally

$$\dot{\mathbf{y}} = \mathbf{B}\mathbf{y}. \tag{2.26}$$

The solution to (2.26) differs from that of (2.23) only because it is distorted by the linear transformation $\mathbf{x} = \mathbf{T}\mathbf{y}$; they are qualitatively identical. (One can be obtained from the other by rescaling and rotation of the axes.) In the following section we study the geometric properties of (2.26) keeping in mind that **B** can take only the forms described in (2.25).

2.4.2 *Phase diagram representation of the solution*

Recall that the origin is the only equilibrium point of (2.26) under our assumption that $|\mathbf{A}| \neq 0$ (i.e., there are no zero roots). Throughout we denote the initial point by (y_1^0, y_2^0) at $t = 0$ and assume that it is not the origin.

From a diagrammatic point of view the following classification into six cases is appropriate.

Case (a)

$$\mathbf{B} = \begin{bmatrix} \lambda & 0 \\ 0 & \lambda \end{bmatrix}.$$

There is a single real root. The solution is $y_1(t) = y_1^0 e^{\lambda t}$, $y_2(t) = y_2^0 e^{\lambda t}$. The ratio $y_2(t)/y_1(t)$ is constant over time, and the trajectories are rays through the origin. This is called a *proper node*. If $\lambda < 0$, it is globally asymptotically stable; if $\lambda > 0$, it is unstable. The stable case is shown in Figure 2.2a.

Case (b)

$$\mathbf{B} = \begin{bmatrix} \lambda_1 & 0 \\ 0 & \lambda_2 \end{bmatrix} \quad \text{and} \quad \lambda_1\lambda_2 > 0.$$

The roots are real, distinct, and of the same sign. The solution is $y_1(t) = y_1^0 e^{\lambda_1 t}$, $y_2(t) = y_2^0 e^{\lambda_2 t}$. The trajectories will be stable if the roots are negative, and unstable if they are positive. The slope of the trajectories is not constant, since $y_2(t)/y_1(t) = (y_2^0/y_1^0)e^{(\lambda_2 - \lambda_1)t}$; thus, depending on the signs of y_2^0, y_1^0, and $\lambda_2 - \lambda_1$, we have different subcases; Figure 2.2b illustrates $0 < \lambda_1 < \lambda_2$, where the ratio goes to infinity. In any event the origin is called an *improper node*.

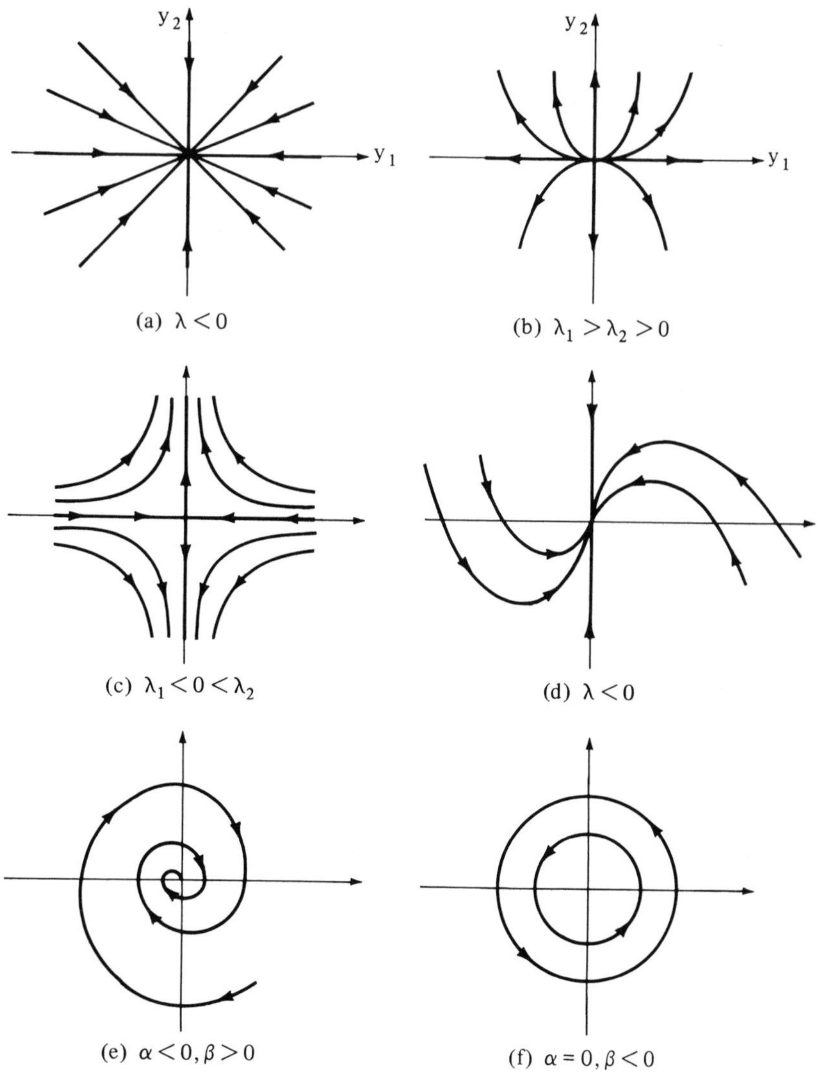

Figure 2.2

Case (c)

$$\mathbf{B} = \begin{bmatrix} \lambda_1 & 0 \\ 0 & \lambda_2 \end{bmatrix} \quad \text{and} \quad \lambda_1\lambda_2 < 0.$$

The roots are real, distinct, and of opposite signs (say, $\lambda_1 < 0 < \lambda_2$). The algebraic solution is the same as in case (b) but, at $t \to \infty$, $y_1(t) = y_1^0 e^{\lambda_1 t}$

goes to 0 and $y_2(t)=y_2^0e^{\lambda_2 t}$ goes to $\pm\infty$ depending on whether y_2^0 is positive or negative, as long as $y_2^0\neq 0$. If $y_2^0=0$, the trajectory is along the horizontal axis and is stable. For any other trajectory, $y_2(t)$ and also $y_2(t)/y_1(t)$ go to infinity. The origin is called a *saddle point* and is depicted in Figure 2.2c. (It is so named by analogy to the saddle-point concept encountered in Section 1.1.5, where it can be verified that the Hessian matrix has characteristic roots of opposite signs, just as the matrix of coefficients does here.) Such points exhibit *conditional stability* in the sense that the solution is stable for some initial points ($(y_1^0, 0)$ here) and unstable for others.

Case (d)

$$\mathbf{B}=\begin{bmatrix}\lambda & 0\\ 1 & \lambda\end{bmatrix}.$$

As in case (a) there is a single real root, but no real matrix **T** can transform **A** into a diagonal matrix and this variant is obtained. The solution includes another t term:

$$y_1(t)=y_1^0e^{\lambda t}, \qquad y_2(t)=(y_2^0+y_1^0t)e^{\lambda t}.$$

Suppose $\lambda<0$; then both $y_1(t)$ and $y_2(t)$ go to the origin as t goes to infinity. If $y_1^0\neq 0$, the ratio $y_2(t)/y_1(t)=(y_2^0/y_1^0)+t$ is not constant but increases to infinity with t, starting from negative values if y_2^0 and y_1^0 differ in sign. The trajectories are as depicted in Figure 2.2d; the origin is called an *improper node,* as in case (b) and is stable since $\lambda<0$. When $\lambda>0$, the trajectories follow the same lines in the opposite direction and are unstable.

Case (e)

$$\mathbf{B}=\begin{bmatrix}\alpha & \beta\\ -\beta & \alpha\end{bmatrix} \quad \text{and} \quad \alpha\neq 0,\ \beta\neq 0.$$

The roots are the complex conjugates $\alpha+i\beta$ and $\alpha-i\beta$ with nonzero real part ($\alpha\neq 0$). The solution is

$$\begin{aligned} y_1(t)&=e^{\alpha t}(y_1^0\cos\beta t+y_2^0\sin\beta t),\\ y_2(t)&=e^{\alpha t}(y_2^0\cos\beta t-y_1^0\sin\beta t). \end{aligned} \tag{2.27}$$

The expressions in parentheses in (2.27) have a periodicity of $2\pi/\beta$, because $\cos\beta[t+2\pi/\beta]=\cos\beta t$ and $\sin\beta[t+2\pi/\beta]=\sin\beta t$. Furthermore, if $\alpha<0$ then the exponential term goes to zero and the trajectory spirals toward the origin; it is called a *stable focus.* If $\alpha>0$ then the trajectories form an *unstable focus* spiraling away from the origin. The stable case is depicted in Figure 2.2e when $\beta>0$, which induces a clockwise movement.

(If β were negative, the movement would be anticlockwise, but still stable as $\alpha<0$.)

Case (f)

$$\mathbf{B}=\begin{bmatrix}0 & \beta\\ -\beta & 0\end{bmatrix} \quad \beta\neq 0.$$

This is much the same as case (e) except that the $e^{\alpha t}$ term is identically 1. Instead of spiraling toward or away from the origin, the trajectories are closed circles. This configuration is called a *center* and is illustrated in Figure 2.2f with $\beta<0$, which results in an anticlockwise movement. Note that for a center the origin is stable but not asymptotically so.

Remark. Some important results emerge from the preceding discussion:

(i) A system such as (2.23) has a stable origin if and only if its characteristic roots have negative real parts.
(ii) A saddle point occurs if and only if the determinant of $\mathbf{A}$ is negative.
(iii) A sufficient condition for instability is that $\operatorname{tr}\mathbf{A}>0$.

2.5 Systems of two nonlinear FODE

In many interesting economic problems, the equations describing the evolution of a system are typically nonlinear. The behavior of the nonlinear system around an equilibrium point (if it exists) can be approximated by that of a linear system, as we now show.

2.5.1 *Local characterization*

Consider the nonlinear system

$$\begin{aligned}\dot{x}_1&=f(x_1,x_2),\\ \dot{x}_2&=g(x_1,x_2),\end{aligned} \tag{2.28}$$

where f and g are assumed to have continuous second-order derivatives. We also assume that system (2.28) possesses an equilibrium point $(\hat{x}_1,\hat{x}_2)$, so that $f(\hat{x}_1,\hat{x}_2)=0$ and $g(\hat{x}_1,\hat{x}_2)=0$. For $\mathbf{x}$ close to $\hat{\mathbf{x}}$ we can take a first-order approximation of f and g to obtain

$$\begin{aligned}f(x_1,x_2)&=f(\hat{x}_1,\hat{x}_2)+f_1(\hat{x}_1,\hat{x}_2)(x_1-\hat{x}_1)+f_2(\hat{x}_1,\hat{x}_2)(x_2-\hat{x}_2),\\ g(x_1,x_2)&=g(\hat{x}_1,\hat{x}_2)+g_1(\hat{x}_1,\hat{x}_2)(x_1-\hat{x}_1)+g_2(\hat{x}_1,\hat{x}_2)(x_2-\hat{x}_2),\end{aligned}$$

but since $f(\hat{x}_1,\hat{x}_2)=0$ and $g(\hat{x}_1,\hat{x}_2)=0$, if we denote f_1, f_2, g_1, and g_2 evaluated at $(\hat{x}_1,\hat{x}_2)$ by a_{11}, a_{12}, a_{21}, and a_{22}, respectively, we can approximate (2.28) by

$$\begin{bmatrix} \dot{x}_1 \\ \dot{x}_2 \end{bmatrix} = \begin{bmatrix} a_{11} & a_{12} \\ a_{21} & a_{22} \end{bmatrix} \begin{bmatrix} x_1 - \hat{x}_1 \\ x_2 - \hat{x}_2 \end{bmatrix}, \tag{2.29}$$

and defining z_i as the deviation $x_i - \hat{x}_i$, we can write (2.29) as

$$\begin{bmatrix} \dot{z}_1 \\ \dot{z}_2 \end{bmatrix} = \begin{bmatrix} a_{11} & a_{12} \\ a_{21} & a_{22} \end{bmatrix} \begin{bmatrix} z_1 \\ z_2 \end{bmatrix}, \tag{2.30}$$

which is exactly the form of (2.23) analyzed in the preceding section. The following theorem formalizes the similarities between (2.28) and (2.30).

Theorem 2.5.1. Assume $a_{11}a_{22} - a_{12}a_{22} \neq 0$. The qualitative behavior of the trajectories of the nonlinear system (2.28) in the neighborhood of the equilibrium point $(\hat{x}_1, \hat{x}_2)$ is the same as that of the trajectories of the corresponding linear system (2.30) around the origin, with the exception that if the origin is a center, then $(\hat{x}_1, \hat{x}_2)$ may be either a center or a focus.

The exception stated in the theorem can be explained as follows: the occurrence of a center hinges on the real parts of the roots of **A** in (2.30) being exactly zero; any perturbation gives rise to a focus. For more details see Lefschetz (1965) or Coddington and Levinson (1955, ch. 15).

2.5.2 *Global characterization and phase diagrams*

To determine the behavior of trajectories away from the equilibrium is a much more difficult task. There is no comprehensive general theory, and precise analysis often requires complicated topological arguments (e.g., Lefschetz, 1965). However, with sufficiently simple systems and with appropriate restrictions on the functions $f(x_1, x_2)$ and $g(x_1, x_2)$, one can often obtain a good qualitative description of the global properties of the solution with the use of phase diagrams. Examples can be found in Hirsch and Smale (1974, ch. 12). Before we illustrate the technique we must describe one type of configuration that is exhibited by nonlinear systems but that could not occur in linear ones: these are *limit cycles.* In such cases there exists a closed curve around the equilibrium point. Any starting point on that curve will remain on it indefinitely, but contrary to a "center" configuration this is the only such closed curve, or loop, in the neighborhood. This curve is called a limit cycle and may be asymptotically stable or unstable from within and/or outside the area it delineates in the sense that trajectories sufficiently close to the loop may either approach asymptotically or move away from it. Several types of limit cycles are depicted in Figure 2.3, where the thicker curve is the limit cycle; (a) is asymptotically stable, (d) is unstable, while (b) and (c) are conditionally stable. It is in general quite difficult to prove or disprove the existence of limit

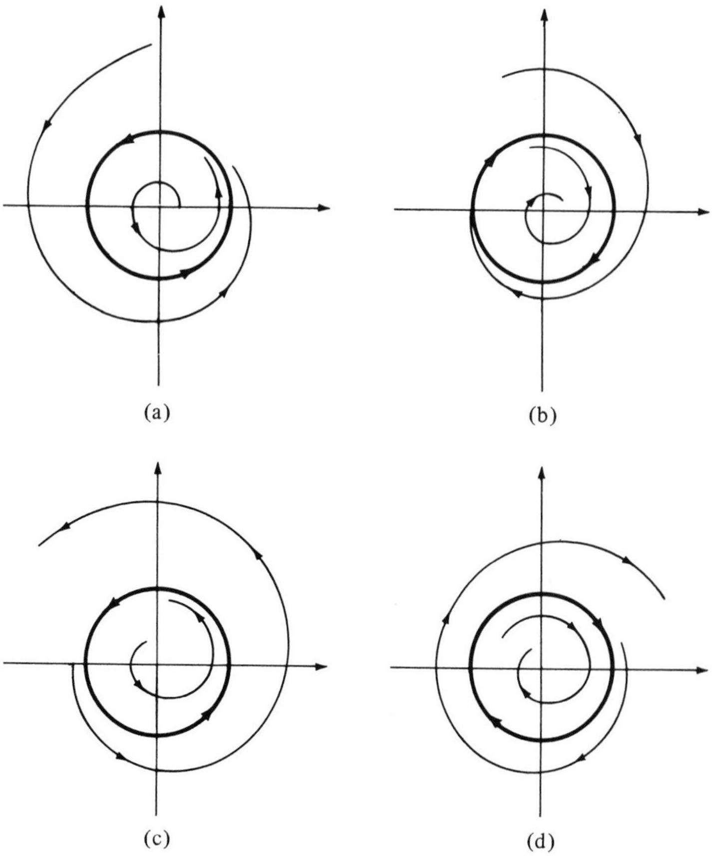

Figure 2.3

cycles. Necessary and sufficient conditions for their existence are available (see, e.g., Coddington and Levinson, 1955, ch. 16). We now turn to some economic examples.

Example 2.5.1: pollution and growth. Let K denote capital stock and P the stock of pollution. Output is $Y=K^{\alpha}$, where $0<\alpha<1$, and savings are a constant proportion of output, so that the rate of growth of capital is

$$\dot{K}=sK^{\alpha}-\delta K=K(sK^{\alpha-1}-\delta), \tag{2.31}$$

where $\delta>0$ is the rate of depreciation of capital. If a capital stock K generates a flow of pollution K^{β}, where $\beta>1$ and the stock of pollution decays at rate $\gamma>0$, the net rate of change in the stock of pollution is

$$\dot{P}=K^{\beta}-\gamma P. \tag{2.32}$$

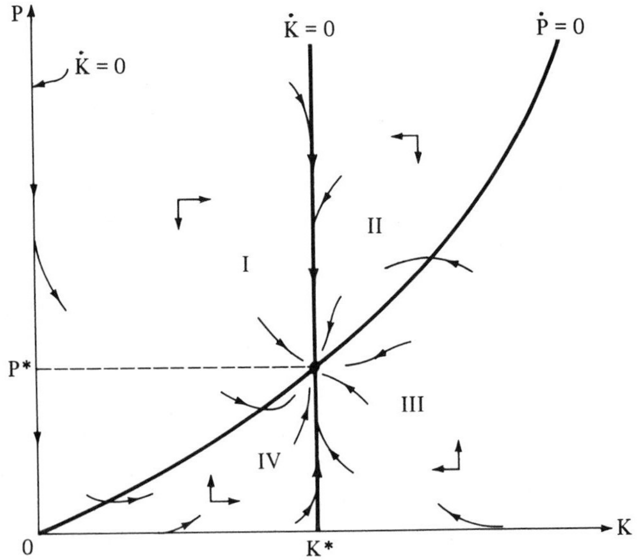

Figure 2.4

The initial conditions are $K(0)=K_0 \geq 0$ and $P(0)=P_0 \geq 0$. We wish to characterize the solution of the system (2.31)–(2.32) in the nonnegative orthant of the (K, P) plane.

Our first task is to determine the locus of points in the (K, P) plane along which $\dot{P}=0$. From (2.32) $\dot{P}=0$ if $P=(1/\gamma)K^{\beta}$. This curve is drawn in Figure 2.4; it is increasing and convex and goes through the origin. Next we turn to the $\dot{K}=0$ locus. From (2.31), $\dot{K}=0$ if $K=0$ or if

$$K=(\delta/s)^{1/(\alpha-1)} \equiv K^*. \tag{2.33}$$

The $\dot{K}=0$ locus thus consists of two vertical lines: $K=0$ and $K=K^*$. There are two points at which both $\dot{K}=0$ and $\dot{P}=0$. The first is the origin $(0,0)$, and the second is at the intersection of the $\dot{P}=0$ locus and $K=K^*$; hence, it is at (K^*, P^*), where $P^*=(1/\gamma)(K^*)^{\beta}$. The locus $\dot{P}=0$ and the line $K=K^*$ (along which $\dot{K}=0$) define four regions, labeled I to IV, in the nonnegative orthant. These regions are called *isosectors,* and the signs of $\dot{K}$ and $\dot{P}$ are uniquely determined inside each isosector. To see this, note that by continuity of the expressions in (2.31) and (2.32) the sign of $\dot{P}$ changes when the trajectories cross the $\dot{P}=0$ locus and that of $\dot{K}$ changes across the $K=K^*$ line. Therefore, above the $\dot{P}=0$ locus, $P>(1/\gamma)K^{\beta}$ implies $\dot{P}<0$, but $\dot{P}>0$ below that locus. From (2.31) and (2.33) it is clear that if $K>K^*$, then $\dot{K}<0$, and if $0<K<K^*$, then $\dot{K}>0$. Therefore, in region I, $\dot{K}>0$ and $\dot{P}<0$. In this isosector we have drawn a horizontal arrow pointing to the right (eastward) to indicate that K increases in this region; similarly, the downward vertical arrow (southward) signifies that P decreases in this region. All trajectories in region I point southeast; they can enter region IV only by crossing the $\dot{P}=0$ locus, and the slope of these trajectories is zero (horizontal) at the crossing, since it is given by $dP/dK=\dot{P}/\dot{K}$, and the numerator is zero on the $\dot{P}=0$ locus. Trajectories cannot enter region II from region I because when K approaches K^*, $\dot{K}$ approaches zero and K^* is not reached in finite time. We turn now to region III, where by examining (2.31) and (2.32) we can ascertain that $\dot{P}>0$ and $\dot{K}<0$, indicated by arrows in Figure 2.4. Here again trajectories cannot move on to region IV but may move on to region II and have a horizontal slope when crossing the $\dot{P}=0$ locus into it. Similar considerations show that trajectories point southwest in region II and northeast in region IV and cannot leave any of those regions once in it. For this reason regions II and IV are called *terminal isosectors* or, more succinctly, *traps*. It is clear that the $K=K^*$ line provides two routes of access to the equilibrium, from above and from below. There exist, however, many others. We have seen that any trajectory in region II (or IV) moves toward the equilibrium and that any trajectory in region III (or I) either goes to the equilibrium or moves on to region II (or IV). Therefore, all trajectories go to the equilibrium, and we expect to have a stable node (proper or improper) at (K^*, P^*). The other equilibrium point, $(0,0)$, is "conditionally stable": only trajectories with the initial condition $K(0)=0$ will converge to it.

Our diagrammatic characterization of (K^*, P^*) as a stable node can be confirmed locally by linearizing the differential equations (2.31) and (2.32) as we did in the general case in (2.30). We obtain

$$\begin{bmatrix}\dot{K}\\ \dot{P}\end{bmatrix}=\begin{bmatrix}\alpha sK^{\alpha-1}-\delta & 0\\ \beta K^{\beta-1} & -\gamma\end{bmatrix}\begin{bmatrix}K-K^*\\ P-P^*\end{bmatrix}. \tag{2.34}$$

The matrix of coefficients in (2.34) must be evaluated at (K^*, P^*), and we have

$$\begin{bmatrix}\dot{K}\\ \dot{P}\end{bmatrix}=\begin{bmatrix}\delta(\alpha-1) & 0\\ \beta(\delta/s)^{(\beta-1)/(\alpha-1)} & -\gamma\end{bmatrix}\begin{bmatrix}K-K^*\\ P-P^*\end{bmatrix}. \tag{2.35}$$

The roots of the characteristic equation of (2.35),

$$\lambda^2-[\delta(\alpha-1)-\gamma]\lambda-\gamma\delta(\alpha-1)=0,$$

are

$$\lambda_1=\delta(\alpha-1) \quad\text{and}\quad \lambda_2=-\gamma,$$

both negative, confirming that we have a stable node (at least around (K^*, P^*)).

Let us consider a more complex example taken from the economics of fisheries.

Example 2.5.2: the fish and the fishermen. Suppose that the stock of fish present in some fishing grounds increases (or decreases) according to the process

$$\dot{R}(t)=R(t)-(R(t))^2-x(t), \tag{2.36}$$

where $R(t)$ is the stock of fish and $x(t)$ the total catch, both at date t. Let $K(t)$ be the tonnage capacity of the fishing fleet. We assume that the relative skillfulness of fish and fishermen results in the following effective catch:

$$x(t)=R(t)[K(t)]^{1/2}. \tag{2.37}$$

We assume that there is free entry and exit into the fleet according to the flow of net returns and specifically that the proportional rate of change in fleet capacity is equal to a multiple of the difference between the average revenue per ton of capacity, px/K (where p is the price of fish), and the fleet average operating cost per ton, c:

$$\dot{K}/K=m(px/K-c). \tag{2.38}$$

We assume for simplicity of exposition that $c=m=p=1$, so that using (2.36)-(2.38) we obtain, after eliminating $x(t)$, the following two-dimensional, nonlinear system of differential equations (we skip the time argument):

$$\dot{K}=K(RK^{-1/2}-1), \tag{2.39}$$

$$\dot{R}=R(1-R-K^{1/2}). \tag{2.40}$$

In order to study the global qualitative properties of the solutions of this system, we construct a phase diagram (Figure 2.5). First, we determine the locus of points where $\dot{R}=0$, using (2.40). This consists of the K axis (along which $R=0$) and the curve $R=1-K^{1/2}$ (defined in the nonnegative orthant for $K\leq 1$). In the region to the right of this curve and above the horizontal axis, $R>1-K^{1/2}$, so that $\dot{R}<0$ by (2.40). To the left of the curve, $\dot{R}>0$. Second, we see from (2.39) that $\dot{K}=0$ along the vertical axis ($K=0$) and the curve $R=K^{1/2}$. In the region above this curve $\dot{K}>K(K^{1/2}K^{-1/2}-1)=0$, and in the region below it $\dot{K}<0$. Since $dR/dK=\dot{R}/\dot{K}$, trajectories crossing the $\dot{K}$ locus have an infinite (vertical) slope at the crossing while trajectories crossing the $\dot{R}=0$ locus have a zero (horizontal) slope there. Equipped with these observations, we are able to delineate four isosectors in the (K,R) plane and the general direction of trajectories in each; these are indicated by right-angled arrows

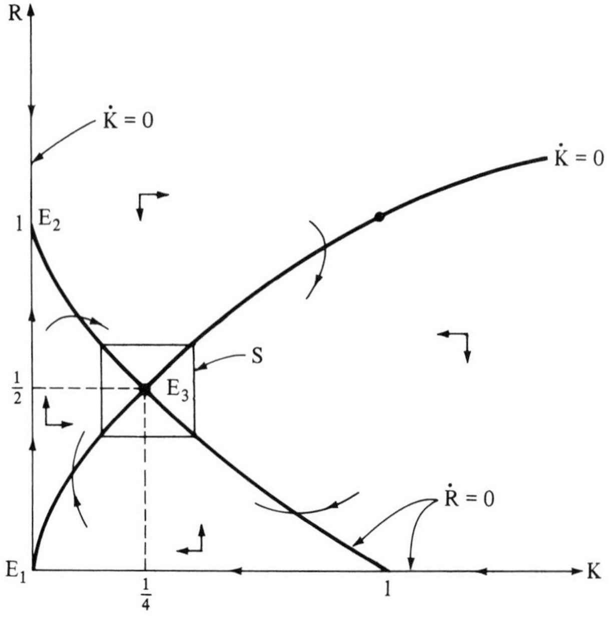

Figure 2.5

in Figure 2.5. Furthermore, the slopes of trajectories when crossing the boundaries of isosectors or touching the axes are also indicated. It is worth noting that this diagram has no "trap"; indeed, trajectories pass freely from one region to the next in a clockwise sequence. There are three equilibrium points: E_1 at $(0,0)$, E_2 at $(0,1)$, and E_3 at $(\frac{1}{4},\frac{1}{2})$. The first equilibrium is conditionally stable: only trajectories starting at $R=0$ converge to it. The second equilibrium is also conditionally stable, for trajectories starting at $K=0$. The local stability of the third equilibrium can be determined by linearizing the system of equations (2.39)–(2.40) around $(\frac{1}{4},\frac{1}{2})$ using the general formula (2.29). We obtain

$$\begin{bmatrix} \dot{K} \\ \dot{R} \end{bmatrix} = \begin{bmatrix} \frac{1}{2}RK^{-1/2}-1 & K^{1/2} \\ -\frac{1}{2}RK^{-1/2} & 1-2R-K^{1/2} \end{bmatrix} \begin{bmatrix} K-\frac{1}{4} \\ R-\frac{1}{2} \end{bmatrix} \tag{2.41}$$

and, evaluating the matrix of coefficients at $(\frac{1}{4},\frac{1}{2})$,

$$\mathbf{A} = \begin{bmatrix} -\frac{1}{2} & \frac{1}{2} \\ -\frac{1}{2} & -\frac{1}{2} \end{bmatrix}. \tag{2.42}$$

$\operatorname{tr}\mathbf{A} = -1$ and $|\mathbf{A}| = \frac{1}{2}$; therefore, the characteristic equation of $\mathbf{A}$ is $\lambda^2+\lambda+\frac{1}{2}=0$ and the roots are complex: $0.5(-1\pm i)$. The real parts are negative, and the equilibrium point E_3 is therefore a locally asymptotically stable focus. This property could not be inferred from the phase diagram by examining the directions of the arrows, for these are consistent with a locally unstable focus or a center.

It is important to realize that we have not proved that any arbitrary trajectory with initial conditions $(K(0), R(0)) > (0,0)$ converges to the equilibrium point E_3. We have only shown this to be true if $(K(0), R(0))$ are in the neighborhood of $(\frac{1}{4}, \frac{1}{2})$ (how close is unspecified). Therefore, the possibility of a limit cycle has not been ruled out. It is possible, however, to do so here in a simple way. We need only consider trajectories within a square of side 1, having the origin as its southwest corner. Because of the symmetry of the two curves $\dot{K}=0$ and $\dot{R}=0$ ($R=K^{1/2}$ and $R=1-K^{1/2}$), starting from any point on one of these curves we can draw a rectangle with a corner on each of the curves (one such rectangle S is drawn in Figure 2.5). For a trajectory beginning at a corner of such a rectangle to return there, it would have to follow the sides, which conflicts with all trajectories pointing inward toward the equilibrium E_3.

Example 2.5.3: competing species. Let X and Y denote the biomass of two species respectively. Assume that the two species compete for food, so that the rate of growth of each species is negatively related to the biomass of the other species. We postulate the functional forms

$$\begin{aligned} \dot{X}/X &= 1-X-2Y, \\ \dot{Y}/Y &= 1-Y-2X, \end{aligned} \quad \text{or} \quad \begin{aligned} \dot{X} &= X(1-X-2Y), \\ \dot{Y} &= Y(1-Y-2X). \end{aligned} \tag{2.43}$$

Thus, $\dot{X}=0$ if $X=0$ or $Y=(1-X)/2$ and $\dot{Y}=0$ if $Y=0$ or $Y=1-2X$. These loci are drawn in Figure 2.6. It is easy to see that there are four equilibria: $(X_1, Y_1)=(0,0)$, $(X_2, Y_2)=(1,0)$, $(X_3, Y_3)=(0,1)$, and $(X_4, Y_4)=(\frac{1}{3}, \frac{1}{3})$. The last one is a saddle point, the first one is an unstable node, and the other two are locally asymptotically stable nodes. The matrix of coefficient of the linearized system is

$$\mathbf{A} = \begin{bmatrix} 1-2X-2Y & -2X \\ -2Y & 1-2Y-2X \end{bmatrix}. \tag{2.44}$$

The reader should check that $\mathbf{A}$ of (2.44), evaluated at (X_4, Y_4), has a negative determinant, thus confirming that the interior equilibrium is a saddle point. The configuration of the other equilibria can be similarly checked.

2.5.3 *A special case: Hamiltonian systems*

Systems of nonlinear differential equations are often very difficult to solve. It is sometimes possible to do so by transforming a system into a higher-

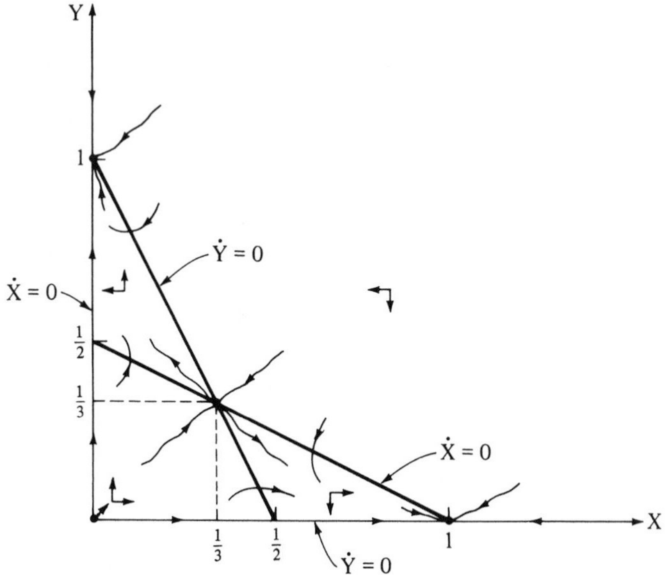

Figure 2.6

order differential equation in one of the variables. In this section we concentrate on a class of systems often encountered in optimal control theory.

Suppose that $H(x_1, x_2)$ is a twice-differentiable function and that we derive a system of two equations in the following way,

$$\begin{aligned} \dot{x}_1 &= \partial H(x_1, x_2)/\partial x_2, \\ \dot{x}_2 &= -\partial H(x_1, x_2)/\partial x_1, \end{aligned} \tag{2.45}$$

where the time argument has been suppressed.

If we differentiate the first equation totally with respect to time, we obtain $\ddot{x}_1$ in terms of $\dot{x}_1$, $\dot{x}_2$, x_1, and x_2; $\dot{x}_2$ can be eliminated using the second equation; and finally an expression for x_2 in terms of $\dot{x}_1$ and x_1 can be extracted from the first equation. This yields $\ddot{x}_1$ in terms of $\dot{x}_1$ and x_1. We now proceed with some examples in which the solution is straightforward.

Example 2.5.4. Let $H = (x_1)^{\alpha_1}(x_2)^{\alpha_2}$; then the system of differential equations is

$$\begin{aligned} \dot{x}_1 &= \alpha_2 (x_1)^{\alpha_1}(x_2)^{\alpha_2 - 1}, \\ \dot{x}_2 &= -\alpha_1 (x_1)^{\alpha_1 - 1}(x_2)^{\alpha_2}. \end{aligned} \tag{2.46}$$

Using the above technique, we obtain

$$\ddot{x}_1 = \alpha_1\alpha_2(x_1)^{\alpha_1-1}(x_2)^{\alpha_2-1}\dot{x}_1 + \alpha_2(\alpha_2-1)(x_1)^{\alpha_1}(x_2)^{\alpha_2-2}\dot{x}_2$$
$$= \alpha_1\alpha_2(x_1)^{\alpha_1-1}(x_2)^{\alpha_2-1}\dot{x}_1 - \alpha_1\alpha_2(\alpha_2-1)x_1^{2\alpha_1-1}x_2^{2\alpha_2-2},$$

and using $x_2 = [\alpha_2(x_1)^{\alpha_1}(\dot{x}_1)^{-1}]^{1/(1-\alpha_2)}$, we obtain after simplification

$$\ddot{x}_1 = \frac{\alpha_1}{\alpha_2}\frac{(\dot{x}_1)^2}{x_1}. \tag{2.47}$$

This is easily solved in two steps:

$$\frac{\ddot{x}_1}{\dot{x}_1} = \frac{\alpha_1}{\alpha_2}\frac{\dot{x}_1}{x_1},$$

$$\ln|\dot{x}_1| = \frac{\alpha_1}{\alpha_2}\ln|x_1| + A,$$

$$|\dot{x}_1| = e^A|x_1|^{\alpha_1/\alpha_2}. \tag{2.48}$$

We must now distinguish several branches of the solution:

$$\left.\begin{aligned} &x_1>0,\ \dot{x}_1>0;\quad \text{thus,}\quad \dot{x}_1 = x_1^{\alpha_1/\alpha_2}e^A,\quad \text{and}\\ &\alpha_1\neq\alpha_2,\quad x_1(t) = (Kt+B)^{\alpha_2/(\alpha_2-\alpha_1)},\\ &\qquad\qquad \text{where } K = (\alpha_2-\alpha_1)e^A/\alpha_2,\\ &\alpha_1=\alpha_2,\quad x_1(t) = \exp(e^At+B). \end{aligned}\right. \tag{2.49a}$$

$$\left.\begin{aligned} &x_1<0,\ \dot{x}_1>0;\quad \text{thus,}\quad \dot{x}_1 = (-x_1)^{\alpha_1/\alpha_2}e^A,\quad \text{and}\\ &\alpha_1\neq\alpha_2,\quad x_1(t) = -(-Kt+B)^{\alpha_2/(\alpha_2-\alpha_1)},\\ &\qquad\qquad \text{where } K = (\alpha_2-\alpha_1)e^A/\alpha_2,\\ &\alpha_1=\alpha_2,\quad x_1(t) = -\exp(-e^At+B). \end{aligned}\right. \tag{2.49b}$$

$$\left.\begin{aligned} &\text{Similarly,}\ \ x_1<0,\ \dot{x}_1<0\ \ \text{yields}\\ &\alpha_1\neq\alpha_2,\quad x_1(t) = -(Kt+B)^{\alpha_2/(\alpha_2-\alpha_1)},\\ &\qquad\qquad \text{where } K = (\alpha_2-\alpha_1)e^A/\alpha_2,\\ &\alpha_1=\alpha_2,\quad x_1(t) = -\exp(e^At+B). \end{aligned}\right. \tag{2.49c}$$

$$\left.\begin{aligned} &x_1>0,\ \dot{x}_1<0\ \ \text{yields}\\ &\alpha_1\neq\alpha_2,\quad x_1(t) = (-Kt+B)^{\alpha_2/(\alpha_2-\alpha_1)},\\ &\qquad\qquad \text{where } K = (\alpha_2-\alpha_1)e^A/\alpha_2,\\ &\alpha_1=\alpha_2,\quad x_1(t) = \exp(-e^At+B). \end{aligned}\right. \tag{2.49d}$$

We can obtain the corresponding solution for x_2 by using the first equation of (2.46). For instance, if $x_1(t) = (Kt+B)^{\alpha_2/(\alpha_2-\alpha_1)}$, we have

$$\dot{x}_1(t) = \frac{\alpha_2}{\alpha_2-\alpha_1}K(Kt+B)^{\alpha_1/(\alpha_2-\alpha_1)};$$

hence, after simplification,

$$x_2(t) = \left(\frac{\alpha_2 - \alpha_1}{K}\right)^{1/(1-\alpha_2)} (Kt+B)^{\alpha_1/(\alpha_2-\alpha_1)},$$

and so on.

Example 2.5.5. Let

$$H = \frac{(x_1)^{\alpha_1}}{\alpha_1} + \frac{(x_2)^{\alpha_2}}{\alpha_2}, \quad \alpha_i \neq 1, 0;$$

then the system is

$$\dot{x}_1 = x_2^{\alpha_2 - 1}, \qquad \dot{x}_2 = -x_1^{\alpha_1 - 1}.$$

Using a similar technique, we quickly obtain

$$\ddot{x}_1 = (1-\alpha_2)(x_1)^{\alpha_1 - 1}(\dot{x}_1)^{(\alpha_2 - 2)/(\alpha_2 - 1)};$$

$$\ddot{x}_1\left(\frac{(\dot{x}_1)^{1/(\alpha_2 - 1)}}{1-\alpha_2}\right) = \dot{x}_1(x_1)^{\alpha_1 - 1}.$$

After integrating, we have

$$\dot{x}_1 = \left[A - \frac{\alpha_2}{\alpha_1}(x_1)^{\alpha_1}\right]^{(\alpha_2 - 1)/\alpha_2}. \tag{2.50}$$

The general solution to this equation cannot be obtained, but we can give the solution in some special cases. As we shall see, it is in implicit form.

(i) Let $\alpha_1 = 0.5$, $\alpha_2 = -0.5$; then (2.50) become $\dot{x}_1 = [A + \sqrt{x_1}\,]^3$. Hence,

$$\int \frac{dx_1}{(A+\sqrt{x_1}\,)^3} = \int dt.$$

This can be integrated using substitution ($y = A + \sqrt{x_1}$) to obtain

$$A(A+\sqrt{x_1}\,)^{-2} - 2(A+\sqrt{x_1}\,)^{-1} = B + t. \tag{2.51}$$

(ii) Let $\alpha_1 = \alpha_2 = \frac{1}{3}$; then (2.50) becomes

$$\dot{x}_1[A-(x_1)^{1/3}]^2 = 1$$

or

$$[A^2 - 2A(x_1)^{1/3} + (x_1)^{2/3}]\dot{x}_1 = 1,$$

$$A^2 x_1 - \tfrac{3}{2}A(x_1)^{4/3} + \tfrac{3}{5}(x_1)^{5/3} = t + B. \tag{2.52}$$

(iii) Let $\alpha_1 = \alpha_2 = 0.5$; then (2.50) becomes $x_1 = [A - \sqrt{x_1}\,]^{-1}$. Hence,

$$Ax_1 - \tfrac{2}{3}(x_1)^{3/2} = t + B. \tag{2.53}$$

As this last example shows, systems of nonlinear differential equations are often complicated to solve.

Appendix

Indefinite integrals

Given any continuous function $f(x)$, one can find a function $F(x)$ such that its derivative is $f(x)$. Such a function is called an antiderivative of $f(x)$. For example, if $f(x)=10x$, then $5x^2+7$ is one of its antiderivatives, and so is $5x^2-3$, or any function of the form $5x^2+C$, where C is a constant. It is clear that all antiderivatives of a given function differ from one another only by a constant. For this reason, it is convenient to define the indefinite integral of a given function $f(x)$ as the general form of its antiderivatives; it always contains an arbitrary constant. The symbol for it is

$$\int f(x)\,dx.$$

Indefinite integrals of some common functions:

$$\int x^n dx=(n+1)^{-1}x^{n+1}+C, \quad n\neq -1, \tag{A1}$$

$$\int x^{-1}dx=\ln|x|+C, \tag{A2}$$

$$\int e^x dx=e^x+C. \tag{A3}$$

Properties of indefinite integrals:

$$\int kf(x)\,dx=k\int f(x)\,dx \quad \text{for any constant } k, \tag{A4}$$

$$\int (f(x)+g(x))\,dx=\int f(x)\,dx+\int g(x)\,dx. \tag{A5}$$

Properties (A4) and (A5), together with the rules of integration by parts and by substitution (described below) are very useful for finding the indefinite integral of sums or products of functions.

Integration by parts:

$$\int u'(x)v(x)\,dx=u(x)v(x)-\int u(x)v'(x)\,dx. \tag{A6}$$

Example. Find the indefinite integral of xe^x. Let $u'(x)=e^x$ and $v(x)=x$. Then $u(x)=e^x$ and $v'(x)=1$. Using (A6), we obtain

$$\int xe^x\,dx = e^x x - \int e^x\,dx = e^x x - e^x + C. \tag{A7}$$

The reader should verify that the derivative of the right-hand side of (A7) is xe^x.

Integration by substitution:

$$\int f(g(x))g'(x)\,dx = \int f(u)\,du, \quad \text{where } u = g(x). \tag{A8}$$

Example. Find the indefinite integral of $2e^{(2x+5)}$. Let $u = 2x+5 \equiv g(x)$ and $f(u) = e^u$. Then $f(g(x)) = e^{(2x+5)}$ and $g'(x) = 2$. Applying formula (A8) yields

$$\int 2e^{(2x+5)}\,dx = \int e^u\,du = e^u + C = e^{(2x+5)} + C.$$

It is easy to see that the derivative of the right-hand side of the preceding equation is $2e^{(2x+5)}$.

Definite integrals

The definite integral

$$\int_a^b f(x)\,dx \tag{A9}$$

is the same as $F(b) - F(a)$, where $F(x)$ is any antiderivative of $f(x)$. In the definite integral (A9), a and b are called the lower and upper limits of the integral. The variable x in (A9) is "mute"; that is, x can be replaced by any other symbol without affecting the value of (A9). For example,

$$\int_a^b f(x)\,dx = \int_a^b f(t)\,dt.$$

Example. Evaluate the following definite integral:

$$\int_1^2 (10x)\,dx.$$

Since $F(x) = 5x^2 + C$, $F(2) - F(1) = 20 - 5 = 15$.

If $F(x)$ is of the form $u(x)v(x)$, then $f(x) = u'(x)v(x) + u(x)v'(x)$, and

$$\begin{aligned} F(b) - F(a) &= u(b)v(b) - u(a)v(a) \\ &= \int_a^b (u'(x)v(x) + u(x)v'(x))\,dx. \end{aligned} \tag{A10}$$

It follows from (A10) that the definite integral counterpart of (A6) is

$$\int_a^b u'(x)v(x)\,dx = u(b)v(b) - u(a)v(a) - \int_a^b u(x)v'(x)\,dx. \tag{A11}$$

The definite integral counterpart of (A8) is

$$\int_a^b f(g(x))g'(x)\,dx = \int_{g(a)}^{g(b)} f(u)\,du. \tag{A12}$$

The derivative of a definite integral with respect to the upper or lower limit of integration

If we define

$$I(a,b) \equiv F(b) - F(a) = \int_a^b f(x)\,dx, \tag{A13}$$

then the derivative $\partial I/\partial a$ can be shown to be equal to $-f(a)$; similarly, $\partial I/\partial b = f(b)$.

The derivative of a definite integral with respect to a parameter

If $f(x,s)$ is continuous in x and its partial derivative with respect to s is defined, then the definite integral

$$I(a,b,s) \equiv \int_a^b f(x,s)\,dx$$

is a function of the parameter s; its partial derivative with respect to s is given by

$$\frac{\partial I}{\partial s} = \int_a^b \frac{\partial f}{\partial s}\,dx. \tag{A14}$$

If a and b are themselves functions of s, then the total derivative dI/ds is

$$\begin{aligned}\frac{dI}{ds} &= \frac{\partial I}{\partial a}\frac{da}{ds} + \frac{\partial I}{\partial b}\frac{db}{ds} + \frac{\partial I}{\partial s} \\ &= -f(a,s)a'(s) + f(b,s)b'(s) + \int_a^b \frac{\partial f(x,s)}{\partial s}\,dx. \end{aligned} \tag{A15}$$

Equation (A15) is called *Leibniz's rule.*

Exercises

1. In a certain economy, capital is accumulated according to the rule $\dot{K}(t) = sF(K(t), L(t)) - \delta K(t)$, where s is the saving ratio, $0 < s < 1$, δ is the rate of

depreciation, $\delta > 0$, and F is the production function. Let $w > 0$ denote the wage rate in a neighboring large country; assume that w is exogenous. This economy's labor force grows or declines according to the difference between the marginal product of labor and w: $\dot{L}(t) = L(t)[F_L(K(t), L(t)) - w]$. Assume that $F(K(t), L(t)) = 4[K(t)L(t)]^{1/4}$.

(a) In the (L, K) space, find the locus of points $\dot{K} = 0$; identify the regions where $\dot{K} > 0$ and $\dot{K} < 0$, respectively.

(b) Find the $\dot{L} = 0$ locus and the regions where $\dot{L} > 0$ and $\dot{L} < 0$.

(c) Is there an equilibrium (L^*, K^*) with $K^* > 0$ and $L^* > 0$? Is it stable?

(d) Assume that $w = 1$, $s = 0.1$, $\delta = 0.4$ and determine K^* and L^*. Linearize the system around the equilibrium point, calculate the characteristic roots, and describe the behavior of the system around the equilibrium.

2. Let $K(t)$ denote capital stock at time t and $P(t)$ be the level of pollution. Output is $Y(t) = (K(t))^\alpha / (1 + P(t))$, where $0 < \alpha < 1$. Savings is a constant fraction of output, so that the rate of change in capital stock is $\dot{K}(t) = sY(t) - \delta K(t)$, where δ is the rate of depreciation and s is the saving ratio, $0 < s < 1$. For simplicity assume that $s = 2\delta$. The pollution level changes according to the formula $\dot{P}(t) = K(t) - P(t)$.

(a) Find the loci $\dot{P} = 0$ and $\dot{K} = 0$ and identify the regions of the (P, K) plane where $\dot{P}$ and $\dot{K}$ have definite signs.

(b) Show that there exists an equilibrium, say (K^*, P^*). Calculate these values and show that they are positive.

(c) Linearize the differential equations around the equilibrium and determine whether the characteristic roots are real or complex if $\delta = 0.4$, $s = 0.8$, and $\alpha = 0.1$. What are the local stability properties of this equilibrium?

3. Suppose that aggregate output adjusts according to the equation $\dot{Y}(t) = G(t) + A - sY(t)$, $0 < s < 1$ and $A > 0$.

(a) If government expenditure varies according to the rule $\dot{G}(t) = 0.5\dot{Y}(t) - G(t)$, determine the path of aggregate output. Is there an equilibrium? Is it stable? Construct a phase diagram in the (Y, G) space.

(b) If, instead, $G(t)$ is determined by $G(t) = G_0 + \alpha \int_0^t (\bar{Y} - Y(\tau))\, d\tau - \beta Y(t)$, where $G_0 > 0$, $\alpha > 0$, $\beta > 0$, and $\bar{Y} > 0$, analyze the consequences of this new policy rule on the path of aggregate output. Construct a phase diagram in the (Y, G) space. (Differentiate $G(t)$ with respect to t.)

4. Let $u(w)$ be the utility of wealth w.

(a) Determine the functional form $u(\cdot)$ for which the coefficient of absolute risk aversion is constant, i.e., $-u''(w)/u'(w) = K$, a positive constant. (*Hint:* The basic variable here is wealth w, not time t.)

(b) Determine the functional form $u(\cdot)$ for which the coefficient of relative risk aversion is constant, i.e., $-wu''(w)/u'(w) = R$, a positive constant.

5. The rate of increase in the price of apples is proportional to excess demand $\dot{P}(t) = k(D(t) - S(t))$. In each of the following two cases determine the general time path of price. Find the equilibrium price; what assumption must you make about the sign of k to ensure that the equilibrium is stable?

(a) $D(t) = A - BP(t)$, $A > 0$, $B > 0$,
$S(t) = -M + NP(t)$, $M > 0$, $N > 0$.

(b) $D(t) = A - B[P(t) - 1]$, $A > 0$, $B > 0$,
$S(t) = A + [P(t) - 1]^3$.

(*Hint:* Let $Q(t) = P(t) - 1$.)

Draw a phase diagram and examine the stability properties of the equilibrium when $D(t) = \exp[-P(t)]$ and $S(t) = [P(t)]^3$.

6. Consider the following predator-prey model:

$$\dot{x}(t) = x(t)[A - By(t) - Mx(t)], \quad x(t) \geq 0,$$

$$\dot{y}(t) = y(t)[Cx(t) - D - Ny(t)], \quad y(t) \geq 0,$$

where A, B, C, D, M, and N are specified positive constants and we assume that $DM < CA$.

(a) Which variable represents the population of predators? Explain the meaning of each equation.

(b) Draw a phase diagram in the (x, y) space. Is there a positively valued equilibrium? If so, linearize the system about the equilibrium point and determine whether the characteristic roots have negative real parts. What can you infer about the stability of the equilibrium?

CHAPTER 3

Introduction to dynamic optimization

This chapter is a very informal attempt to motivate the exposition of the dynamic optimization methods that take up the remainder of the book. We use a simple dynamic macroeconomic model to introduce several important concepts through numerical examples. Let y, C, and I be aggregate income, consumption, and investment, respectively; then a simple macroeconomic model of income determination would be

$$C = cy, \tag{3.1}$$

$$C + I = y. \tag{3.2}$$

The first equation is a simple consumption function, while the second is the equilibrium condition. Given an exogenous value for I, the equilibrium values of y and C can be determined as long as y does not exceed the full employment level Y, which we now define. Full-employment income Y depends on the level of capital stock, s, through a production function

$$y \leq Y = f(s). \tag{3.3}$$

Suppose that investment equals the full-employment level of savings, $I = (1-c)Y$; then the economy will be at full employment. To obtain a growth model we formally equate the net rate of change in capital stock to investment

$$\dot{s} = I. \tag{3.4}$$

Then equations (3.1)–(3.4), with y replaced by Y, constitute a simple descriptive growth model under full employment, with all variables evaluated at the same time t and the time argument suppressed for simplicity of notation.

In such a model we no longer solve for the static equilibrium values but determine the functional form of the variables in terms of t. For this we reduce (3.1)–(3.4) to a differential equation in s: (3.1)–(3.3) yield $I = Y - C = (1-c)Y = (1-c)f(s)$, and with (3.4)

$$\dot{s} = (1-c)f(s). \tag{3.5}$$

This equation can be solved for s, provided that the functional form $f(s)$ is specified and given some initial condition, $s(0) = s_0$, say.

3.1 Optimal borrowing

In this section we entertain the possibility of borrowing to augment the initial stock of capital. Let the size of the loan be L, with continuously compounded interest at the rate r; the amount to be repaid at the expiry date of the loan, say T, is Le^{rT}. We use the production function $Y = s^{\alpha}/(1-\alpha)$, $0<\alpha<1$, and the initial condition $s(0)=s_0$; letting the price of capital be 1 at any time, it is now possible to begin with s_0+L units of capital; L is to be optimally chosen to maximize the amount of capital available at time T, after the loan has been repaid. The differential equation (3.5) takes the form

$$\dot{s} = (1-c)s^{\alpha}/(1-\alpha), \tag{3.6}$$

which yields

$$\int (1-\alpha)s^{-\alpha}\,ds = \int (1-c)\,dt,$$

$$s^{1-\alpha} = (1-c)t + A. \tag{3.7}$$

If L is borrowed, the initial condition is

$$(s_0+L)^{1-\alpha} = A. \tag{3.8}$$

Hence,

$$s(t) = [(1-c)t + (s_0+L)^{1-\alpha}]^{1/(1-\alpha)} \tag{3.9}$$

is the solution for $0 \le t < T$. At time T the loan is repaid and what remains is

$$[(1-c)T + (s_0+L)^{1-\alpha}]^{1/(1-\alpha)} - Le^{rT}. \tag{3.10}$$

Maximizing (3.10) with respect to L yields the first-order condition

$$[(1-c)T + (s_0+L)^{1-\alpha}]^{\alpha/(1-\alpha)}(s_0+L)^{-\alpha} = e^{rT},$$

and raising each side to the power $(1-\alpha)/\alpha$, we get

$$(1-c)T\cdot(s_0+L)^{\alpha-1} = e^{rT(1-\alpha)/\alpha} - 1,$$

and finally

$$L^* = \left[\frac{(1-c)T}{e^{rT(1-\alpha)/\alpha}-1}\right]^{1/(1-\alpha)} - s_0. \tag{3.11}$$

This is assumed to be a maximum. (The exact conditions to ensure this can be derived but are intricate.) Note that L could be negative. Substituting (3.11) into (3.10) yields the value of the remaining stock after repayment:

$$s^*(T) = e^{rT}[s_0 + [(1-c)T]^{1/(1-\alpha)}[e^{rT(1-\alpha)/\alpha}-1]^{-\alpha/(1-\alpha)}].$$

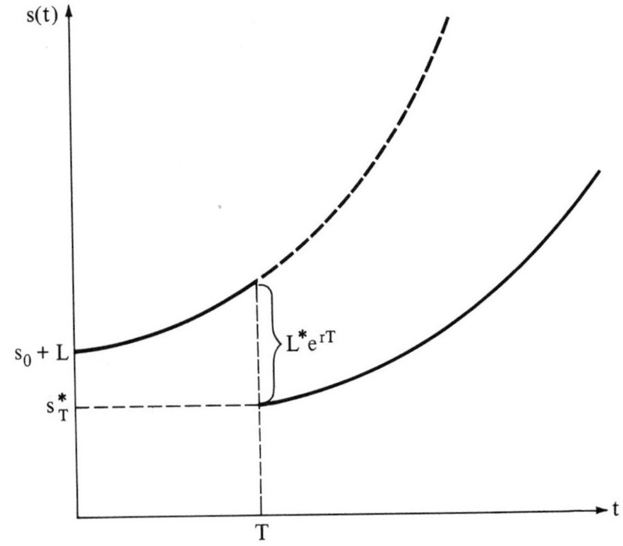

Figure 3.1

In order to obtain the solution for $t > T$, we must not simply deduct L^*e^{rT} from (3.9). Instead, we must use $s^*(T)$ as the new boundary condition and determine anew the constant A in (3.7). This implies that we jump onto a new trajectory at time T, as illustrated in Figure 3.1. This emphasizes the importance of correctly specifying boundary conditions.

3.2 Fiscal policy

Here we add a government to the preceding model and give it the power to tax; the government's tax revenue is invested, so that it is a case of forced savings. The model is

$$C = c(1-\theta)Y, \tag{3.12}$$

$$G = \theta Y, \tag{3.13}$$

$$Y = s^{\alpha}/(1-\alpha), \tag{3.14}$$

$$C + I + G = Y, \tag{3.15}$$

$$\dot{s} = I + G, \tag{3.16}$$

where θ is the tax rate, $(1-\theta)Y$ is disposable income, and G is government revenue, or public investment, and I is private investment. For ease of calculations we take $\alpha = 0.5$ and obtain the differential equation

$$\dot{s} = 2s^{1/2}(1-c+c\theta), \tag{3.17}$$

which is solved to yield

$$s(t) = [(1-c+c\theta)t + s_0^{1/2}]^2, \tag{3.18}$$

where $s(0) = s_0$.

Suppose that the government's objective is to maximize total consumption over some horizon $[0, T]$. Its problem is then to choose θ so as to maximize

$$W = \int_0^T C(t)\,dt = \int_0^T 2c(1-\theta)[(1-c+c\theta)t + s_0^{1/2}]\,dt$$
$$= c(1-\theta)[(1-c+c\theta)T^2 + 2s_0^{1/2}T]. \tag{3.19}$$

The optimum value of θ is found by setting $dW/d\theta = 0$, which yields

$$\theta = 1 - \frac{2s_0^{1/2} + T}{2cT} = 1 - \frac{2s_0^{1/2}T^{-1} + 1}{2c}. \tag{3.20}$$

(It is easy to verify that $d^2W/d\theta^2 = -2c^2T^2 < 0$.)

We now wish to illustrate the effect of the length of the time horizon (the duration of the political mandate of the government?) on the fiscal policy parameter θ. From (3.20) it is clear that θ increases with T, to approach the value of $(c-0.5)/c$. For concreteness we set $s_0 = 1$ and $c = 0.75$ and let T take on various values. We have, from (3.20), $\theta = (T-4)/3T$, $Y(t) = 2[(1-c+c\theta)t + s_0^{1/2}] = t(T-2)/T + 2$, and $\dot{s} = (1-c+c\theta)Y = Y(T-2)/2T$ after substitution.

If $T = 5$, we have $\theta = \frac{1}{15}$ and $Y(t) = 0.6t + 2$; government policy increases net investment. If $T = 3$, we have $\theta = -\frac{1}{9}$ and $Y(t) = t/3 + 2$; government policy decreases net investment, but this is still positive since $I + G = \dot{s} = \frac{1}{6}Y$. For a very long horizon, $T \to \infty$, $\theta \to \frac{1}{3}$ and $Y(t)$ becomes $t + 2$. This is the largest rate of tax and the fastest growth rate of income. These results are understandable. Savings decreases current consumption in favor of investment in capital, which will yield larger income, hence consumption, in the future. The more future periods are taken into account, the stronger is the incentive to save. The length of the planning horizon can thus have a drastic effect on policies. Finally, if $T = 1$, we have $\theta = -1$ and $Y(t) = -t + 2$. In this case the incentive to save is reversed so strongly that net investment is negative ($\dot{s} = -Y$) and the growth rate of income is negative. This raises the issue of whether it is possible to use old capital stock to generate current consumption - the jargon for it is *reversibility of investment*.

3.3 Suboptimal consumption path

We revert to a model without a government sector and choose $f(s) = 4s$; $s(0) = 1$. Then equation (3.5) is

$$\dot{s}=4(1-c)s, \quad s(0)=1. \tag{3.21}$$

Our objective is to maximize utility over some horizon $\int_0^T U(C(t))\,dt$. Taking $U(C(t))=\ln C(t)$ and $T=1$, we must maximize

$$V=\int_0^1 \ln C(t)\,dt \tag{3.22}$$

subject to (3.21), where $C(t)=4cs(t)$.

Solving (3.21) yields $s(t)=\exp(4(1-c)t)$ and $C(t)=4c\exp(4(1-c)t)$. When this is substituted in (3.22), we must choose c to maximize

$$V=\int_0^1 [\ln 4c+4(1-c)t]\,dt, \tag{3.23}$$

$$V=\ln 4c+2(1-c). \tag{3.24}$$

The maximum is at $c=0.5$, and we have $s^*(t)=e^{2t}$, $C^*(t)=2e^{2t}$, and $V^*=\ln(2)+1\simeq 1.69$.

Let us now pause and reflect on the "optimality" of this procedure – in particular, the fact that the propensity to consume, c, was held constant throughout the horizon. Letting c vary over time could only improve the integral maximand, if it had any effect. In that sense the above solution is suboptimal. With a variable propensity to consume, our problem would be to choose $c(t)$ to maximize

$$\int_0^1 \ln[4c(t)s(t)]\,dt \tag{3.25}$$

subject to

$$\dot{s}(t)=4(1-c(t))s(t) \tag{3.26}$$

and boundary conditions on s, where $c(t)$ is an unknown function of time. Clearly, then, it is not possible to solve equation (3.26), since we do not know the form of $c(t)$ – this is precisely what we seek. The elementary calculus techniques used in this chapter are of no help in solving the problem of (3.25)–(3.26). This is indeed a simple example of what we call a *control problem;* the solution of such problems is the subject of the next chapter.

3.4 Discounting and depreciation in continuous-time models

The concept of an interest rate is essential to dynamic economic models. It is well understood in a financial context. Denoting the rate of interest per period by r, we can define the present value P of an amount of dollars A to be paid T periods hence as the number of dollars that, if deposited today, would grow into \$$A$ if left to compound interest for T periods.

Formally,

$$P(1+r)^T = A \quad \text{or} \quad P = A(1+r)^{-T}. \tag{3.27}$$

This assumes that interest is compounded each period, that is, reinvested after each period. If interest is in fact compounded, say, n times during a period, we then have nT subperiods, each bearing an interest rate of r/n. Then the present value of A is

$$P = A(1+r/n)^{-nT}. \tag{3.28}$$

As compounding takes place more and more frequently, n increases without bound, and recalling that $\lim_{x\to\infty}(1+x^{-1})^x = e$, we have $\lim_{n/r\to\infty} P = \lim_{n/r\to\infty} A[(1+r/n)^{n/r}]^{-rT} = Ae^{-rT}$. Therefore, as interest is continuously compounded, the present value formula becomes

$$P = Ae^{-rT}, \tag{3.29}$$

where r is the interest rate per period and T the number of periods.

In continuous-time models, that is, when the time variable is real-valued, equation (3.29) provides a convenient way of calculating present values. Note that using the exponential discounting formula does not presume that interest is actually compounded at every instant; we need only assume that the calculation can be made when T is a real number and that r is the effective interest rate calculated on the basis of continuous compounding. To see this, note that an interest rate i compounded once per period is equivalent to an interest rate $r = \ln(1+i)$ compounded continuously since it follows that $e^r = 1+i$ and $Ae^{-rT} = A(1+i)^{-T}$.

A positive real interest rate can be observed in most economies operating near full employment; hence, its existence is not often an issue. Furthermore, we can give it a neat theoretical justification by the argument that production processes take time and that the use of a productive asset – or money to buy it – for some length of time must attract an economic rent. Thus, if our objective is expressed in money terms, it seems appropriate to use a discount factor. For instance, if the flow of profit at time t is $\pi(c(t),t)$, where c is a policy variable, its present value would be $e^{-rt}\pi(c(t),t)$ and the total present value of profits over the horizon $[0,T]$, $\int_0^T e^{-rt}\pi(c(t),t)\,dt$.

We use discounting for other criteria as well. For instance, if the objective is a level of utility, we often discount it using the same exponential formula, but the rate of discount is now a subjective one reflecting the individual's or the planner's relative valuation of present over future enjoyment. For example, if the level of utility at instant t is given by $u(c(t),t)$, we discount and aggregate this much as we would a monetary reward to obtain the criterion $\int_0^T e^{-\delta t}u(c(t),t)\,dt$. Here δ is the subjective rate of discount. Such welfare or utility criteria are used throughout the literature on dynamic economic optimization. There are several criticisms of this. The first, a moral one, is that it is wrong for the current generation to attach less value to the enjoyment of future generations and that δ should be zero. A second criticism is that of the summation of utility levels across time, which is implicit in the integral formulation of the criterion. Whereas it is reasonable to assume that money flows can be added, there is less justification for adding flows of utility. One remarkable feature of the exponential form of the discount factor is that it is the only one that ensures that the planner will be able to formulate a plan that he will actually wish to follow, even if offered the opportunity to change it at a later date. This is the issue of consistency that will be briefly discussed in Chapter 4.

For now we turn to the model of the preceding section in order to illustrate what our intuition tells us is true: the introduction of a positive rate of discount will lead to a relatively higher consumption early in the horizon and a relatively lower consumption later, reflecting the planner's "impatience." Equation (3.23) is modified by the introduction of a rate of discount $\delta > 0$.

$$\begin{aligned} V &= \int_0^1 [\ln 4c + 4(1-c)t] e^{-\delta t}\, dt \\ &= \ln 4c \left[-\frac{1}{\delta} e^{-\delta t} \right]_0^1 + 4(1-c) \left[-\frac{t}{\delta} e^{-\delta t} - \frac{e^{-\delta t}}{\delta^2} \right]_0^1 \\ &= (\ln 4c)\left(\frac{1-e^{-\delta}}{\delta} \right) + \frac{4(1-c)}{\delta^2} \left[1 - e^{-\delta}(\delta+1) \right]. \end{aligned} \tag{3.30}$$

To maximize V in (3.30), set

$$\frac{\partial V}{\partial c} = \left(\frac{1-e^{-\delta}}{\delta} \right) c^{-1} - \frac{4}{\delta^2}(1 - e^{-\delta}(\delta+1)) = 0,$$

from which

$$c = \frac{\delta(1-e^{-\delta})}{4(1-e^{-\delta}(1+\delta))}. \tag{3.31}$$

One can show that $\lim_{\delta \to 0+} c = 0.5$, but positive values of δ will yield higher c values. As an illustration, $\delta = 0.5$ yields $c = 0.5453$ – hence $\hat{C}(t) \simeq 2.172 e^{1.1819t}$ instead of $C^*(t) = 2e^{2t}$ without discounting – and we verify that $\hat{C}(0) = 2.172 > C^*(0)$, while $\hat{C}(1) = 13.39 < 14.78 = C^*(1)$.

Depreciation

We have seen that if an amount of money P is invested with continuous compounding at the interest rate r, it will grow into Pe^{rt} after t units of time. This would be the case for any stock of goods growing at a constant rate r; if at time zero there is $s(0)$ of it, this will grow into $s(t)=s(0)e^{rt}$ after t periods. Clearly, if the stock decays or depreciates instead of growing, the stock will decrease as $s(t)=s(0)e^{-mt}$, where $m>0$ is the rate of depreciation. This is the exponential decay typical of radioactive materials. We can express this phenomenon as a differential equation. The time derivative is

$$\dot{s}(t)=s(0)(-me^{-mt})=-ms(0)e^{-mt}=-ms(t),$$

so that

$$\dot{s}=-ms \tag{3.32}$$

represents depreciation at the constant rate m, and

$$\dot{s}=rs \tag{3.33}$$

represents growth at the constant rate r. Equation (3.33) is a simple instance of equation (3.5); rs is seen as the new amount generated at each instant (the "interest").

In a more general fashion s could generate $f(s)$ at each instant and also depreciate at the constant rate m; this would yield

$$\dot{s}=f(s)-ms. \tag{3.34}$$

These and other forms will be used throughout this book to describe the evolution of dynamic systems.

Exercises

1. Reconsider the optimal borrowing problem of Section 3.1. Suppose that you now wish to choose both the amount of the loan $\hat{L}$ and the expiry date $\hat{T}$ to maximize the present value of the capital available after repayment (i.e., the expression in (3.10) multiplied by e^{-rT}). Derive the first-order conditions. Obtain values for $\hat{T}$ and $\hat{L}$ when $\alpha=0.5$, $c=0.75$, $s_0=0$, and $r=10\%$; verify that your answers satisfy equation (3.11). Does $\hat{T}$ depend on c?
2. A bottle of wine costs \$3.00 now. Its future sale value at time t is given by $V(t)=3.00+\sqrt{t}$. The storage cost per unit of time is \$0.10, and the prevailing interest rate is 10%. You wish to choose a date for selling the bottle that maximizes the present value of profit.

   Calculate the total discounted value of storage cost from time 0 to time t. Express the present value of profit in terms of t. Find the optimal year of sale and the value of profits. Redo the calculations with interest rates of 5% and 20%. How does the rate of interest affect the optimal time of sale? Comment.
3. Reconsider the fiscal policy problem of Section 3.2. Find the form of W when $\alpha=\frac{2}{3}$, $s_0=1$, and $c=0.75$. Determine the value of θ that maximizes W and evaluate it for various values of T. Does it increase with T? What is the lowest upper bound on θ?

4. Reconsider the consumption path problem of Section 3.3. Suppose now that the planner has for horizon the time interval $[0, 3]$. He has more latitude in the choice of the propensity to consume in the sense that he can choose one value, c_1, during the time interval $[0, 1]$ and another value, c_2, during the time interval $[1, 3]$. Show that the general solution to $\dot{s} = 4(1-c)s$ is $s(t) = A\exp(4(1-c)t)$, where A depends on initial conditions and is valid on intervals where c remains constant. Let $V_1 = \int_0^1 \ln C(t)\,dt$ and $V_2 = \int_1^3 \ln C(t)\,dt$. Calculate V_1 as a function of c_1; calculate $s(1)$; use this initial condition to determine $C(t)$ over $[1, 3]$ and calculate V_2. Find c_1 and c_2 that maximize $V_1 + V_2$. Is the value of c_1 you obtain different from 0.5 as in the example of Section 3.3? Can you explain why?

   Attempt now a more elaborate exercise in which the horizon $[0, 4]$ is split into four intervals of length 1. On each interval the propensity to consume is constant. The task is to choose c_1, c_2, c_3, and c_4 to maximize the total utility (logarithm) of consumption on the whole interval. Let $s_1(t), \ldots, s_4(t)$ be the time paths of capital over the four intervals. Determine their exact forms by using the boundary conditions. (The first one depends only on c_1; the last one depends on c_1, c_2, c_3, and c_4.) Calculate $V_i = \int_{i-1}^{i} \ln C(t)\,dt$, $i = 1, 2, 3, 4$, and find the values of c_1, c_2, c_3, and c_4 that maximize $\sum_{i=1}^{4} V_i$. Can you detect a pattern in the c values? Can you rationalize it?

5. Modify the basic growth model adding depreciation as $\dot{s} = I - ms$. Using $f(s) = 4s$, express $C(t)$ when the propensity to consume is constant. Find c that maximizes $V = \int_0^1 \ln C(t)\,dt$. How is the value of consumption affected by the rate of depreciation?

CHAPTER 4

The maximum principle

In this chapter we present a first account of optimal control theory. The maximum principle is the central result of the theory. (It was originally developed by Pontryagin and his associates; see Pontryagin et al., 1962.) To help the reader become thoroughly acquainted with it, we proceed with the analysis of a simple case, without paying undue attention to some technical regularity conditions. (These and other matters will be dealt with in Chapter 6.)

4.1 A simple control problem

Consider a dynamic system – for instance, a moving spaceship or an economy. Some variables can be identified that describe the state of the system: they are called *state variables* – for instance, the distance of the spaceship from earth or the stock of goods present in the economy. The rate of change over time in the value of a state variable may depend on the value of that variable, time itself, or some other variables, which can be controlled at any time by the operator of the system. These other variables are called *control variables* – for instance, the pitch of the motor or the flow of goods consumed at any instant. The equations describing the rate of change in the state variables are usually differential equations, as discussed in Chapter 2. Once values are chosen for the control variables (at each date), the rates of change in the values of the state variables are thus determined at any time, and given the initial value for the state variables, so are all future values. For instance, the pitch of the spaceship engine determines its speed and hence its distance from earth once its initial position is known; the consumption path of the economy determines net investment and hence capital stock accumulation over time. The object of controlling a system is usually to contribute to a given objective. For instance, the values of all the relevant variables determine the fuel consumption of the spaceship at any time, and the objective is to minimize total fuel consumption so that some destination is reached within a given time period. Similarly, the values of consumption, capital stock, and time may determine the welfare of the community at each instant, and the objective is to maximize total welfare over a fixed time horizon, given specific values of the stock at the beginning and the end.

A salient feature of optimal control problems that emerges from the foregoing discussion is that it is necessary to choose a value for the control variable (or variables) at each instant; when, as is usually the case, time is taken to be real-valued, there are infinitely many values of the control to be chosen. Another way of putting this is to say that we must find a functional form over some time interval, which the control variable is to follow. Thus, the problem appears to be far more difficult than those encountered in static optimization. Fortunately, the maximum principle provides a framework that makes these problems amenable to solution.

We now formally define a simple optimal control problem and state the maximum principle for it. For all t, find $c(t)$ that maximizes

$$V=\int_0^T v(s(t),c(t),t)\,dt \tag{4.1}$$

subject to

$$\dot{s}=f(s(t),c(t),t) \tag{4.2}$$

and

$$s(0)=s_0, \quad s(T)=s_T, \tag{4.3}$$

where $s(t)$ is the state variable, $\dot{s}(t)$ is the rate of change of the state variable with respect to time, $c(t)$ is the control variable, and t denotes the date; the interval $[0, T]$ is the planning horizon, s_0 and s_T are the values the state variable must take on at the boundaries. The values of T, s_0, and s_T are exogenously specified. (For simplicity we assume throughout this chapter that $c(t)$ is unconstrained; this assumption will be relaxed in Chapter 6.)

If a functional form is chosen for $c(t)$ over $[0,T]$, the differential equation (4.2) together with the boundary conditions (4.3) will determine $s(t)$ uniquely over $[0,T]$; this in turn will yield a value for the integral V in (4.1). The problem is to choose $c(t)$ to yield the largest possible V. In this chapter we assume that an optimal control exists, is unique, and is differentiable with respect to time. This is a very restrictive assumption; see the final remark of Section 4.6 for details.

The necessary conditions that constitute the maximum principle are most conveniently stated after some auxiliary variables, akin to multipliers, have been introduced; with the state variable $s(t)$ is associated an auxiliary variable called a *costate variable* denoted by $\pi(t)$. We define, at each instant, a new function called a *Hamiltonian,* similar to a Lagrangean. The Hamiltonian for the problem defined in (4.1)–(4.3) is

$$H(s(t),c(t),\pi(t),t)\equiv v(s(t),c(t),t)+\pi(t)f(s(t),c(t),t). \tag{4.4}$$

Theorem 4.1.1: the maximum principle. An optimal solution to the above problem is a triplet $(s(t), c(t), \pi(t))$ and must satisfy the following conditions:

(i) $c(t)$ maximizes $H(s(t), c(t), \pi(t), t)$, that is,

$$\frac{\partial H}{\partial c(t)} = 0; \tag{4.5}$$

and

(ii) the state and costate variables satisfy a pair of differential equations,

$$\dot{s}(t) = \frac{\partial H}{\partial \pi(t)}, \tag{4.6}$$

$$\dot{\pi}(t) = -\frac{\partial H}{\partial s(t)}, \tag{4.7}$$

with boundary conditions as in (4.3).

Using the definition of the Hamiltonian in (4.4), equations (4.5)–(4.7) can be expanded as

$$\frac{\partial v}{\partial c(t)} + \pi(t)\frac{\partial f}{\partial c(t)} = 0, \tag{4.8}$$

$$\dot{s}(t) = f(s(t), c(t), t), \tag{4.9}$$

$$\dot{\pi}(t) = -\frac{\partial v}{\partial s(t)} - \pi(t)\frac{\partial f}{\partial s(t)}, \tag{4.10}$$

with $s(0) = s_0$ and $s(T) = s_T$.

Therefore, the optimal triplet is a solution of equations (4.8)–(4.10). These consist of two differential equations and an algebraic equation, often called the first-order condition since it optimally selects the control. Before attempting to apply the maximum principle to a specific problem, we will show how easy it is to derive it in a discrete analog of the problem presented in this section.

4.2 Derivation of the maximum principle in discrete time

For a thorough grasp of important results such as the maximum principle, it is essential to work through and assimilate a heuristic proof. We present a proof for the continuous-time case in Section 4.6. For now we are content to derive a proof in the simpler case where time is a discrete variable: the horizon consists of T periods, $t = 1, 2, \ldots, T$, instead of a continuous interval, as in the preceding section. Thus, it is a constrained maximum problem with a special recursive structure.

An optimal control problem in discrete time

Find $c(1), c(2), \ldots, c(T)$ that maximize

$$V=\sum_{t=1}^{T} v(s(t), c(t)) \tag{4.11}$$

subject to

$$s(t+1)-s(t)=f(s(t), c(t)), \quad t=1,2,\ldots,T, \tag{4.12}$$

$$s(1)=s_1, \quad s(T+1)=s_{T+1}. \tag{4.13}$$

This is the discrete analog of the problem formulated in (4.1)–(4.3). The difference equation (4.12) that describes how the state variable changes from one period to the next replaces the differential equation (4.2) that described the change at each instant. The independent time argument has been suppressed here to simplify the notation; its inclusion does not affect this derivation. The symbols $s(t)$ and $c(t)$ denote, respectively, the values of the state variable and control variable at the beginning of period t; thus, specifying a value for $s(T+1)$ is the same as requiring the state variable to take on this value at the end of period T. We are free to choose any values for $c(t)$ and $s(t)$ to maximize (4.11) as long as the constraints (4.12) and (4.13) are satisfied. To this end we substitute (4.13) into (4.12) and assign a multiplier to each constraint. The Lagrangean of the problem is

$$\begin{aligned}
L=\;& v(s_1, c(1))+v(s(2), c(2))+\cdots+v(s(t), c(t))+\cdots \\
&+v(s(T-1), c(T-1))+v(s(T), c(T)) \\
&+\pi(1)[s_1+f(s_1, c(1))-s(2)]+\pi(2)[s(2)+f(s(2), c(2))-s(3)] \\
&+\cdots+\pi(t-1)[s(t-1)+f(s(t-1), c(t-1))-s(t)] \\
&+\pi(t)[s(t)+f(s(t), c(t))-s(t+1)]+\cdots \\
&+\pi(T-2)[s(T-2)+f(s(T-2), c(T-2))-s(T-1)] \\
&+\pi(T-1)[s(T-1)+f(s(T-1), c(T-1))-s(T)] \\
&+\pi(T)[s(T)+f(s(T), c(T))-s_{T+1}].
\end{aligned}$$

The first-order necessary conditions are obtained by partially differentiating with respect to all free $c(t)$, $s(t)$, and $\pi(t)$:

$$\begin{aligned}
v_{c(1)}+\pi(1)f_{c(1)}&=0, \\
&\cdots \\
v_{c(t)}+\pi(t)f_{c(t)}&=0, \\
&\cdots \\
v_{c(T-1)}+\pi(T-1)f_{c(T-1)}&=0, \\
v_{c(T)}+\pi(T)f_{c(T)}&=0;
\end{aligned} \tag{4.14}$$

$$\begin{aligned} v_{s(2)} - \pi(1) + \pi(2) + \pi(2) f_{s(2)} &= 0, \\ &\dots \\ v_{s(t)} - \pi(t-1) + \pi(t) + \pi(t) f_{s(t)} &= 0, \\ &\dots \\ v_{s(T-1)} - \pi(T-2) + \pi(T-1) + \pi(T-1) f_{s(T-1)} &= 0, \\ v_{s(T)} - \pi(T-1) + \pi(T) + \pi(T) f_{s(T)} &= 0; \end{aligned} \tag{4.15}$$

$$\begin{aligned} s_1 + f(s_1, c(1)) - s(2) &= 0, \\ &\dots \\ s(t) + f(s(t), c(t)) - s(t+1) &= 0, \\ &\dots \\ s(T) + f(s(T), c(T)) - s_{T+1} &= 0. \end{aligned} \tag{4.16}$$

These three sets of equations can be written more compactly as

$$v_{c(t)} + \pi(t) f_{c(t)} = 0, \quad t = 1, 2, \dots, T, \tag{4.17}$$

$$\pi(t) - \pi(t-1) = -v_{s(t)} - \pi(t) f_{s(t)}, \quad t = 2, 3, \dots, T, \tag{4.18}$$

$$s(t+1) - s(t) = f(s(t), c(t)), \quad t = 1, 2, \dots, T, \tag{4.19}$$

with

$$s(1) = s_1 \quad \text{and} \quad s(T+1) = s_{T+1}.$$

If we define a new function,

$$H(s(t), c(t), \pi(t)) \equiv v(s(t), c(t)) + \pi(t) f(s(t), c(t)), \tag{4.20}$$

then these necessary conditions can be expressed as

$$\frac{\partial H}{\partial c(t)} = 0, \quad t = 1, 2, \dots, T, \tag{4.21}$$

$$\pi(t) - \pi(t-1) = -\frac{\partial H}{\partial s(t)}, \quad t = 2, 3, \dots, T, \tag{4.22}$$

$$s(t+1) - s(t) = \frac{\partial H}{\partial \pi(t)}, \quad t = 1, 2, \dots, T. \tag{4.23}$$

It is obvious that the expressions in (4.20)–(4.23) are the discrete counterparts of the Hamiltonian and maximum principle of (4.4) and (4.5)–(4.7). It is instructive to apply our result to a simple example.

Example 4.2.1. Find $c(t)$ that maximizes

$$V = \sum_{t=1}^{3} \ln c(t)$$

subject to

$$s(t+1) - s(t) = 0.1 s(t) - c(t), \quad t = 1, 2, 3,$$

$$s(1) = 1, \quad s(4) = 1.21.$$

We form the Hamiltonian

$$H(s(t), c(t), \pi(t)) = \ln c(t) + \pi(t)[0.1s(t) - c(t)].$$

Then (4.21) yields

$$\frac{1}{c(t)} - \pi(t) = 0, \quad t = 1, 2, 3; \tag{4.24}$$

(4.22) yields

$$\pi(t) - \pi(t-1) = -0.1\pi(t), \quad t = 2, 3; \tag{4.25}$$

and (4.23) yields

$$s(t+1) - s(t) = 0.1s(t) - c(t), \quad t = 1, 2, 3. \tag{4.26}$$

We can use (4.24) to eliminate the costate variables to get

$$c(t) = 1.1c(t-1), \quad t = 2, 3. \tag{4.27}$$

Using (4.26), (4.27), and the initial condition, we can proceed recursively:

$$\begin{aligned} s(2) &= (1.1)(1) - c(1), \\ s(3) &= (1.1)s(2) - c(2) = 1.21 - (1.1)c(1) - (1.1)c(1), \\ s(4) &= (1.1)s(3) - c(3) = 1.331 - 2(1.21)c(1) - (1.21)c(1), \\ s(4) &= 1.331 - 3(1.21)c(1). \end{aligned} \tag{4.28}$$

In order to meet the terminal condition $s(4) = 1.21$ we must therefore choose $c(1) = 0.0333\ldots.$ Substituting it into the above equations we easily obtain all values of $s(t)$, $c(t)$, and $\pi(t)$:

$$c(1) = 0.0333, \quad c(2) = 0.0367, \quad c(3) = 0.0403;$$

$$\pi(1) = 30, \quad \pi(2) = 27.27, \quad \pi(3) = 24.79;$$

$$s(2) = 1.0667, \quad s(3) = 1.1367, \quad s(4) = 1.21.$$

It is interesting that in equation (4.28) we obtained the relationship between the terminal value of the state variable and the initial value of the control, when an optimal path is followed. Hence, there is a different optimal path for each terminal condition. It is worth remarking that for some terminal conditions no feasible solution exists: since the control must remain positive, we get from (4.28) $s(4) < 1.331$.

The procedure followed in solving this example by means of the discrete maximum principle involved several steps: (i) eliminating one of the variables (the costate in this case) by making use of the first-order condition; (ii) solving two difference equations (one for the state and one for the control); (iii) making use of the boundary conditions to determine the initial value of the control; (iv) substituting this value in the solutions of the difference equations and the first-order condition to obtain the optimal values of all variables.

A similar solution procedure will be followed when we apply the maximum principle in continuous time in the next section.

4.3 Numerical solution of an optimal control problem in continuous time

In Section 3.3 we presented a simple dynamic choice model. In that problem the optimal policy was determined by choosing the value of the consumption/output ratio: it was a parameter that remained constant for the whole horizon. Although it was pointed out that this was unduly restrictive, more flexibility was shown to yield an insoluble problem.

In this section we reconsider this problem. It will be possible to choose a consumption/output ratio that varies optimally over time through the use of the maximum principle.

4.3.1 *Optimal consumption*

The problem, set up in an optimal control format, is to find $c(t)$ that maximizes

$$V=\int_0^1 \ln[c(t)4s(t)]\,dt \tag{4.29}$$

subject to

$$\dot{s}(t)=4s(t)(1-c(t)) \tag{4.30}$$

with

$$s(0)=1, \quad s(1)=e^2. \tag{4.31}$$

The reason (4.31) includes a terminal condition is to make this problem comparable with that of Section 3.3, in which this was the terminal value of the stock.

The Hamiltonian of the problem is

$$H(s(t),c(t),\pi(t))=\ln 4+\ln c(t)+\ln s(t)+\pi(t)[4s(t)(1-c(t))],$$

and applying the maximum principle yields the following first-order condition and two differential equations:

$$\frac{\partial H}{\partial c(t)}=\frac{1}{c(t)}-4\pi(t)s(t)=0, \tag{4.32}$$

$$\dot{\pi}(t)=-\frac{\partial H}{\partial s(t)}=-\frac{1}{s(t)}-4(1-c(t))\pi(t), \tag{4.33}$$

$$\dot{s}(t)=\frac{\partial H}{\partial \pi(t)}=4s(t)(1-c(t)). \tag{4.34}$$

For notational simplicity we suppress the time argument while deriving the solution. From (4.32), $c=1/(4\pi s)$, and substituting it in the other equations we obtain a pair of differential equations in π and s:

$$\dot{\pi}=-\frac{1}{s}-4\pi\left(1-\frac{1}{4\pi s}\right)$$

and

$$\dot{s}=4s\left(1-\frac{1}{4\pi s}\right).$$

These simplify to

$$\dot{\pi}=-4\pi,$$

$$\dot{s}=4s-(1/\pi).$$

The first differential equation immediately yields

$$\pi(t)=\pi(0)e^{-4t}, \tag{4.35}$$

which we substitute into the second equation,

$$\dot{s}=4s-e^{4t}/\pi(0).$$

This is most easily solved by passing all s terms on the left-hand side and multiplying through by the integrating factor e^{-4t}:

$$\dot{s}e^{-4t}-4se^{-4t}=-1/\pi(0).$$

The left-hand side is the derivative of se^{-4t}, and integration yields the general solution

$$se^{-4t}=-t/\pi(0)+A.$$

We use (4.31) to determine $\pi(0)$ and A:

$$s(0)=1;\quad \text{hence,}\quad 1=A;$$

$$s(1)=e^{2};\ \text{hence,}\quad e^{-2}=-1/\pi_0+1;$$

thus, $\pi_0\simeq 1.156$.

We have the solution

$$s(t)=e^{4t}-0.865te^{4t}. \tag{4.36}$$

Substituting (4.35) and (4.36) into (4.32), we obtain

$$c(t)=\frac{1}{4.624-4t}. \tag{4.37}$$

Therefore, it is optimal for the consumption/output ratio to increase in the manner described by (4.37). In order to show convincingly that this solution is preferred to that of Section 3.3, we calculate the optimal value of V:

$$V=\int_0^1 \ln(4c(t)s(t))\,dt=\int_0^1 \ln\frac{1}{\pi(t)}\,dt \quad \text{(by (4.32))}$$
$$=\int_0^1 [\ln(0.865)+4t]\,dt=[t\ln 0.865+2t^2]_0^1.$$

Therefore, $V\simeq 1.855$, which is larger than $V=1.69$ obtained previously.

Before leaving this example it is useful to reflect on one hidden assumption. Substituting $t=1$ into (4.37) we find that at that time the consumption/output ratio is 1.662; therefore, for some time the amount consumed exceeds the amount produced, $\dot{s}$ is negative, and this implicitly assumes that it is possible to eat into the capital stock. Our purpose is not to argue for or against this assumption, as already mentioned in Section 3.2, but to note its importance. Note that $\dot{s}$ could become negative because the control was free of all constraints. Had we wished to restrict it to values between 0 and 1, we would have needed more general results; the constrained control problem and other extensions are the subject of Chapter 6.

To conclude this section we shall solve a slightly more general version of this model.

4.3.2 *Optimal consumption with discounting*

Here the problem is to find $c(t)$ that maximizes

$$V=\int_0^T e^{-\delta t}\ln c(t)\,dt \tag{4.38}$$

subject to

$$\dot{s}(t)=rs(t)-c(t), \tag{4.39}$$
$$s(0)=s_0, \quad s(T)=s_T. \tag{4.40}$$

In this formulation $c(t)$ denotes the consumption flow itself and not the consumption/output ratio. Moreover, the logarithmic utility function is discounted at the rate $\delta>0$. The rate of interest (r) and the boundary values of the state variable (s_0 and s_T) are exogenously specified.

The Hamiltonian of the problem is

$$H(s(t),c(t),t)=e^{-\delta t}\ln c(t)+\pi(t)[rs(t)-c(t)].$$

Applying the maximum principle yields

$$\frac{\partial H}{\partial c(t)}=e^{-\delta t}\frac{1}{c(t)}-\pi(t)=0, \tag{4.41}$$

$$\dot{\pi}(t) = -\frac{\partial H}{\partial s(t)} = -r\pi(t), \tag{4.42}$$

$$\dot{s}(t) = \frac{\partial H}{\partial \pi(t)} = rs(t) - c(t). \tag{4.43}$$

When the time argument is omitted, these necessary conditions appear less cumbersome:

$$e^{-\delta t} = \pi c,$$
$$\dot{\pi} = -r\pi,$$
$$\dot{s} = rs - c.$$

It is obvious how to proceed: we can use the second equation to get the general solution for π, substitute it in the first equation to get c, and substitute c into the last equation to get the general solution for s. The boundary conditions will determine the two constants of integration:

$$\pi(t) = \pi(0)e^{-rt}, \tag{4.44}$$
$$c(t) = [\pi(0)]^{-1}e^{(r-\delta)t}, \tag{4.45}$$
$$\dot{s} - sr = -[\pi(0)]^{-1}e^{(r-\delta)t},$$
$$\dot{s}e^{-rt} - re^{-rt}s = -[\pi(0)]^{-1}e^{-\delta t},$$
$$se^{-rt} = e^{-\delta t}/\delta\pi(0) + A,$$
$$s(t) = e^{(r-\delta)t}/\delta\pi(0) + Ae^{rt}. \tag{4.46}$$

Substituting (4.40) into (4.46), we get

$$s(0) = s_0; \quad \text{hence,} \quad s_0 = A + \frac{1}{\delta\pi(0)};$$

$$s(T) = s_T; \quad \text{hence,} \quad s_T = Ae^{rT} + \frac{e^{(r-\delta)T}}{\delta\pi(0)}.$$

Solving these two equations yields

$$\frac{1}{\pi(0)} = \frac{s_0 - s_T e^{-rT}}{(1-e^{-\delta T})/\delta}, \tag{4.47}$$

$$A = \frac{s_T e^{-rT} - s_0 e^{-\delta T}}{1 - e^{-\delta T}}. \tag{4.48}$$

Since $c(t)$ has the sign of $\pi(t)$, which is the same as that of $\pi(0)$, we must require that $s_0 - s_T e^{-rT} > 0$, or $s_T < s_0 e^{rT}$, for otherwise $c(t)$ would be negative and the logarithm undefined. Upon reflection this is only the requirement that the final value of the stock not exceed the level to which the initial stock would have naturally grown had its growth not been curtailed by consumption.

From (4.45) we see that the flow of consumption varies exponentially; whether it increases or decreases depends on the relative values of r and δ. If the rate of interest r exceeds the consumer's own rate of discount δ, she tends to postpone consumption; thus, her consumption flow increases over time. From (4.48) it is obvious that the sign of A cannot be ascertained in general; hence, the path of $s(t)$ may or may not be monotone. Given specific values for the parameters T, r, δ, s_0, and s_T, the paths of all variables can be precisely determined. The reader is invited to verify that the derivative of $c(0)$ with respect to δ has the sign of $[e^{-\delta T}-(1+\delta T)]>0$, confirming that a higher subjective discount rate induces higher consumption at the beginning of the horizon, just as in the example of Section 3.4.

4.4 Phase diagram analysis of optimal control problems

In the simple type of optimal control problem that is the subject of this chapter, the maximum principle yields a first-order condition and a pair of differential equations. Nevertheless, these are often difficult to solve analytically. If all functional forms and relevant parameter values were specified, it would be possible to use numerical methods to derive the solution. This may be useful in physics or engineering; however, since our ultimate purpose in using control theory is to gain insight into the dynamic behavior of economic models, we often deal with problems involving unspecified functional forms. For example, often for the sake of generality, we do not wish to specify a particular form of the utility function (e.g., logarithmic) or a particular form of the production function (e.g., Cobb–Douglas). In such cases the explicit solution of the differential equations is impossible. The best we can hope for is a qualitative characterization of the optimal solution, as was often the case in static optimization problems in economics. This seems a formidable task. Fortunately, we have just the device needed: the representation of the solution on a phase diagram as described in Section 2.5. We will be able to partition the phase space into regions in which we know whether the variables increase or decrease over time. Further analysis will yield restrictions on the shape of the trajectories that are candidates for the optimal path. As with most qualitative tools this is not a perfect device. In many cases it will require some ingenuity to pinpoint exactly the optimal trajectory. Nonetheless, it provides a structure for detailed qualitative analysis.

Initially we shall illustrate the technique with a numerical example.

Example 4.4.1. Consider the problem of finding $c(t)$ to maximize

$$V=\int_0^T e^{-0.05t}[\ln c(t)]\,dt$$

subject to

$$\dot{s}(t) = 2[s(t)]^{0.5} - c(t),$$
$$s(0) = s_0, \quad s(T) = s_T.$$

We may interpret $c(t)$ as the consumption flow and $s(t)$ as the stock of capital; $2[s(t)]^{0.5}$ is the output produced with capital stock $s(t)$, and there is no depreciation.

The Hamiltonian of the problem is

$$H(s(t), c(t), \pi(t), t) = e^{-0.05t} \ln c(t) + \pi(t)[2(s(t))^{0.5} - c(t)],$$

and applying the maximum principle yields

$$\frac{\partial H}{\partial c(t)} = e^{-0.05t}[c(t)]^{-1} - \pi(t) = 0,$$
$$\dot{\pi}(t) = -\frac{\partial H}{\partial s(t)} = -\pi(t)[s(t)]^{-0.5},$$
$$\dot{s}(t) = \frac{\partial H}{\partial \pi(t)} = 2[s(t)]^{0.5} - c(t).$$

Skipping the time argument we must solve

$$e^{-0.05t}c^{-1} = \pi, \tag{4.49}$$
$$\dot{\pi} = -\pi s^{-0.5}, \tag{4.50}$$
$$\dot{s} = 2s^{0.5} - c. \tag{4.51}$$

Let us use the first-order condition (4.49) to eliminate c and obtain the pair of differential equations

$$\dot{\pi} = -\pi s^{-0.5}, \tag{4.52}$$
$$\dot{s} = 2s^{0.5} - e^{-0.05t}\pi^{-1}. \tag{4.53}$$

These differential equations involve not only π and s but also an independent exponential time trend. This makes it impossible to draw a phase diagram in the (s, π) space because the locus of points at which $\dot{s} = 0$ is not well defined, since it depends on t. There is a way out of this difficulty and we shall present it later in this section.

For now let us look at an alternative pair of differential equations, one in s and one in c. Although we do not have a differential equation in c, we can obtain one by totally differentiating the first-order condition (4.49) with respect to time:

$$\begin{aligned} -0.05e^{-0.05t}c^{-1} - e^{-0.05t}c^{-2}\dot{c} &= \dot{\pi} \\ &= -\pi s^{-0.5} \qquad \text{by (4.50)} \\ &= -e^{-0.05t}c^{-1}s^{-0.5} \qquad \text{by (4.49).} \end{aligned}$$

Multiplying through by $e^{0.05t}c^2$ yields

$$-0.05c-\dot{c}=-cs^{-0.5},$$

from which we obtain

$$\dot{c}=c(s^{-0.5}-0.05). \tag{4.54}$$

Equation (4.54) is a differential equation in c involving c and s only; together with (4.51) they can be analyzed with a phase diagram in the (s,c) space. The procedure used to derive (4.54) is used so often that it is worth spelling it out. First, we totally differentiated the first-order condition and obtained an equation involving a $\dot{c}$ term and a $\dot{\pi}$ term. We eliminated the latter using the differential equation in π. This in turn introduced a π term, which was eliminated by using the original first-order condition again. Finally, some algebraic manipulation yielded a simpler form. This procedure is often useful in solving simple problems, but it sometimes fails. In that event we will attempt to devise another way out of the difficulty.

We follow the procedure outlined in Section 2.5.2. The main task is to use equations (4.51) and (4.54) to partition to (s,c) space into regions in which the respective signs of $\dot{s}$ and $\dot{c}$ are known. The first step is to obtain the loci of points where $\dot{s}=0$ and $\dot{c}=0$. The curve for $\dot{s}=2s^{0.5}-c=0$ is the graph of the function $c=2s^{0.5}$. This is a concave and increasing function of s. The curve goes through the origin, where it has an infinitely large slope. The other equation, $\dot{c}=c(s^{-0.5}-0.05)=0$, defines two straight lines: $c=0$ and $s^{-0.5}=0.05$, or $s=400$. These "critical loci" are drawn in Figure 4.1. They define four regions (or isosectors), which are labeled I–IV. The expressions defining $\dot{s}$ and $\dot{c}$ are continuous in the positive orthant (check (4.51) and (4.54)); therefore, $\dot{s}$ and $\dot{c}$ can change sign only when we cross over one of the above critical loci. In order to ascertain the sign of $\dot{s}$ and $\dot{c}$ in any one of the four regions, it is sufficient to evaluate these signs at an arbitrary point of the region.

In region I, c is less than $2s^{0.5}$; hence, $\dot{s}>0$ by (4.51). Since $c>0$ and $s<400$, (4.54) implies $\dot{c}>0$. In region II we still have $c<2s^{0.5}$; hence, $\dot{s}>0$, and now that $s>400$ we have $\dot{c}<0$. Region III is above the $c=2s^{0.5}$ curve, so that $\dot{s}<0$ and $s>400$ implies $\dot{c}<0$. In region IV, $\dot{s}<0$ and $\dot{c}>0$. We could have come to this conclusion by noting that at any point below the $c=2s^{0.5}$ curve $\dot{s}>0$, and that $\dot{c}<0$ to the right of the $s=400$ line. The signs of $\dot{s}$ and $\dot{c}$ are represented in each region of Figure 4.1 by two small perpendicular arrows. The horizontal arrow indicates an increase or a decrease in s, while the vertical one refers to changes in c, since the s axis is horizontal and the c axis vertical. Since optimal trajectories must obey equations (4.51) and (4.54), they must follow the direction indicated by the pair of arrows in each region. A curve indicates the direction of

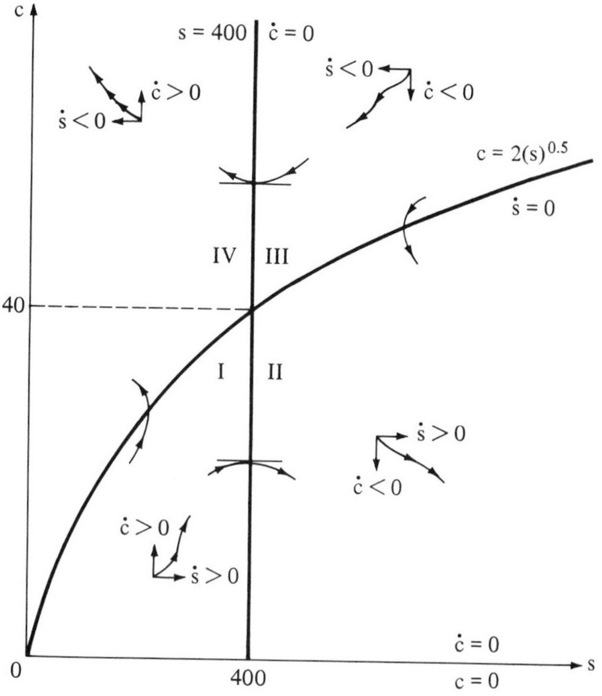

Figure 4.1

admissible trajectories in each region, with the arrows on the curve denoting movement as time passes.

We must gather another piece of information before we can draw the general shape of trajectories. The slope of trajectories in the (s, c) space can be obtained from the relationship $dc/ds = (dc/dt)(dt/ds) = \dot{c}/\dot{s}$. Hence, when a trajectory goes through a locus where $\dot{c} = 0$, it has slope zero, and when it goes through a locus where $\dot{s} = 0$, it has an infinite slope. This information, along with the direction of trajectories in each region, allows us to draw the shape of trajectories when crossing a critical locus from one region to the next; this is done in Figure 4.1 in all four cases. The general solution to (4.51) and (4.54) is a family of trajectories. The knowledge of the boundary conditions will determine the specific solution. We did not assign specific values to s_0 and s_T in order to discuss the influence of the boundary conditions on the choice of the optimal trajectory. To this end we represent a few possible trajectories in Figure 4.2. The positions of trajectories relative to the equilibrium point E ($c = 40$, $s = 400$)

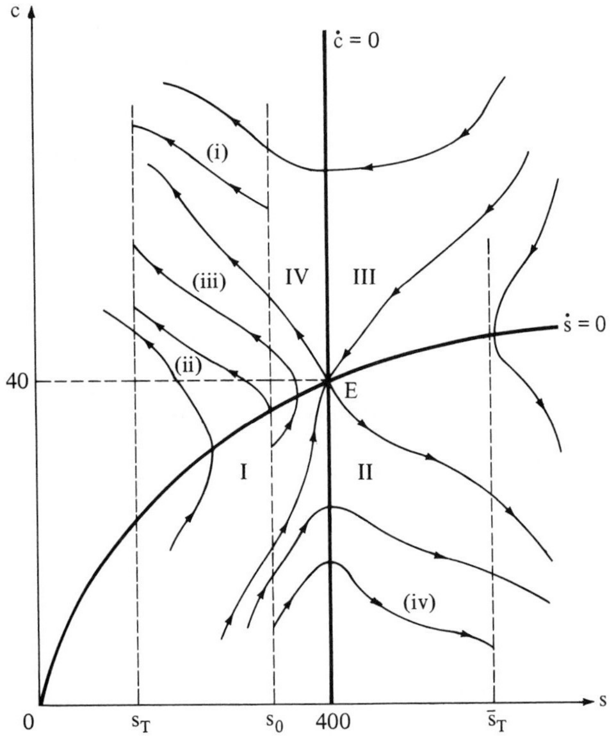

Figure 4.2

are of interest. In region I the trajectories go up and to the right; if one reaches the $\dot{s} = 0$ locus, it turns left; if one reaches the $\dot{c} = 0$ locus, it turns down. We know from the theory of differential equations that trajectories cover the whole space; therefore, there is one trajectory in region I that reaches the equilibrium point E. We refer to it as a *stable path*. Note that because $\dot{s}$ and $\dot{c}$ become arbitrarily small near E, it would take an infinite amount of time to reach equilibrium.

Similar reasoning reveals the existence of a downward stable path in region III, while regions II and IV each possess an unstable path leading away from E, identifying E as a saddle point. We can check the saddle-point property locally by linearizing the system (4.51), (4.54) around E. The matrix of coefficients obtained is

$$\begin{bmatrix} 0.05 & -1 \\ -0.0025 & 0 \end{bmatrix},$$

which has a negative determinant corresponding to a saddle point, as noted at the end of Section 2.4. The arms of the saddle point and some other trajectories are drawn in Figure 4.2.

Suppose now that $s(0)=s_0$ and $s(T)=s_T$, as indicated in Figure 4.2. Vertical dashed lines have been drawn through those values to emphasize the fact that although boundary values are specified for the state variable, all values for the control are to be optimally chosen. Thus, the optimal trajectory may begin anywhere along the $s=s_0$ line and must end somewhere on the $s=s_T$ line. The optimal path will be within regions I and IV. Indeed, there is no way to reach region III beginning at s_0, and were the path to enter region II, it would never reach s_T. The optimal path may be wholly in region IV, as are (i) and (ii), or begin in I, as does (iii). Since our objective is to maximize the integral of $[\ln c(t)]e^{-0.05t}$, it would seem that we should select the higher trajectory. However, to travel the length of a path such as (i) takes some fixed amount of time. Since the time horizon has been specified exogenously as $[0,T]$, we must select the path that will go from s_0 to s_T in exactly T units of time. The further this path is from the $\dot{s}=0$ locus, the larger $\dot{s}$ is and the faster s changes. Hence, if T is small, the optimal path will be a high path, such as (i); the larger T, the further down we go until we reach path (ii). This is, of all paths wholly within region IV, the one that takes the longest time to go from s_0 to s_T. If T is still larger, we must select as optimal a path such as (iii) that begins in region I. No matter how large T is, we shall always be able to select an appropriate path, because we can choose one that goes arbitrarily close to E before turning left into region IV, and in the neighborhood of E, movement along the path would be very slow. The effect of T on the choice of an optimal path can be explained in simple terms. Recall that c is consumption and s a stock of capital. You have been given T periods to eat into your capital from s_0 to s_T. The shorter the time you have, the higher the rate of consumption you will be able to afford. If you must plan for a long enough time, you will find it optimal to begin with a level of consumption low enough that you accumulate capital initially. In any event, consumption increases monotonically through time.

Other boundary conditions would yield different paths, based on analyses similar to the one just presented. To take one more instance, suppose that the initial stock is still s_0 but the terminal stock must now be $\bar{s}_T$. We would follow a path such as (iv). The length of the planning horizon would determine exactly which one, since the higher the path, the closer it is to the $\dot{s}=0$ locus, around which s increases very slowly. Note that in this case s always increases throughout and c goes through a peak. Note also that if s_0 and T are small enough and $\bar{s}_T$ large enough so that the differential equation $\dot{s}=2s^{0.5}$ has no solution satisfying these boundary conditions, then the control problem has no feasible solution, let alone an optimal one. If there is no way to satisfy the boundary requirements even by starving ($c=0$), there can be no optimal way to solve the problem.

This completes the analysis of the (s,c) phase diagram for this problem. In some cases it may be useful, necessary, or the only feasible choice to conduct the phase diagrammatic analysis in the (state, costate) plane. Recall that we attempted this task but encountered difficulties because of an independent time term in the costate differential equation. We now show how to deal with this problem by a change of variable. We had to deal with equations (4.52) and (4.53):

$$\dot{\pi} = -\pi s^{-0.5}, \tag{4.52}$$

$$\dot{s} = 2s^{0.5} - e^{-0.05t}\pi^{-1}. \tag{4.53}$$

Define

$$\psi(t) \equiv e^{0.05t}\pi(t). \tag{4.55}$$

Taking the time derivative, we have

$$\dot{\psi} = 0.05e^{0.05t}\pi + e^{0.05t}\dot{\pi}.$$

Substituting (4.52) and (4.55) into the above equation yields

$$\dot{\psi} = 0.05\psi - e^{0.05t}\pi s^{-0.5}$$
$$= 0.05\psi - \psi s^{-0.5}.$$

We now have a pair of differential equations in ψ and s that are autonomous, that is, free of an independent time argument:

$$\dot{\psi} = \psi(0.05 - s^{-0.5}), \tag{4.56}$$

$$\dot{s} = 2s^{0.5} - \psi^{-1}. \tag{4.57}$$

There is a way to obtain these equations directly; it will be explained at the end of this section.

The phase diagram in the (s,ψ) space can be constructed in the same manner as the preceding one. The critical loci are

$$\dot{\psi} = \psi(0.05 - s^{-0.5}) = 0, \quad \text{or} \quad \psi = 0 \text{ and } s = 400,$$
$$\dot{s} = 2s^{0.5} - \psi^{-1} = 0, \quad \text{or} \quad \psi = 0.5s^{-0.5}.$$

The first locus is a pair of straight lines, while the second one is a rectangular hyperbola with both axes as asymptotes; these are drawn in Figure 4.3. We shall again restrict our attention to the positive orthant since it is obvious from the optimality condition (4.49) that π (hence ψ) cannot be negative.

From (4.57) we see that $\dot{s} < 0$ below the hyperbola and $\dot{s} > 0$ above it. Equation (4.56) implies that $\dot{\psi} < 0$ to the left of $s = 400$ and $\dot{\psi} > 0$ to the

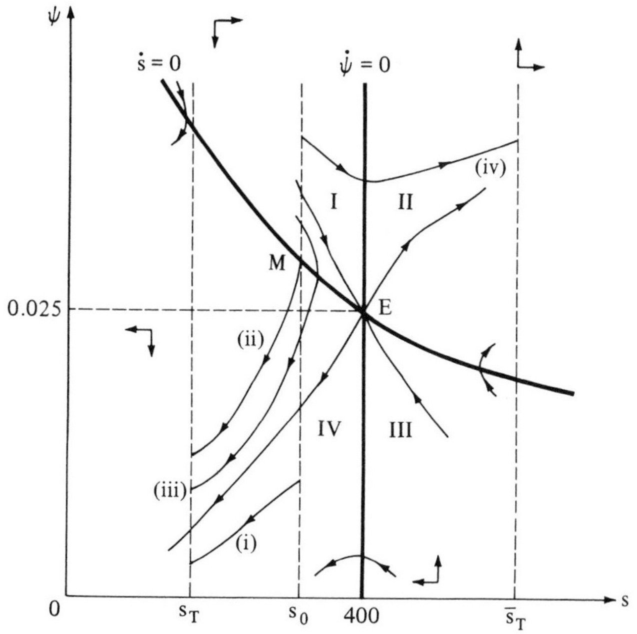

Figure 4.3

right. These signs together with the knowledge that the trajectories have a zero slope when crossing the $\dot{\psi} = 0$ locus and an infinite slope when crossing the $\dot{s} = 0$ locus enable us to draw the general shape of the trajectories in Figure 4.3. There are again four regions in the phase space. We have numbered them in a way that is consistent with the numbering in Figure 4.2. In fact it is very important to grasp the correspondence between the two diagrams. The first-order condition (4.49) rewritten after the change of variable is $c^{-1} = \psi$. Therefore, whereas the horizontal axes in Figures 4.2 and 4.3 bear the same variable, the vertical axes bear variables that are the reciprocals of one another. Thus, region I of Figure 4.3, where s increases and ψ decreases, is indeed the mirror image of region I in Figure 4.2, where both c and s increase. Similar correspondences apply between other regions, as can readily be checked. With boundary values s_0 and s_T, a short planning horizon will lead to an optimal path such as (i). The most time-consuming monotone path is (ii). If more time is available, a path such as (iii) would be selected. If boundary values were s_0 and $\bar{s}_T$, a path such as (iv) would be chosen. Paths (i)–(iv) in Figure 4.3 correspond to like-numbered paths in Figure 4.2.

This completes our analysis by phase diagrams of Example 4.4.1. We have been able to obtain a fairly precise qualitative description of the optimal solution. Note that if a numerical solution were required, the insights gained from the phase diagram analysis would be very useful in tracking it down. Our next task in this section is to introduce a far more general growth model from which the previous examples were drawn. It will be shown that the phase diagram analysis is no more complicated in the general case than it was in the example. Hence, the technique is more powerful than this example and those of Section 2.5 may have led us to believe.

Find $c(t)$ to maximize

$$V=\int_0^T e^{-\delta t}u(c(t))\,dt$$

subject to

$$\dot{s}(t)=F(s(t))-c(t),$$
$$s(0)=s_0, \quad s(T)=s_T,$$

where we assume $u'>0$, $u''<0$, $u'(0)=\infty$, $F(0)=0$, $F'>0$, $F''<0$, and $F'(+\infty)<\delta<F'(0)$. The assumptions that the utility function $u(\cdot)$ and the production function $F(\cdot)$ are increasing and concave are standard ones. The additional assumption that marginal utility at the origin is infinite guarantees that consumption is always positive along an optimal path. We shall see that the condition $F'(+\infty)<\delta<F'(0)$ is necessary and sufficient to guarantee the existence of an equilibrium.

The Hamiltonian of the problem is

$$H(s,c,\pi,t)=e^{-\delta t}u(c)+\pi[F(s)-c],$$

and applying the maximum principle yields

$$\frac{\partial H}{\partial c}=e^{-\delta t}u'(c)-\pi=0, \tag{4.58}$$

$$\dot{\pi}=-\frac{\partial H}{\partial s}=-\pi F'(s), \tag{4.59}$$

$$\dot{s}=\frac{\partial H}{\partial \pi}=F(s)-c. \tag{4.60}$$

To obtain a differential equation in c we totally differentiate (4.58) with respect to time:

$$\begin{aligned}-\delta e^{-\delta t}u'(c)+e^{-\delta t}u''(c)\dot{c}&=\dot{\pi}\\ &=-\pi F'(s) \qquad \text{by (4.59)}\\ &=-e^{-\delta t}u'(c)F'(s) \quad \text{by (4.58)}.\end{aligned}$$

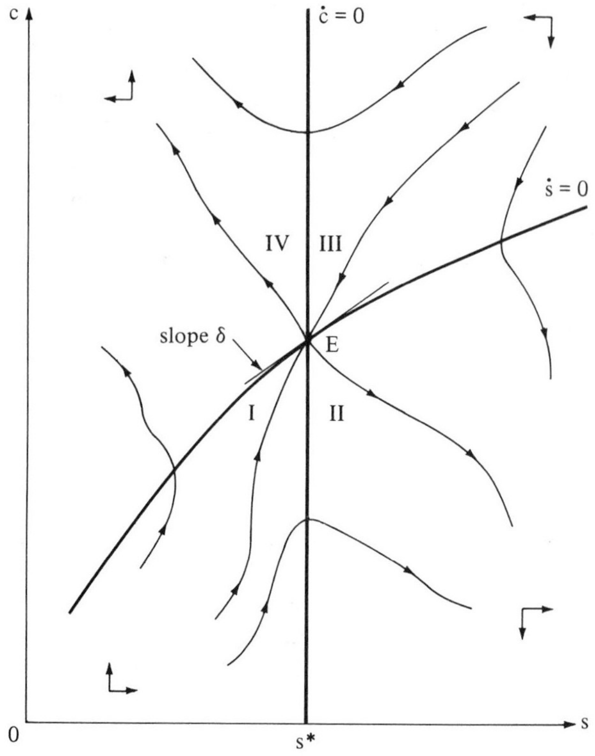

Figure 4.4

Multiplying through by $e^{\delta t}/u''(c)$ yields

$$\dot{c} = \frac{u'(c)}{u''(c)}[\delta - F'(s)]. \tag{4.61}$$

Equations (4.60) and (4.61) can be used to draw a phase diagram.

The $\dot{s} = 0$ locus is $c = F(s)$, an increasing concave curve going through the origin. The $\dot{c} = 0$ locus is a straight line, $s = s^*$, where s^* is the number defined by $F'(s^*) = \delta$. The assumption $F'(+\infty) < \delta < F'(0)$ ensures that s^* exists. These loci are drawn in Figure 4.4. The observations that $\dot{s} > 0$ below the $\dot{s} = 0$ locus and $\dot{s} < 0$ above it, and that $\dot{c}$ is negative to the right of s^* and positive to the left of it, enable us to draw in each region arrows indicating the directions followed by trajectories. Moreover, as the trajectories have slope zero and infinity, respectively, when crossing a $\dot{c} = 0$ and an $\dot{s} = 0$ locus, the general shapes can be drawn. We note the existence of two stable paths in regions I and III and two unstable paths in regions II and IV. The equilibrium value of s, namely, s^*, is determined by finding the point on the $\dot{s}=0$ locus that has slope δ. The similarities between Figures 4.4 and 4.2 are so obvious as to require no comments; indeed, we need not have drawn a new diagram, except that s^* now replaces 400. We can prove the existence of a local saddle point by linearizing (4.60) and (4.61) around the equilibrium. The matrix of coefficients is

$$\begin{bmatrix} \delta & -1 \\ -F''(s^*)u'(c^*)/u''(c^*) & 0 \end{bmatrix}.$$

Its determinant is negative, confirming that the equilibrium is a saddle point. We now simply derive the differential equations that we would need for a diagram in the (s, ψ) space. Define

$$\psi(t) = e^{\delta t}\pi(t). \tag{4.62}$$

Differentiating (4.62) with respect to time yields

$$\dot{\psi} = \delta e^{\delta t}\pi + e^{\delta t}\dot{\pi} = \delta e^{\delta t}\pi - e^{\delta t}\pi F'(s)$$

(by (4.59)). Hence.

$$\dot{\psi} = \psi(\delta - F'(s)). \tag{4.63}$$

This equation is the general form corresponding to (4.56) and defines the $\dot{\psi}=0$ locus as two straight lines of equations $\psi=0$ and $s=s^*$; it can be used to draw a diagram in the (s, ψ) space. Note, however, that the other equation, $\dot{s}=F(s)-c$, includes c. In Example 4.4.1 we eliminated c with the aid of the first-order condition; this, in the present case, takes the form

$$u'(c) = \psi, \tag{4.64}$$

which is obtained from (4.58) and (4.62). We had assumed $u''<0$; therefore, u' is strictly decreasing and possesses an inverse function, which is also decreasing. Hence, the first-order condition can be expressed as

$$c = c(\psi), \quad dc/d\psi < 0, \tag{4.65}$$

where $c(\cdot)$ is the inverse function of $u'(\cdot)$.

Substitution into (4.60) yields

$$\dot{s} = F(s) - c(\psi), \tag{4.66}$$

which can be used in conjunction with (4.63) to draw the phase diagram in the (s, ψ) space. The derivation of the $\dot{s}=0$ locus is the only new aspect of the procedure. Its equation is $F(s)=c(\psi)$. Since c is the inverse of the function u', applying u' to each side we get $u'(F(s)) = u'(c(\psi)) = \psi$. This gives ψ as a function of s. Note that since u' is decreasing and F is

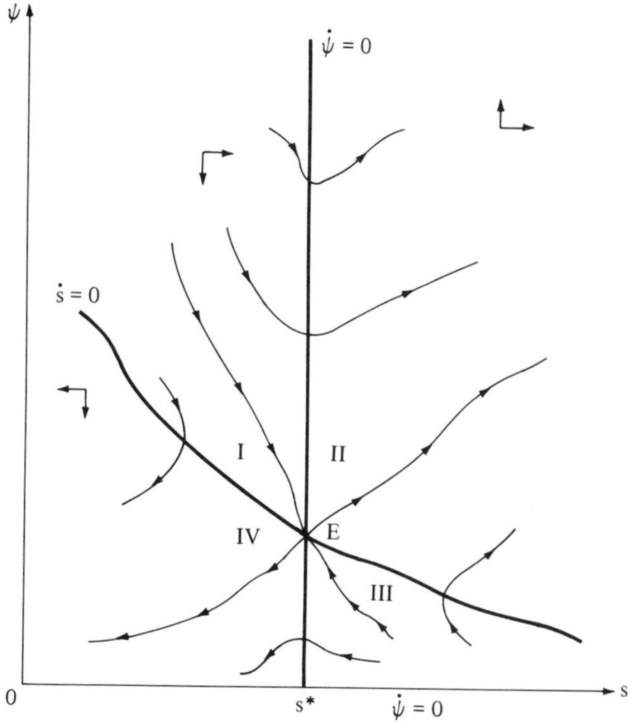

Figure 4.5

increasing, ψ is here a decreasing function of s. Furthermore, we know that both u' and F are positively valued and defined for all positive values of their arguments. This implies that the graph of $\psi = u'(F(s))$ is downward sloping and intersects the line $s = s^*$ but not the line $\psi = 0$. Hence, this point of intersection is the unique equilibrium. Points above the graph of $\psi = u'(F(s))$ are such that $\psi > u'(F(s))$; hence, $c(\psi) < F(s)$ (because $c(\cdot)$ is a decreasing function of ψ); therefore, $\dot{s} > 0$ at such points. Recalling that F' is decreasing since $F'' < 0$, (4.63) implies that $\dot{\psi} > 0$ at points to the right of the $s = s^*$ line. Figure 4.5 is the phase diagram. The only feature of the phase diagram in Figure 4.3 which is not necessarily present in Figure 4.5 is that the $\dot{s} = 0$ locus need not be an asymptote to both axes. This would be guaranteed here by the special assumptions $F(0) = 0$, $F(\infty) = \infty$, $u'(\infty) = 0$. To see this recall that the $\dot{s} = 0$ locus can be represented in the form $\psi = u'(F(s))$; if s is close to zero, so is $F(s)$; hence, $u'(F(s))$, which is ψ, approaches infinity. Conversely, if s is arbitrarily large, (4.66) and $\dot{s}=0$ imply that $F(s)$ and c are too; then our special assumptions imply that u', hence ψ, approaches zero. Nonetheless, all the important features of the solution, given specific values for T, s_0, and s_T, are common to the two problems, whether or not the above special assumptions are made. We have established our contention that phase diagram analysis is a powerful technique that can be used on models without specified functional forms.

This concludes our introduction to phase diagram analysis. Before we move on to the next section, it is convenient to include here an alternative statement of the maximum principle that yields equations of the type (4.63) directly. This procedure applies to control problems in which the only independent time term appears in the discount factor in the maximand; these are an important class of problems. They are called *autonomous problems,* of which a simple prototype is as follows: find $c(t)$ to maximize

$$V=\int_0^T e^{-\delta t}u(s(t),c(t))\,dt \tag{4.67}$$

subject to

$$\dot{s}(t)=f(s(t),c(t)) \tag{4.68a}$$

and

$$s(0)=s_0, \quad s(T)=s_T. \tag{4.68b}$$

The reason we call such problems "autonomous" is that they yield an autonomous system of differential equations (see Section 2.2) to characterize the solution, as we now demonstrate. Instead of introducing the costate variable $\pi(t)$ and the Hamiltonian

$$H(s(t),c(t),\pi(t),t)=e^{-\delta t}u(s(t),c(t))+\pi(t)f(s(t),c(t)),$$

we introduce the costate variable $\psi(t)$ and the Hamiltonian

$$\tilde{H}(s(t),c(t),\psi(t))=u(s(t),c(t))+\psi(t)f(s(t),c(t)).$$

It is still true that $\psi(t)=e^{\delta t}\pi(t)$ and consequently

$$\tilde{H}(s(t),c(t),\psi(t))=e^{\delta t}H(s(t),c(t),\pi(t),t).$$

The economic interpretation of the maximum principle is taken up in the next section, but it is already apparent that if $\tilde{H}$ and ψ reflect current values, then H and π reflect discounted or present values (see Section 3.4). $\tilde{H}$ and $\psi(t)$ are called the *current value Hamiltonian* and *current value costate,* respectively. The maximum principle, when applied to $\tilde{H}$ of an autonomous problem, is stated as follows:

$$\frac{\partial \tilde{H}}{\partial c(t)}=\frac{\partial u}{\partial c(t)}+\psi(t)\frac{\partial f}{\partial c(t)}=0, \tag{4.69}$$

$$\dot{s}(t) = \frac{\partial \tilde{H}}{\partial \psi(t)} = f(s(t), c(t)), \tag{4.70}$$

$$\dot{\psi}(t) = -\frac{\partial \tilde{H}}{\partial s(t)} + \delta\psi(t) = -\frac{\partial u}{\partial s(t)} - \psi(t)\frac{\partial f}{\partial s(t)} + \delta\psi(t). \tag{4.71}$$

Condition (4.68b), of course, remains the relevant boundary condition. Special attention is drawn to the additional term $\delta\psi(t)$ in (4.71). One can easily check that (4.71) yields (4.63) of the preceding problem directly.

It is an instructive exercise to verify that (4.69)–(4.71) are equivalent to the equations obtained by applying the maximum principle to H in the way defined in Section 4.1. We display these equations here for easy reference:

$$\frac{\partial H}{\partial c(t)} = e^{-\delta t}\frac{\partial u}{\partial c(t)} + \pi(t)\frac{\partial f}{\partial c(t)} = 0, \tag{4.69$'$}$$

$$\dot{s}(t) = \frac{\partial H}{\partial \pi(t)} = f(s(t), c(t)), \tag{4.70$'$}$$

$$\dot{\pi}(t) = -\frac{\partial H}{\partial s(t)} = -e^{-\delta t}\frac{\partial u}{\partial s(t)} - \pi(t)\frac{\partial f}{\partial s(t)}. \tag{4.71$'$}$$

Since $\pi(t) = e^{-\delta t}\psi(t)$, we see that (4.69) and (4.69′) are equivalent. Equations (4.70) and (4.70′) are identical. To prove the equivalence of (4.71) and (4.71′), we totally differentiate the identity $\pi(t) = e^{-\delta t}\psi(t)$ to get

$$\dot{\pi}(t) = -\delta e^{-\delta t}\psi(t) + e^{-\delta t}\dot{\psi}(t).$$

We substitute this result into (4.71′) to obtain

$$-\delta e^{-\delta t}\psi(t) + e^{-\delta t}\dot{\psi}(t) = -e^{-\delta t}\frac{\partial u}{\partial s(t)} - \pi(t)\frac{\partial f}{\partial s(t)}.$$

Passing the first term on to the right-hand side and multiplying by $e^{\delta t}$ yields

$$\dot{\psi}(t) = -\frac{\partial u}{\partial s(t)} - \psi(t)\frac{\partial f}{\partial s(t)} + \delta\psi(t),$$

which is (4.71).

There are no independent time terms in (4.69)–(4.71); hence, phase diagrams obtained from these equations will not exhibit an $\dot{s} = 0$ locus that shifts over time. It is therefore recommended that for autonomous problems the maximum principle be stated using the current-value Hamiltonian. The results presented in Section 4.1 nonetheless apply to all problems.

Remark. Before concluding this section we must comment on the special form of the discount factor used in problem (4.67); it is the usual exponential discount factor introduced in Section 3.4. Had we instead used the more general formulation

$$\max \int_0^T \alpha(t)u(s(t), c(t))\,dt,$$

we might have encountered a serious problem. Suppose that the individual when allowed to recalculate her optimal policy at a later date, say $\theta > 0$, used the discount factor $\alpha(t-\theta)$. She would do this if the discount factor reflected the weight attached to the utility flow at time t, not by virtue of its calendar time but merely because of its distance from the planning date (e.g., the case of "impatience" or "pure time preference"). Let $(s^*(t), c^*(t))$ be the plan that solves the maximization problem at time $t_0 = 0$ subject to $\dot{s}(t) = f(s(t), c(t))$ and $s(0) = s_0$, $s(T) = s_T$. At time θ, the individual would want to solve

$$\max \int_\theta^T \alpha(t-\theta)u(s(t), c(t))\,dt$$

subject to $\dot{s}(t) = f(s(t), c(t))$ and $s(\theta) = s^*(\theta)$, $s(T) = s_T$. It can be shown (see Strotz, 1955–6; Pollak, 1968) that the only form of the discount factor which ensures that the solution of this problem coincides with $(s^*(t), c^*(t))$ over $[\theta, T]$ is the one adopted in (4.67), namely, $\alpha(t) = e^{-\delta t}$ (or equivalently $\alpha(t) = A^t$). The difficulty with any other form of discount factor is that the individual would persistently wish to change her optimal plan, thereby rendering planning rather meaningless. This problem is known as a problem of *dynamic inconsistency;* it is perhaps one of the reasons for the almost universal use of the exponential discount factor.

4.5 Economic interpretation of the maximum principle

In this section we show that the maximum principle can be given an appealing economic interpretation that gives us further insight into the optimality of dynamic choice. We deal with the general control problem introduced in Section 4.1. It is restated here:

$$V^*(s_0, s_T, 0, T) = \max_{c(t)} \int_0^T v(s(t), c(t), t)\,dt, \tag{4.72}$$

subject to

$$\dot{s}(t) = f(s(t), c(t), t),$$
$$s(0) = s_0, \quad s(T) = s_T.$$

We shall refer to s as a stock of capital and to c as a flow of consumption; v is the instantaneous value function, V^* the maximum value function, and f the growth function of stock.

The Hamiltonian of this problem is

$$H(s, c, \pi, t) = v(s, c, t) + \pi f(s, c, t), \tag{4.73}$$

where π is the costate variable and again we omit the time argument. The maximum principle yields the following conditions:

(i) $c^*(t)$ maximizes H at each t; hence,

$$\frac{\partial H}{\partial c} = v_c + \pi^* f_c = 0. \tag{4.74}$$

(ii) $s^*(t)$ and $\pi^*(t)$ satisfy the pair of differential equations

$$\dot{\pi}^*(t) = -\frac{\partial H}{\partial s} = -v_s - \pi^* f_s, \tag{4.75}$$

$$\dot{s}^*(t) = \frac{\partial H}{\partial \pi} = f(s^*, c^*, t). \tag{4.76}$$

Optimal values are denoted by an asterisk, and all functions are evaluated along the optimal path.

4.5.1 *Costate variables as prices*

We have defined the meaning of all variables but the costate, and now we turn to this important task. Let us evaluate the derivatives[1] of the maximum value function V^* of (4.72) with respect to s_0 and s_T. Since $\dot{s} = f(s^*, c^*, t)$ at all time, it is true that

$$V^* = \int_0^T [v(s^*, c^*, t) + \pi f(s^*, c^*, t) - \pi \dot{s}^*]\, dt$$

for any *arbitrary* function $\pi(t)$. Note that the above expression is similar to a Lagrangean in a static problem; indeed, our argument here parallels the one used to interpret Lagrange multipliers. First we need to transform the above expression. We know from the method of integration by parts that $\int \pi \dot{s}^* dt = \pi s^* - \int \dot{\pi} s^* dt$. Therefore,

[1] Here we assume that V^* is a differentiable function of s_0. In some abnormal cases, the derivative $\partial V^*/\partial s_0$ may not exist at some value of s_0. For example, consider the problem

$$\text{maximize} \int_0^T \dot{s}(t)\, dt \quad (= s(T) - s(0))$$

subject to $\dot{s}(t) = c(t)s(t)$, $1 \le c(t) \le 2$, $s(0) = s_0$, $s(T)$ free, T fixed. Without the need to use the maximum principle, it is easy to see that if $s_0 > 0$, then $c(t) = 2$ and $V^* = s_0(e^{2T} - 1)$; if $s_0 < 0$, then $c(t) = 1$ and $V^* = (e^T - 1)s_0$; if $s_0 = 0$, then $V^* = 0$. Hence, $\partial V^*/\partial s_0$ does not exist at $s_0 = 0$.

$$V^* = \int_0^T [v(s^*, c^*, t) + \pi f(s^*, c^*, t) + \dot{\pi} s^*]\, dt - [\pi s^*]_0^T$$

$$= \int_0^T [H(s^*, c^*, \pi, t) + \dot{\pi} s^*]\, dt - \pi(T) s_T + \pi(0) s_0. \tag{4.77}$$

The derivative of V^* with respect to s_0 is

$$\frac{\partial V^*}{\partial s_0} = \int_0^T \left[(H_s)\frac{ds^*}{ds_0} + (H_c)\frac{dc^*}{ds_0} + (H_\pi)\frac{d\pi}{ds_0} + (H_t)\frac{dt}{ds_0} + \dot{\pi}\frac{ds^*}{ds_0} + s^*\frac{d\dot{\pi}}{ds_0} \right] dt + \pi(0), \tag{4.78}$$

where subscripts also denote partial derivatives. The value of s_0 does not affect the independent time trend; thus, $dt/ds_0 = 0$. Furthermore, since π is an arbitrarily chosen function of time, $d\pi/ds_0 = 0$ and $d\dot{\pi}/ds_0 = 0$ also. Thus, (4.78) reduces to

$$\frac{\partial V^*}{\partial s_0} = \int_0^T \left[(H_s + \dot{\pi})\frac{ds^*}{ds_0} + (H_c)\frac{dc^*}{ds_0} \right] dt + \pi(0).$$

This is true for any function $\pi(t)$; but suppose that we select the optimal path $\pi^*(t)$ obtained from the maximum principle equations (4.74)–(4.76). Then $H_s + \dot{\pi}^* = 0$ and $H_c = 0$, and we have

$$\frac{\partial V^*}{\partial s_0} = \pi^*(0), \tag{4.79}$$

where $\pi^*(0)$ is the optimal value of the costate variable at time zero. An identical argument and almost identical calculations yield

$$\frac{\partial V^*}{\partial s_T} = -\pi^*(T), \tag{4.80}$$

where $\pi^*(T)$ is the optimal value of the costate variable at time T.

The meaning of the results (4.79) and (4.80) is clear. A marginal increase in the initial stock would contribute $\pi^*(0)$ per unit to the total value obtainable over the horizon. Hence, $\pi^*(0)$ is the worth, or imputed value, of one unit of initial stock. The meaning of this costate variable parallels that of the multipliers and dual variables encountered in static optimization. Similarly, requiring that one reach terminal time with more capital would deduct $\pi^*(T)$ per unit from V^*, thus identifying $\pi^*(T)$ as the imputed value of stock at time T. The difference in signs between (4.79) and (4.80) stems from the fact that we are endowed with s_0 at time 0, whereas we must relinquish s_T at time T.

These results can be generalized to interpret $\pi^*(\theta)$ as the imputed value of stock at any instant θ along the optimal path (see Léonard, 1987, for more details). It is necessary to alter the format of the control problem, because we cannot prove that $\partial V^*/\partial s^*(\theta) = \pi^*(\theta)$ since $s^*(\theta)$ is not an exogenously specified parameter but is optimally chosen. We must formalize the notion that at some time θ in the interval $(0, T)$, a small amount of capital stock is suddenly added to the existing stock. The rate of change of capital stock is now defined by

$$\dot{s} = f(s, c, t) + a(t), \tag{4.81}$$

where

$$a(t) = \begin{cases} 0, & 0 \le t < \theta, \\ \alpha\epsilon^{-1}, & \theta \le t < \theta + \epsilon, \\ 0, & \theta + \epsilon \le t \le T. \end{cases}$$

The number ϵ is arbitrarily small and positive. The injection of capital takes place abruptly during the interval $[\theta, \theta + \epsilon)$; the smaller ϵ, the more abrupt the injection. At the limit $(\epsilon \to 0)$ it mimicks the addition of α units of capital at time θ, since

$$\int_\theta^{\theta+\epsilon} a(t)\,dt = \int_\theta^{\theta+\epsilon} \alpha\epsilon^{-1}\,dt = \alpha[\epsilon^{-1}t]_\theta^{\theta+\epsilon} = \alpha.$$

The Hamiltonian become

$$H(s, c, \pi, t, a) = v(s, c, t) + \pi(f(s, c, t) + a),$$

and (4.76) is replaced by (4.81). The same calculations as those used to interpret $\pi^*(0)$ and $\pi^*(T)$ yield an equation similar to (4.78) but including $a(t)$:

$$V^* = \int_0^T [H(s^*, c^*, \pi, t, a) + \dot{\pi}s^*]\,dt - \pi(T)s_T + \pi(0)s_0,$$

where $\pi(t)$ is arbitrarily chosen.

For ϵ sufficiently small, the derivative of V^* with respect to α will be an adequate approximation of the rate of increase in V^* per unit of increase in capital at time θ, which we shall as usual interpret as the worth of a unit of capital stock at that time. Recalling that $dt/d\alpha$, $d\pi/d\alpha$, and $d\dot{\pi}/d\alpha$ vanish as previously argued, we have

$$\frac{\partial V^*}{\partial \alpha} = \int_0^T \left[H_s \frac{ds^*}{d\alpha} + H_c \frac{dc^*}{d\alpha} + \dot{\pi} \frac{ds^*}{d\alpha} + H_a \frac{da}{d\alpha} \right] dt.$$

Selecting the optimal trajectory $\pi^*(t)$ and using (4.74), (4.75), and (4.81), this reduces to

$$\begin{aligned} \frac{\partial V^*}{\partial \alpha} &= \int_0^T \left[\pi^*(t) \frac{da}{d\alpha} \right] dt \\ &= \int_\theta^{\theta+\epsilon} \pi^*(t)\epsilon^{-1}\,dt. \end{aligned}$$

Letting $\Pi^*(t)=\int \pi^*(t)\,dt$, we can calculate

$$\lim_{\epsilon\to 0}\frac{\partial V}{\partial \alpha}=\lim_{\epsilon\to 0}\left[\frac{\Pi^*(\theta+\epsilon)-\Pi^*(\theta)}{\epsilon}\right]=\pi^*(\theta).$$

Therefore, we shall interpret the optimal value of the costate variable at any time θ as the imputed value of a unit of stock at that time. Clearly, the costate variable is the dynamic analog of the static multipliers, and we shall again use the words "imputed value," "worth," "shadow price," or simply "value" or "price" interchangeably to refer to them. The units in which $\pi(t)$ is expressed are "value units" (as for V^*) per unit of stock at that time. Now that the meaning of each variable is understood, we can turn to the maximum principle itself.

Recall that $v(s,c,t)$ is some instantaneous value function and that $f(s,c,t)$ describes the growth of s. The Hamiltonian function

$$H(s,c,\pi,t)=v(s,c,t)+\pi f(s,c,t)$$

is the sum of the instantaneous value plus the value of the instantaneous growth of s. One could call it a dynamic value function because it also takes into account the effect of current stock and control on the size and valuation of future stock.

Choosing $c(t)$ to maximize the Hamiltonian at each instant of time takes into account the immediate effect of $c(t)$ as well as its effect at all future dates. Therefore, maximizing H at each instant t yields a dynamic optimum at that time. The links between these optima are provided by the differential equations $\dot{s}=f(s,c,t)$ and $\dot{\pi}=-\partial H/\partial s$.

That is the general structure of the maximum principle. Let us now examine each of the equations (4.73)–(4.76) in detail.

4.5.2 *The maximum principle as an economic program*

First we will take the point of view of a central planner, alone responsible for solving the control problem. The Hamiltonian of (4.73) represents the nonmyopic value function at time t. When the control is chosen to maximize that function, (4.74) must hold; it states

$$\frac{\partial v}{\partial c(t)}+\pi(t)\frac{\partial f}{\partial c(t)}=0.$$

The first term is the marginal contribution of consumption to the instantaneous value function, that is, to the current flow of value. The second term is the product of the value of stock at time t and the marginal effect of consumption on the rate of growth of stock at that time; hence, it is the marginal contribution of current consumption to future value, through its effect on the growth of stock. Together these two terms account for all marginal benefits (and/or costs) of consumption, and (4.74) is seen as the counterpart of the familiar first-order condition of a static model of choice.

Of the two differential equations the second one ($\dot{s}=f(s,c,t)$) is already understood. The first one, (4.75), requires some thought. It states

$$\frac{\partial v}{\partial s(t)}+\pi(t)\frac{\partial f}{\partial s(t)}=-\dot{\pi}(t).$$

The first term is the marginal effect of stock on the instantaneous value function, that is, on the current flow of value. The second term is the product of the value of stock at time t by the marginal effect of the level of stock on its own rate of growth; hence, it reflects the marginal impact of current stock on future value. With regard to the term on the right-hand side, since $\dot{\pi}(t)$ is the rate of change in the value of stock, we can interpret $-\dot{\pi}(t)$ as the rate at which the value of stock depreciates. Then the equation states that along the optimal path the value of a unit of stock should be depreciated at a rate equal to its net (or combined) marginal contribution: current ($\partial v/\partial s(t)$) and future ($\pi(t)\partial f/\partial s(t)$). This rate of depreciation may, of course, be negative, but then the "contribution" of stock would itself be a deduction. This interpretation is appropriate for a central planner who seeks the optimal pricing of stock over time.

Let us now imagine the reasoning of an individual agent to whom the price of stock is exogenously given. The Hamiltonian of (4.73) represents the net benefit accruing to the agent at instant t since it takes into account the current flow of value v and the value of the growth of stock at instant t. Hence, (4.74) states that at each instant the agent chooses the control variable so as to maximize net benefit. There is no need for the agent to take into account the whole planning problem. Indeed, we may view agents at different times as distinct individuals who simply maximize the Hamiltonian at one instant. Their knowledge of π and their myopic optimizing behavior at each instant are sufficient to guide the economy along the optimal path.[2] This interpretation brings to light once again the decentralizing role of prices.

While the agent at each instant t need choose only the control, it is instructive to reconsider equation (4.75) with the myopic agent interpretation in mind. It can be written as $\partial v/\partial s(t)+\pi(t)\partial f/\partial s(t)+\dot{\pi}(t)=0$. The

[2] It may be worth noting an implicit restriction of this interpretation. Suppose that, for some time, $f(s,c,t)>0$ while $\pi(t)<0$; then the stock grows at a positive rate but its value is negative (it is a "bad," or a nuisance, for whatever reason). If free disposal is a possibility, the agent should, of course, discard this spurt of unwanted growth. This is not allowed and the free-disposal assumption (often made in general equilibrium theory) does not apply to this problem. This remark also applies when the central planner interpretation is adopted but is perhaps more natural in that context.

interpretation of the first two terms is similar to that of the central planner: they account for the benefits (and/or costs) of holding the marginal unit of stock through its effect on current and future value. The last term, however, is now interpreted as the speculative gain made by the agent while holding a unit of stock at instant t, recalling that the value of stock is seen to vary exogenously by the agent. Setting this exhaustive sum of marginal benefits (and/or costs) equal to zero is necessary to determine the optimal number of units of stock to be held by the agent at instant t. Therefore, it is as if the agent were following an optimal investment plan at each instant.

Finally, we can show that if there are several economic agents who sequentially share the responsibility for decision making, the optimal path will still be followed, provided that each agent respects appropriate boundary conditions on the levels of stock. Consider an agent responsible for the subhorizon $[t_0, t_1]$, where $0 \leq t_0 \leq t_1 < T$, and with boundary conditions $s(t_0) = s^*(t_0)$; $s(t_1) = s^*(t_1)$, where $(s^*(t), c^*(t), \pi^*(t))$ denotes the optimal solution to the global problem (i.e., the one with horizon $[0, T]$). Thus, the boundary values of stock in this problem are on the optimal path of the global problem. We want to show that the optimal solutions of the two problems coincide over the subhorizon $[t_0, t_1]$.[3] Suppose that the optimal controls differ from one another in the two problems; since both of them take the stock from $s^*(t_0)$ to $s^*(t_1)$, one of them must yield a higher value for the maximand, a contradiction (we assume uniqueness of the optimal solution). But if the controls are identical, the stock variables that both satisfy the differential equation (4.76) and have identical endpoints must follow the same path. If we take the interpretation of economic agents for whom the price path is exogenously specified, nothing further need be said. If we choose to consider the agent as central planner for the interval $[t_0, t_1]$, it is easy to see that the pair of differential equations (4.75) and (4.76) along with two boundary conditions, $s(t_0) = s^*(t_0)$ and $s(t_1) = s^*(t_1)$, will yield the same solution in this subproblem as in the global problem once the controls are identically chosen. Thus, we can transform the aggregate global problem into a sequence of subproblems with appropriate boundary conditions.

[3] Here we assume that the problem of inconsistent planning, referred to at the end of Section 4.4, does not arise, either because $v(s, c, t)$ takes the form $e^{-\delta t}u(s, c)$ or because the independent time argument in v does not represent pure time preference and hence at time t_0 the individual maximizes the integral

$$\int_{t_0}^{T} v(s(t), c(t), t)\, dt,$$

and not

$$\int_{t_0}^{T} v(s(t), c(t), t - t_0)\, dt.$$

4.5.3 *Optimal growth*

We now illustrate our economic interpretation of the maximum principle with a more specific problem. It is identical to the "general" growth model presented at the end of Section 4.4 in all respects but one: the rate of change in stock is now

$$\dot{s} = F(s(t)) - ms(t) - c(t), \quad m > 0.$$

To begin with, we restate the problem and describe it in economic terms. Consider an economy with a single capital resource that can be used to produce a single good according to some technology and can also be consumed. If we denote the stock of capital at time t by $s(t)$, gross production is $F(s(t))$, where F is the production function. This stock of capital depreciates naturally over time at the exponential rate $m > 0$ (see Section 3.4). Thus, if neither gross production nor consumption were to take place, the path of capital would be described by $\dot{s}(t) = -ms(t)$, whose solution is $s(t) = s_0 e^{-mt}$, where s_0 is the specified value of $s(0)$. However, as we have seen, production does take place, and the stock of capital is also depleted by a flow of consumption $c(t)$ at time t. Thus, the rate of change of capital stock is described by

$$\dot{s}(t) = F(s(t)) - ms(t) - c(t).$$

Suppose that a central planner is in charge of charting the path of the economy over some time horizon $[0, T]$. He knows that at time 0 there are s_0 units of capital available and that he must reach time T with s_T units of the stock. Meanwhile, he has chosen as his objective the maximization of the total discounted utility over the planning horizon. At any instant t, utility depends on the flow of consumption $c(t)$ only; hence, it can be formulated as $u(c(t))$. This implicitly assumes that utility at any instant is not directly dependent of the stock of capital at that time (no Scrooge element). The planner attaches less utility to future consumption than to current consumption, and this is reflected by a subjective exponential rate of discount $\delta > 0$.

This means that the discounted utility of consuming c at time t is given by $e^{-\delta t}u(c)$. In a continuous-time model the sum of utilities takes the form of an integral, and the planner's objective is to maximize

$$V = \int_0^T e^{-\delta t} u(c(t))\, dt,$$

subject to

$$\dot{s}(t) = F(s(t)) - ms(t) - c(t),$$
$$s(0) = s_0, \quad s(T) = s_T.$$

We assume that $u'>0$, $u''<0$, $u'(0)=\infty$, $F(0)=0$, $F'>0$, $F''<0$, and $F'(\infty)<\delta+m<F'(0)$. These assumptions are identical with those made in Example 4.4.1, save for the last one, which now incorporates the rate of depreciation m as well as the rate of discount. Its purpose is again to guarantee the existence of an equilibrium. The production and utility functions are strictly concave and increasing. This implies that more consumption (resp. capital) yields more utility (resp. output) but does so at a decreasing marginal rate. A zero capital stock yields no output. Marginal utility becomes infinite as consumption approaches zero.

The Hamiltonian of this problem is

$$H(s,c,\pi,t)=e^{-\delta t}u(c(t))+\pi(t)[F(s(t))-ms(t)-c(t)],$$

and the necessary conditions are (skipping arguments)

$$\frac{\partial H}{\partial c}=e^{-\delta t}u'(c)-\pi=0, \tag{4.82}$$

$$\dot{\pi}=-\frac{\partial H}{\partial s}=-\pi[F'(s)-m], \tag{4.83}$$

$$\dot{s}=\frac{\partial H}{\partial \pi}=F(s)-ms-c. \tag{4.84}$$

$\pi(t)$ is the present value of capital at instant t. (Note that value is measured in units of utility.) To reach a dynamic maximum the Hamiltonian is maximized at each instant; the influence of the consumption flow on the growth of capital is taken into account. This maximization is characterized by (4.82), which equates at each instant the discounted marginal utility of consumption to the present value of capital. This reflects the fact that consumption is simply subtracted from the growth in capital stock. Equation (4.84) reflects the optimality of the path of capital. The net marginal physical product is the gross marginal physical product less the rate of depreciation. Hence, equation (4.83) prescribes that the rate of change in the value of stock plus its net marginal value product sum to zero. This sum represents the net marginal gain of a would-be investor holding s units of capital and facing the price path π. If it were to be negative or positive, the investor would be holding too much or too little capital, respectively. We can derive further insights into the nature of the optimal solution by examining the consumption path. To this end we totally differentiate (4.82) with respect to time and use it with (4.83) to obtain $\dot{c}$:

$$\begin{aligned}-\delta e^{-\delta t}u'(c)+e^{-\delta t}u''(c)\dot{c}&=\dot{\pi}\\ &=\pi[m-F'(s)]\\ &=e^{-\delta t}u'(c)[m-F'(s)],\end{aligned}$$

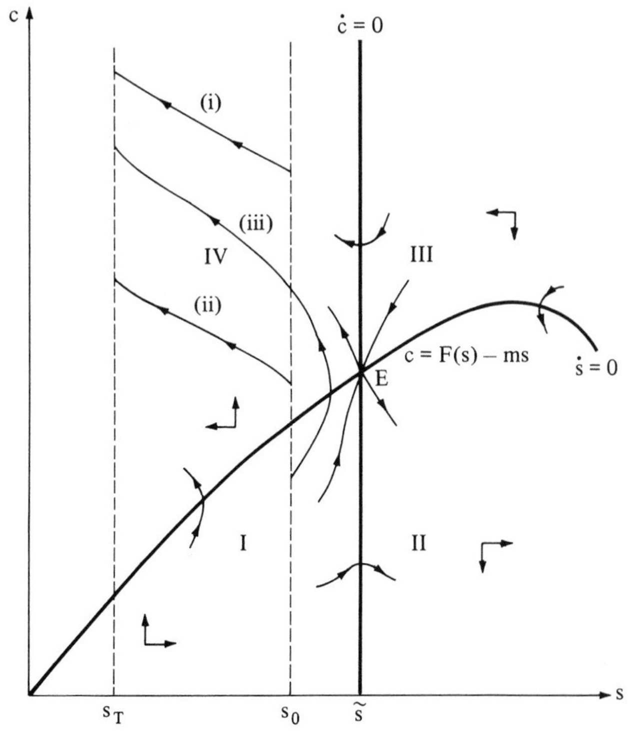

Figure 4.6

$$\dot{c} = -\frac{u'(c)}{u''(c)}[(F'(s) - m) - \delta]. \tag{4.85}$$

Equation (4.85) indicates that the optimal flow of consumption will increase (resp. decrease) if and only if the net marginal product of capital is larger (resp. smaller) than the subjective rate of discount. It pays to have a higher future consumption relative to present consumption if the cost of postponing it is smaller than the gains it provides through its marginal effect on net capital return, where the cost of postponing consumption evaluated at the margin is indicated by the discount rate δ.

To conclude this section we now briefly present a phase diagram analysis of this problem in the (s, c) plane, using equations (4.84) and (4.85). Figure 4.6 is the phase diagram; it is similar to Figure 4.4 except that the $\dot{s} = 0$ locus is now $c = F(s) - ms$ and thus may have a maximum. (It does if $F'(\infty) < m$; this is the case depicted in Figure 4.6.) The construction of the diagram is left to the reader as an exercise. Three optimal paths are illustrated for various lengths of the planning horizon T. If T is very small, a high path such as (i) is chosen; for larger T, (ii) may be optimal. Along both these paths capital decreases monotonically. If T is even larger, we may, as in case (iii), go through an initial phase when capital stock is increasing. This diagram will be referred to in Section 6.4.

4.6 Necessity and sufficiency of the maximum principle

In this section we provide a heuristic derivation of the maximum principle as a set of necessary conditions for optimality, and we also define a class of problems for which the maximum principle is sufficient for optimality.

Consider the control problem stated at the beginning of Section 4.5 with the Hamiltonian given by (4.73). Denote the optimal solution by $(s^*(t), c^*(t), \pi^*(t))$, and let $H^* \equiv v(s^*(t), c^*(t), t) + \pi^*(t) f(s^*(t), c^*(t), t)$. We assume for simplicity that all variables are continuously differentiable functions of time. Given the equation $\dot{s}(t) = f(s(t), c(t), t)$, consider the class of functions $c(t)$ that drives $s(0) = s_0$ to $s(T) = s_T$. This class of functions is denoted by A. Clearly, $c^*(t)$ is a member of A. Given any arbitrary function $\bar{\alpha}(t)$, let us construct a family of functions[4] $\alpha(t, \epsilon)$, where ϵ is a parameter such that $\alpha(t, 0) = \bar{\alpha}(t)$ and $c^*(t) + \epsilon\alpha(t, \epsilon)$ is a member of A. Suppose now that we deviate from the optimal trajectory by using $c(t) = c^*(t) + \epsilon\alpha(t, \epsilon)$. The resulting state variable path (through (4.76)) is denoted by $s(t, \epsilon)$ since it depends on the value of ϵ; it obeys the original boundary conditions. Since $c^*(t)$ is the optimal control, V must be maximized when $\epsilon = 0$. We first manipulate the expression for V and then evaluate its derivative with respect to ϵ when $\epsilon = 0$:

$$\begin{aligned} V &= \int_0^T v(s(t, \epsilon), c^* + \epsilon\alpha(t, \epsilon), t)\, dt \\ &= \int_0^T [v(s(t, \epsilon), c^* + \epsilon\alpha(t, \epsilon), t) + \pi(t) f(s(t, \epsilon), c^* + \epsilon\alpha(t, \epsilon), t) \\ &\qquad - \pi(t)\dot{s}(t, \epsilon)]\, dt, \end{aligned}$$

for any arbitrary function $\pi(t)$. Proceeding, let

$$H(t, \epsilon) \equiv v(s(t, \epsilon), c^* + \epsilon\alpha(t, \epsilon), t) + \pi(t) f(s(t, \epsilon), c^* + \epsilon\alpha(t, \epsilon), t)$$

and

$$\begin{aligned} V &= \int_0^T [H(t, \epsilon) - \pi(t)\dot{s}(t, \epsilon)]\, dt \\ &= \int_0^T [H(t, \epsilon) + \dot{\pi}(t) s(t, \epsilon)]\, dt - \pi(T) s_T + \pi(0) s_0. \end{aligned}$$

[4] The existence of such families of functions and their differentiability with respect to ϵ are simply assumed here. This assumption would, of course, be innocuous if s_T were free, and not exogenously specified.

Then

$$\left.\frac{dV}{d\epsilon}\right|_{\epsilon=0}=\int_0^T\left[\left.\frac{\partial H}{\partial c}\right|_{\epsilon=0}\alpha(t,0)+\left(\left.\frac{\partial H}{\partial s}\right|_{\epsilon=0}+\dot{\pi}(t)\right)\frac{ds(t,\epsilon)}{d\epsilon}\right]dt.$$

It is necessary for a maximum that the derivative $dV/d\epsilon$ vanish when evaluated at $\epsilon=0$. This must be true for any arbitrarily chosen pair of functions $\bar{\alpha}(t)$ and $\pi(t)$. Therefore, it must be true, in particular, for a $\pi(t)$ function satisfying the differential equation $\dot{\pi}(t)+(\partial H/\partial s)|_{\epsilon=0}=0$, and for this particular choice of the function $\pi(t)$ we have, for any arbitrarily chosen $\bar{\alpha}(t)$,

$$\left.\frac{dV}{d\epsilon}\right|_{\epsilon=0}=\int_0^T\left.\frac{\partial H}{\partial c}\right|_{\epsilon=0}\bar{\alpha}(t)\,dt=0.$$

This implies that $(\partial H/\partial c)|_{\epsilon=0}$ vanishes at each instant, for otherwise it would be possible to find a function $\bar{\alpha}(t)$ such that the above integral does not vanish. Therefore, collecting the results, the necessary conditions for optimality are

$$\frac{\partial H}{\partial c}=0, \qquad \dot{\pi}+\frac{\partial H}{\partial s}=0,$$

and, of course,

$$\dot{s}=f(s,c,t),$$

with $s(T)=s_T$ and $s(0)=s_0$. We recognize here the maximum principle stated in more detail in equations (4.74)–(4.76).

To obtain sufficient conditions consider now a class of optimal control problems for which the Hamiltonian $H(s,c,\pi,t)$ is concave in (s,c). We will shortly state conditions guaranteeing this. For simplicity of notation, we will suppress all arguments of functions, and let an asterisk denote optimality when superscripted to variables or functions; the absence of an asterisk denotes any other feasible solution (in particular, s satisfies the boundary conditions). We manipulate the expression for V in a manner similar to that used to derive the maximum principle:

$$\begin{aligned}V^*-V&=\int_0^T(v^*-v)\,dt\\&=\int_0^T[(H^*-\pi^*\dot{s}^*)-(H-\pi^*\dot{s})]\,dt\\&=\int_0^T(H^*+\dot{\pi}^*s^*)-(H+\dot{\pi}^*s)\,dt \quad \text{(by integration by parts)}\\&=\int_0^T[H^*-H+\dot{\pi}^*(s^*-s)]\,dt\end{aligned}$$

$$\geq \int_0^T [(s^*-s)H_s^* + (c^*-c)H_c^* + \dot{\pi}^*(s^*-s)]\,dt$$

(by concavity of the Hamiltonian)

$$= \int_0^T [(s^*-s)(H_s^* + \dot{\pi}^*) + (c^*-c)H_c^*]\,dt$$

$$= 0$$

if the starred solution satisfies the maximum principle. It follows that if H is concave in (s,c), then any solution satisfying the maximum principle yields a value V^* at least as high as the value V yielded by any feasible solution. Note that if H is strictly concave in (s,c), then $V^* > V$ and the optimal solution must be unique. We now state the result formally.

Theorem 4.6.1: sufficiency. Consider the problem defined in (4.1)–(4.3). Suppose that the maximum principle of (4.5)–(4.7) yields the solution $(s^*(t), c^*(t), \pi^*(t))$. If the Hamiltonian of (4.4), with $\pi(t) = \pi^*(t)$, is concave (resp. strictly concave) in (s,c) jointly, then $(s^*(t), c^*(t), \pi^*(t))$ is an optimal solution (resp. the only optimal solution) to the above problem.

This sufficiency result depends on the concavity of H, which may be difficult to ascertain without obtaining an explicit solution. This is why we now provide a set of conditions that ensure the concavity of H.

Theorem 4.6.2. If v is concave in (s,c), and either $\pi \geq 0$ and f is concave in (s,c) or $\pi \leq 0$ and f is convex in (s,c), then H is concave in (s,c) and the necessary conditions provided by the maximum principle are also sufficient for an optimal solution. Furthermore, it can be shown (see Léonard, 1981) that if $v_s > 0$ (resp. $v_s < 0$) along the optimal path then $\pi > 0$ (resp. $\pi < 0$).

In some cases where Theorems 4.61 and 4.62 fail, there is yet another result that might apply. First we must define a new concept and restate the maximum principle.

Definition 4.6.1. For the problem of (4.1)–(4.3), let

$$H^0(s(t), \pi(t), t) \equiv \max_{c(t)} H(s(t), c(t), \pi(t), t), \tag{4.86}$$

where $H(s(t), c(t), \pi(t), t)$ is as in (4.4). We call H^0 the *maximized Hamiltonian.*

Note that (4.86) implicitly defines c as a function of s, π, and t – through equation (4.5).

We can now restate the maximum principle.

Theorem 4.6.3: necessity. Consider the problem of (4.1)–(4.3) and let (c, s, π) be an optimal solution; then c satisfies (4.5) and π and s are solutions to

$$\dot{\pi}(t) = -\partial H^0/\partial s(t) \quad \text{and} \quad \dot{s}(t) = \partial H^0/\partial \pi(t), \tag{4.87}$$

where H^0 is defined in (4.86).

Theorem 4.6.4: sufficiency. Suppose $(s^*(t), c^*(t), \pi^*(t))$ is a solution to (4.3), (4.5), and (4.87). If the maximized Hamiltonian of (4.86) is concave (resp. strictly concave) in $s(t)$ when $\pi(t) = \pi^*(t)$, then $(s^*(t), c^*(t), \pi^*(t))$ constitutes an optimal solution (resp. the unique optimal solution) to the problem of (4.1)–(4.3).

The above statement of the maximum principle using H^0 in Theorem 4.6.3 is equivalent to (4.5)–(4.7), and the two may be used interchangeably. Note that attention must still be paid to the properties of H in terms of c (given s and π) to ensure that (4.86) indeed defines a maximum. The proof proceeds along the same lines as that of Theorem 4.6.1, leaving out terms specifically referring to c and using H^0 instead of H. We now illustrate Theorems 4.6.3 and 4.6.4 with some examples.

Example 4.6.1. We want to maximize $\int_0^T 4c(s)^{1/2}e^{-\delta t}\,dt$ subject to $\dot{s} = -(c)^4$, and s_0 and s_T are exogenously specified and positive. The Hamiltonian $H = 4c(s)^{1/2}e^{-\delta t} - \pi(c)^4$ is clearly not concave in (c, s) jointly. However, it is concave in c alone (assuming $\pi > 0$), and the first-order condition, $\partial H/\partial c = 0$, yields a maximum at

$$c = (s)^{1/6}(\pi)^{1/3}e^{-\delta t/3}. \tag{4.88}$$

[From this we can show $\pi > 0$: s must be positive or the maximand is undefined and a negative value of c is never chosen, since it makes both $\dot{s}$ and the maximand negative; hence, π must be positive.] We find the maximized Hamiltonian by substituting the optimal c value into H:

$$\begin{aligned} H^0 &= 4(s)^{1/6}(\pi)^{-1/3}e^{-\delta t/3}(s)^{1/2}e^{-\delta t} - \pi(s)^{4/6}(\pi)^{-4/3}e^{-4\delta t/3} \\ &= 3e^{-4\delta t/3}(\pi)^{-1/3}(s)^{2/3}, \end{aligned}$$

which is evidently strictly concave in s.

The alternative form of the maximum principle is

$$\dot{\pi} = -\partial H^0/\partial s = -2e^{-4\delta t/3}(\pi)^{-1/3}(s)^{-1/3}, \tag{4.89a}$$

$$\dot{s} = \partial H^0/\partial \pi = -e^{-4\delta t/3}(\pi)^{-4/3}(s)^{2/3}, \tag{4.89b}$$

with c given by equation (4.88). The reader is invited to verify that substituting $\partial H/\partial c = 0$ (i.e., (4.88)) into $\dot{s} = \partial H/\partial \pi$ and $\dot{\pi} = -\partial H/\partial s$ yields equations (4.89).

Example 4.6.2. Here we illustrate the remark made after Theorem 4.6.4. Consider a stock whose growth is enhanced by its own size and the effort made to tend it. Utility depends on the size of the stock and is negatively related to effort. We maximize $\int_0^T (s)^{1/2}(c)^{-1}e^{-\delta t}\,dt$ subject to $\dot{s} = (s)^{1/2}c$, and s_0 and s_T are exogenously specified. The Hamiltonian is $H = (s)^{1/2}(c)^{-1}e^{-\delta t} + \pi c(s)^{1/2}$. The first-order condition $\partial H/\partial c = 0$ yields $(c)^{-2} = \pi e^{\delta t}$ and $H^0 = 2e^{-\delta t/2}(\pi)^{1/2}s^{1/2}$, which is clearly concave in s. However, the first-order condition did not select a maximum over c but indeed a global minimum since the Hamiltonian H is clearly convex in c. In fact, this Hamiltonian has no maximum when no (positive) lower and upper bounds have been imposed on c from the outset.

Remark. Before leaving this section, recall that we have assumed throughout that the variables are continuously differentiable functions of time. It is an attractive feature of optimal control theory that it can deal with a wider class of solutions. All that is needed is that the control variable be piecewise-continuous. This means that it is acceptable to have a control function $c(t)$ that exhibits some jump discontinuity at a finite number of points. It follows that $\dot{s}$ and $\dot{\pi}$ themselves are piecewise-continuous while s and π are continuous and piecewise-differentiable. A more formal statement of the maximum principle is postponed until Chapter 6, and the special features of problems exhibiting such discontinuities are the subject of Chapter 8.

Exercises

1. Applying the maximum principle, find $c(t)$ to maximize $V = \int_0^T \ln(c(t))\,dt$ subject to $\dot{s}(t) = -c(t)$, $s(0) = s_0$, $s(T) = s_T$, where T, s_0, and s_T are specified positive constants, with $s_T < s_0$. Show that the optimal control is constant over the horizon and that the optimal path of the state variable is a linear function of time.
2. Repeat exercise 1 with $\int_0^T (c(t))^\alpha\,dt$ as the maximand $(0 < \alpha < 1)$. Is the optimal control different in this case? Can you show that c would be constant with $\int_0^T U(c(t))\,dt$ as the maximand, where U is any strictly increasing and strictly concave function?
3. Applying the maximum principle, find $c(t)$ to maximize $V = \int_0^T e^{-t}\ln(c(t))\,dt$ subject to $\dot{s}(t) = -c(t)$, $s(0) = s_0$, $s(T) = s_T$, where T, s_0, and s_T are specified positive constants, with $s_T < s_0$. (*Hint:* Obtain the general solution to the costate differential equation, substitute for the control variable in the state variable differential equation, and use the boundary conditions to determine the constants of integration.) Is the control constant over time? Contrast this with the result of exercise 1.
4. Replace the maximand in exercise 3 by $\int_0^T e^{-t}(c(t))^\alpha\,dt$ and repeat the exercise $(0 < \alpha < 1)$. Does the optimal control depend on the exact form of the maximand? Contrast this with the results of exercise 2.

5. Reconsider the control problem of exercise 1 now with the modified state differential equation $\dot{s}(t) = -s(t) - c(t)$. Solve this problem. What restriction must be placed on T, s_0, and s_T to ensure that the optimal control is positive? (Without this a solution cannot exist.)
6. Consider a modified version of the problem of exercise 5. Find $c(t)$ to maximize $V = \int_0^T [(c(t))^{1+\gamma}/(1+\gamma)]\,dt$ subject to $\dot{s}(t) = -s(t) - c(t)$, $s(T) = s_T$, $s(0) = s_0$, where $-\gamma$, T, s_0, and s_T are specified positive constants. Apply the maximum principle and solve the problem.
7. In this exercise you will formulate a control problem and apply the maximum principle. It concerns the optimal management of a mineral spring. You must pay particular attention to the units in which the variables are expressed. Your aim is to maximize the total present value of profits over the planning horizon. The spring flows at a constant rate. Outflow may be sold immediately or stocked in a natural reservoir at no cost; however, the reservoir leaks and a constant proportion of current stocks is lost per unit of time. We count volumes of water in megaliters (L) and time in days (d). The following notation is used (units are indicated in parentheses):

$c(t)$	quantity sold at time t (L/d),
$P(c(t))$	profit, at time t, from the sale of $c(t)$ units at time t (\$/d),
R	constant rate of outflow of the spring (L/d),
$s(t)$	level of stocks at time t (L),
α	constant proportion of stocks that leaks at time t (d^{-1}),
$e^{-\delta t}$	discount factor, i.e., the present value of \$1 at time t (a pure number).

   R, δ, and α are fixed positive constants. The time horizon $[0,T]$ is also fixed, as are the boundary conditions $s(0) = s_0$ and $s(T) = s_T$.

   Express the rate of change in the levels of stocks, $\dot{s}(t)$ (in L/d) in terms of the other variables. Express the present value of profits over $[0,T]$, an integral (in \$). Set up the control problem. Which is the state variable and which is the control? Let the costate variable be $\pi(t)$. Write down the Hamiltonian and apply the maximum principle (you will have three equations).
8. Consider the mineral spring problem of exercise 7 with the following data: $P(c) = 10c - c^2$, $R = 15$, $\alpha = 1$, $\delta = 0.5$. Show that the application of the maximum principle yields $10 - 2c = \pi e^{t/2}$, $\dot{\pi} = \pi$, and $\dot{s} = 15 - s - c$. Solve the differential equation in π - you will need to use an arbitrary constant of integration, say A. Obtain the optimal path for c - it depends on A. Use this to solve the differential equation in s - you will need another constant of integration, say B. Use the initial condition $s(0) = 50$ to eliminate B.

   Now let $T = \ln 2$ and $s(T) = 30$; determine the constant A and the exact paths of c, s, and π. Is c constant over time?

   This time let $T = \ln 5$ and $s(T) = 0$; determine A and the exact paths of c, s, and π; calculate $c(0)$ and $c(T)$. Can you plot a rough path in the (c,s) plane?
9. Repeat exercise 8 with the following data: $P(c) = 50c - c^2$, $R = 15$, $\alpha = 1$, $\delta = 0.5$, $s(0) = 30$. Show that the maximum principle yields $50 - 2c = \pi e^{t/2}$, $\dot{\pi} = \pi$, and $\dot{s} = 15 - s - c$. Determine the exact paths of c, s, and π for the two terminal conditions ($T = \ln 2$, $s(T) = 10$) and ($T = \ln 5$, $s(T) = 0$); plot the latter solution in the (c,s) plane.

Suppose now that you want to determine the value of T, say $\bar{T}$, that would lead to $c(\bar{T})=15$ with boundary condition $s(\bar{T})=0$. Determine the approximate value of $\bar{T}$.

Now choose a value $\bar{\bar{T}}$ larger than $\bar{T}$; use the boundary condition $s(\bar{\bar{T}})=0$ to determine the constant of integration. Show that there is a problem with this outcome. (*Hint:* Calculate $s(t)$ for t very close to $\bar{\bar{T}}$, but smaller.)

10. Consider the problem of choosing $c(t)$ to maximize

$$V=\int_0^T e^{-\delta t}\left[\frac{[c(t)]^{(1+\alpha)}}{1+\alpha}\right]dt$$

subject to $\dot{s}(t)=rs(t)-c(t)$ with $s(0)=s_0$ and $s(T)=s_T$, where T, δ, $-\alpha$, r, s_0, and s_T are specified positive constants. Apply the maximum principle. Obtain a differential equation for c and solve it. Use the result to solve the differential equation in s. Use the boundary conditions on s to determine the constants of integration. Show that if $s_T=s_0$, $c(t)$ is always positive.

11. Use the problem of exercise 1 to derive autonomous differential equations for c and s in which $\dot{c}$ and $\dot{s}$ depend only on the values of c and s. Draw a phase diagram in the (c,s) plane for arbitrary values of T, s_0, and s_T ($s_0>s_T>0$). How does the optimal trajectory change if a larger T value is selected? Show that the results are qualitatively identical if an arbitrary U function is selected as in exercise 2.

12. Use the problem of exercise 3 to derive autonomous differential equations for c and s as in exercise 11 and draw a phase diagram in the positive quadrant of the (c,s) plane. Show that the optimal trajectory is a straight line of slope 1. How does the trajectory change when a larger T value is selected? Repeat the exercise with the problem of exercise 4. Show that the optimal trajectory has slope $(1-\alpha)^{-1}$. Does the exact value of α change the general shape of the optimal trajectory?

13. Consider the problem of exercise 5. Apply the maximum principle and derive autonomous differential equations for the state and the control variable (i.e., $\dot{c}$ and $\dot{s}$ in terms of c and s only). Show that the loci of $\dot{c}=0$ and $\dot{s}=0$ are out of bounds if c and s are positive. Draw the phase diagram. Derive autonomous differential equations for the state and costate (π) variables. Show that the loci of $\dot{s}=0$ and $\dot{\pi}=0$ are out of bounds if c and s are positive. Modify now the state equation to $\dot{s}(t)=s(t)-c(t)$. Draw the phase diagrams in the (c,s) plane and the (π,s) plane. Show that c always increases over time but s is not always monotone. Let $s_0=s_T>0$, and draw an optimal trajectory for various values of T. Can the value of T have a qualitative effect on the trajectory of s?

14. Repeat exercise 13 for the problem of exercise 6.

15. For the problem of exercise 10 apply the maximum principle to the current-value Hamiltonian (denote the costate variable by ψ). Obtain a system of autonomous differential equations for c and s and draw a phase diagram in the (c,s) plane. Do the same for ψ and s. Attempt the same exercise with the maximand $\int_0^T e^{-\delta t}u(c(t))\,dt$, where $u'>0$, $u''<0$. Do you get qualitatively distinct phase diagrams when using $u(c)$ instead of $(c)^{1+\alpha}/(1+\alpha)$?

16. Draw the phase diagrams in the (c, s) plane for exercises 8 and 9. In exercise 7, assume that $P(c)$ reaches a maximum at $c^* > 0$. First assume $c^* > R$ and draw the phase diagram in the (c, s) space; then assume $c^* < R$ and repeat.
17. Consider the problem of choosing $c(t)$ to maximize $V = \int_0^T \ln(c(t))\,dt$ subject to $\dot{s}(t) = 10s(t) - 0.1(s(t))^2 - c(t)$ with $s(0) = s_0$, $s(T) = s_T$, where T, s_0, and s_T are specified positive constants. Use the maximum principle to derive a system of autonomous equations for c and s; draw the phase diagram in the positive quadrant of the (c, s) plane. Show that there is a steady state at $(c = 250, s = 50)$. Linearize the differential equations at this point and formally prove that it is a saddle point. Make sure that the trajectories in your diagram confirm this. Carry out a similar exercise for the state–costate pair. After the diagrams have been completed, attempt to pinpoint the optimal trajectory for selected allocations of T, s_0, and s_T as in Section 4.4.
18. Repeat exercise 17 when discounting and depreciation are included, namely, choose $c(t)$ to maximize $V = \int_0^T e^{-t} \ln(c(t))\,dt$ subject to $\dot{s}(t) = 10s(t) - 0.1(s(t))^2 - 2s(t) - c(t)$. Do the phase diagrams differ qualitatively from those of exercise 17?
19. Suppose that the problems of exercises 1 and 3 are concerned with the optimal consumption of a nonrenewable resource where the utility of consumption is the logarithm of consumption; in exercise 3 utility is discounted. What does the costate variable represent? What does the maximum principle indicate? Contrast the two problems. In exercise 5 a new feature was added: this is now a naturally decaying resource. What does this imply for the costate variable and the consumption path?
20. Repeat exercise 19 for the problems of exercises 2, 4, and 6 ($1 + \gamma$ is equivalent to α).
21. Consider the mineral spring problem of exercise 7. Give an economic interpretation of the costate variable. Discuss the conditions obtained from the maximum principle and comment on the trajectories identified in exercise 16. What does c^* represent?
22. You are exploiting a renewable resource; its stock at time t is $s(t)$; its natural rate of growth is $g(s(t))$; the flow of harvesting is denoted by $c(t)$ (this is subtracted from $\dot{s}(t)$). Furthermore, you can boost the growth of the resource by spending $\$x(t)$; this adds $b(x(t))$ to $\dot{s}(t)$. The revenue from $c(t)$ units harvested at time t is $R(c(t))$ and the cost $x(t)$ is subtracted from it. The rate of interest is δ. Formulate the problem of maximizing the present value of profits from time 0 to time T, subject to the growth equation. The values of s at times 0 and T are specified. Assume that R, g, and b are concave. Show that the maximum principle yields necessary and sufficient conditions for a maximum. Obtain these; interpret the costate variable and all the optimality conditions.

CHAPTER 5

The calculus of variations and dynamic programming

Any introductory treatment of optimal control theory would be incomplete without explicit mention of its predecessor, the calculus of variations, and the parallel development of dynamic programming. The calculus of variations owes much to the eighteenth-century mathematician Euler, but many developments and refinements were made in the following centuries. Optimal control theory, developed by Pontryagin and his co-workers in the late 1950s, may be regarded as a generalization of the calculus of variations: not only is its field of applicability broadened, but the general problem is approached from a fresh and more insightful viewpoint.

Dynamic programming was developed by Bellman, also in the late 1950s. It was designed primarily to deal with optimization problems in discrete time, but Bellman's famous "principle of optimality" also applies to continuous-time problems, where the Hamilton–Jacobi–Bellman equation plays a crucial role.

In this chapter, we examine the connection of optimal control theory with the calculus of variations and dynamic programming. We illustrate how all three approaches lead to the same solution and comment on their relative usefulness in analytical economics.

5.1 The calculus of variations

In Chapter 4, we studied the problem of finding $(s(t), c(t))$ that maximizes

$$\int_0^T v(s(t), c(t), t)\, dt \tag{5.1}$$

subject to

$$\dot{s}(t) = f(s(t), c(t), t), \tag{5.2}$$

$$s(0) = s_0, \tag{5.3}$$

$$s(T) = s_T. \tag{5.4}$$

Let us assume that (5.2) can be inverted to yield

$$c(t) = \phi(s(t), \dot{s}(t), t). \tag{5.5}$$

Substitute (5.5) into $v(s, c, t)$ to obtain

$$v(s(t), c(t), t) = v(s(t), \phi(s(t), \dot{s}(t), t), t). \tag{5.6}$$

The right-hand side of (5.6) is a function of $s(t)$, $\dot{s}(t)$, and t. We give this function a name, $F(s(t), \dot{s}(t), t)$. Problem (5.1) becomes: find $s(t)$ and hence $\dot{s}(t)$ that maximize

$$\int_0^T F(s(t), \dot{s}(t), t)\, dt \tag{5.7}$$

subject to

$$s(0) = s_0, \tag{5.8}$$

$$s(T) = s_T. \tag{5.9}$$

This problem is written in the format of the calculus of variations, in which the choice variable is the rate of change of the state variable. We assume that F possesses continuous second-order partial derivatives. In optimal control problems we require only that the function $s(t)$ be piecewise-differentiable (i.e., it is allowed to have kinks at isolated values of t); this means that $\dot{s}(t)$ is piecewise-continuous (it may have jump discontinuities at isolated values of t). In the calculus of variations, at least in its early development, attention is restricted to the class of functions $s(t)$ that have continuous second-order derivatives for all t in $[0, T]$. This class of functions is denoted by $C^2[0, T]$. In the remainder of this section we present a necessary condition for a maximum and show how it can be used to derive the solution. We first state an important result.

Theorem 5.1.1: Euler's equation. Assume that there exists a function $s^*(t)$ in $C^2[0, T]$ that maximizes (5.7), that is,

$$\int_0^T F(s^*(t), \dot{s}^*(t), t)\, dt \geq \int_0^T F(s(t), \dot{s}(t), t)\, dt \tag{5.10}$$

for all $s(t)$ in $C^2[0, T]$ subject to (5.8) and (5.9). Then $s^*(t)$ must satisfy the equation

$$F_s - \frac{d}{dt}(F_{\dot{s}}) = 0. \tag{5.11}$$

Equation (5.11) can be written more fully as

$$F_s = F_{\dot{s}s}\frac{ds}{dt} + F_{\dot{s}\dot{s}}\frac{d\dot{s}}{dt} + F_{\dot{s}t} \quad \text{(Euler's equation)}. \tag{5.12}$$

Remark. Note that $d\dot{s}/dt$ is the second derivative of $s(t)$, making (5.12) a second-order differential equation. Equation (5.11) assumes the existence of this derivative.

We now proceed to prove the necessity of (5.11) by means of a perturbation argument. Any function $s(t)$ satisfying (5.8) and (5.9) must satisfy the condition

$$s(t)=s^*(t)+\epsilon g(t) \tag{5.13}$$

for some $\epsilon>0$ and some $g(t)$ in $C^2[0,T]$ with the properties $g(0)=0=g(T)$ (otherwise conditions (5.8) and (5.9) would not be satisfied). Let us define the difference between the right-hand side and the left-hand side of (5.10) as $D(\epsilon)$:

$$D(\epsilon)=\int_0^T F(s^*+\epsilon g(t),\dot{s}^*+\epsilon g'(t),t)\,dt-\int_0^T F(s^*,\dot{s}^*,t)\,dt. \tag{5.14}$$

By definition $D(\epsilon)$ is nonpositive and attains its maximum at $\epsilon=0$ (since when $\epsilon=0$, both integrals are identical). Since $D(\epsilon)$ is continuously differentiable, the fact that it attains a maximum at $\epsilon=0$ implies that $D'(0)=0$. Let us use (5.14) to evaluate this derivative:

$$D'(\epsilon)=\int_0^T [F_s g(t)+F_{\dot{s}}g'(t)]\,dt. \tag{5.15}$$

Evaluated at $\epsilon=0$, and set to zero, (5.15) gives

$$-\int_0^T F_s(s^*,\dot{s}^*,t)g(t)\,dt=\int_0^T F_{\dot{s}}(s^*,\dot{s}^*,t)g'(t)\,dt. \tag{5.16}$$

Integrating by parts the right-hand side of (5.16),

$$\int_0^T F_{\dot{s}}g'(t)\,dt=[F_{\dot{s}}g(t)]\Big|_0^T-\int_0^T \frac{g(t)d(F_{\dot{s}})}{dt}\,dt. \tag{5.17}$$

Since $g(T)=g(0)=0$, equations (5.17) and (5.16) yield

$$\int_0^T\left(F_s-\frac{d}{dt}F_{\dot{s}}\right)g(t)\,dt=0. \tag{5.18}$$

Equation (5.18) must hold for any function $g(t)$ (subject only to $g(0)=g(T)=0$). This is possible only if

$$F_s-\frac{d}{dt}F_{\dot{s}}=0,$$

where the derivatives are evaluated at $(s^*(t),\dot{s}^*(t),t)$. This completes the proof of the necessity of Euler's equation.

Our main reasons for not devoting a great deal of space to the calculus of variations are, first, that optimal control theory is capable of dealing with a wider class of problems and, second, that the necessary conditions obtained from the calculus of variations contain no new information, since they can be readily derived from the maximum principle. This we now proceed to demonstrate.

Problem (5.7) can be transformed into the standard form of the control problem (4.1). Let $c(t)$ be a control variable, and let

$$\dot{s}(t) = c(t).$$

Then (5.7) becomes

$$\text{Maximize } \int_0^T F(s(t), c(t), t)\, dt$$

subject to

$$\dot{s} = c, \quad s(0) = s_0, \quad s(T) = s_T.$$

Applying the necessary conditions (4.8)–(4.10) to this problem, we have

$$F_c + \pi = 0, \tag{5.19}$$

$$\dot{s} = c, \tag{5.20}$$

$$\dot{\pi} = -F_s. \tag{5.21}$$

Since it has been assumed that $s(t)$ is twice differentiable, we can differentiate (5.19) with respect to time:

$$-\dot{\pi}(t) = \frac{d}{dt} F_c. \tag{5.22}$$

Euler's equation (5.11) can thus be obtained from (5.21) and (5.22). (Recall that F_c is the same as $F_{\dot{s}}$.)

Example 5.1.1. To illustrate the use of Euler's equation, let us return to problem (4.38) of Chapter 4. In order to formulate this problem in the calculus of variations format, we use (4.39) to eliminate $c(t)$ from the maximand in (4.38). We need to find $s(t)$ and hence $\dot{s}(t)$ that maximize

$$\int_0^T e^{-\delta t} \ln(rs - \dot{s})\, dt \tag{5.23}$$

subject to

$$s(0) = s_0 \quad \text{and} \quad s(T) = s_T. \tag{5.24}$$

Applying Euler's equation to (5.23),

$$e^{-\delta t} r (rs - \dot{s})^{-1} = \frac{d}{dt}(-e^{-\delta t}(rs - \dot{s})^{-1})$$

$$= \delta e^{-\delta t}(rs - \dot{s})^{-1} + e^{-\delta t}(rs - \dot{s})^{-2}(r\dot{s} - \ddot{s}).$$

This yields a second-order linear differential equation,

$$\ddot{s}-(2r-\delta)\dot{s}+(r-\delta)rs=0. \tag{5.25}$$

We can transform this equation into a system of two first-order linear differential equations by defining $z=\dot{s}$; hence, $\dot{z}=\ddot{s}$ can be obtained from (5.25) and we have

$$\begin{pmatrix}\dot{z}\\ \dot{s}\end{pmatrix}=\begin{pmatrix}2r-\delta & -r(r-\delta)\\ 1 & 0\end{pmatrix}\begin{pmatrix}z\\ s\end{pmatrix}.$$

The characteristic roots of this matrix are r and $(r-\delta)$. Therefore, the solution is

$$s(t)=Ae^{rt}+Be^{(r-\delta)t}, \tag{5.26}$$

where A and B are constants to be determined with the help of (5.24):

$$s_0=A+B,$$
$$s_T=Ae^{rT}+Be^{(r-\delta)T}.$$

Solving for A and B,

$$A=(s_Te^{-rT}-s_0e^{-\delta T})/(1-e^{-\delta T}), \tag{5.27}$$
$$B=(s_0-s_Te^{-rT})/(1-e^{-\delta T}). \tag{5.28}$$

It is easily seen that equations (5.26), (5.27), and (5.28) are equivalent to (4.46), (4.47), and (4.48).

This concludes our brief introduction to the calculus of variations. (For a thorough treatment of economic growth using the calculus of variations see Hadley and Kemp, 1971.) We will not discuss this topic again, because it is more economical to approach continuous-time optimization problems using optimal control theory. The latter is a more unified, more elegant, and more systematic body of knowledge that contains all of the results of the calculus of variations as special cases.

5.2 Dynamic programming: discrete-time, finite-horizon problems

In Section 4.2, we introduced an optimization problem in discrete time and stated the discrete-time maximum principle. An alternative method of solving this type of problem is the dynamic programming approach, which successfully exploits the *recursive nature* of problem (4.11). To explain this, we first rewrite (4.11) and (4.12) in a more convenient form for our present purpose. The problem is to find $c(1), c(2), \ldots, c(T)$ that maximize

$$V=\sum_{t=1}^{T} v_t(s(t),c(t)) \tag{5.29}$$

subject to

$$s(t+1)=h_t(s(t),c(t)), \quad t=1,2,\ldots,T, \tag{5.30}$$

$$s(1)=s_1, \quad s(T+1)=\bar{s}, \tag{5.31}$$

where s_1 and $\bar{s}$ are fixed exogenously.

The usual dynamic programming terminology is as follows: In (5.29), $v_t(s(t),c(t))$ is the *net benefit* at time t. Equation (5.30) is called the *transition equation,* and $h_t(s(t),c(t))$ is called the *transition function* at t. The subscript t in v_t and h_t indicates that these functions may depend on t.

Problem (5.29), subject to (5.30) and (5.31), is a discrete version of a control problem and as such has the two fundamental properties of separability and additivity over time periods. More precisely,

(i) for any t, the functions v_t and h_t depend on t and on the current values of the state and control variables, but not on their past or future values;

(ii) the maximand V is the *sum* of the net benefit functions.

Using these two properties, Bellman (1957) enunciates an important theorem about the nature of any optimal solution of problem (5.29). This theorem is known as the *principle of optimality.* Roughly speaking, it says that an optimal policy has the property that at any stage t, the remaining decisions $c^*(t), c^*(t+1), \ldots, c^*(T)$ must be optimal with regard to the current state $s^*(t)$, which results from the initial state s_1 and the earlier decisions $c^*(1), \ldots, c^*(t-1)$. This property is obviously sufficient for optimality since we require it to hold for all t: when we put $t=1$, we have the definitions of an optimal policy. Furthermore, the property is also necessary, since any deviation from the optimal policy, even in the last period, is clearly suboptimal. It was left to Bellman's genius to transform this rather trite, nearly tautological observation into an efficient method of solution. We now state the result formally.

Theorem 5.2.1: principle of optimality. $c^*(1), c^*(2), \ldots, c^*(T)$ is an optimal solution to the problem (5.29)–(5.31) if and only if $c^*(t), c^*(t+1), \ldots, c^*(T)$ solve the following problem for $t=1,\ldots,T$:

$$\text{Maximize } R_t=\sum_{\tau=t}^{T} v_\tau(s(\tau),c(\tau)) \tag{5.32}$$

subject to

$$s(\tau+1)=h_\tau(s(\tau),c(\tau)), \quad \tau=t,t+1,t+2,\ldots,T, \tag{5.33}$$

$$s(t) = s^*(t), \tag{5.34a}$$

$$s(T+1) = \bar{s}. \tag{5.34b}$$

This theorem implies that the values of $c^*(t), c^*(t+1), \ldots, c^*(T)$ can be determined from a set of equations (namely, the necessary conditions of problem (5.32)) that do not contain past values of the control and state variables: the knowledge of $c^*(t-1), c^*(t-2), \ldots, c^*(1), s^*(t-1), s^*(t-2)$, and so on is irrelevant, and only t and $s^*(t)$ matter. This follows the fundamental observation made earlier that the value of the state variables at some time summarizes *all* the relevant information about the system at that time.

Clearly, the principle of optimality relies on properties (i) and (ii) stated earlier. This principle does not apply to problems that cannot be put in the form (5.32), as the following example illustrates:

$$\text{Maximize} \quad [c(1)c(2)]^{1/2} + [c(2)c(3)]^{1/2} + [c(3)]^{1/2}$$

subject to

$$s(t+1) = s(t) - c(t), \quad t = 1, 2, 3,$$

$$s(1) = s_1 \text{ fixed}, \quad s(4) = s_4 \text{ fixed}.$$

A little reflection will establish that in this example, for given $s^*(2)$, we cannot determine $c^*(2)$ and $c^*(3)$ if we do not know $c^*(1)$ and s_1; in other words, $c^*(1)$, $c^*(2)$, $c^*(3)$, $s^*(2)$, and $s^*(3)$ must be determined simultaneously.

We can prove the principle of optimality formally by establishing a contradiction. Thus, let $c^*(1), c^*(2), \ldots, c^*(t-1), c^*(t), \ldots, c^*(T)$ be an optimal solution to (5.29) and suppose that $c^*(t), c^*(t+1), \ldots, c^*(T)$ did not yield a maximum for (5.32) for given $s^*(t)$. Let $\hat{c}(t), \hat{c}(t+1), \hat{c}(t+2), \ldots, \hat{c}(T)$ be an optimal solution of (5.32). Then the path $c^*(1)$, $c^*(2)$, $\ldots$, $c^*(t-1), \hat{c}(t), \ldots, \hat{c}(T)$ would yield a higher value V for (5.29) than the path $c^*(1), c^*(2), \ldots, c^*(t-1), c^*(t), \ldots, c^*(T)$. This would contradict the hypothesis that the latter is an optimal solution of (5.29). This completes the proof of the principle of optimality.

Bellman's principle of optimality gives rise to an important equation called the *functional recurrence equation,* which is the key to the dynamic programming method of solution. Let $V_t(s(t))$ denote the maximum value of R_t in (5.32) for given $s(t)$. We call this the *return function*. The principle of optimality implies that

$$V_t(s(t)) = \max_{c(t)} [v_t(s(t), c(t)) + V_{t+1}(s(t+1))], \tag{5.35a}$$

subject to

$$s(t+1)=h_t(s(t),c(t)), \tag{5.35b}$$

$$s(t) \text{ given} \quad \text{and} \quad s(T+1)=\bar{s}. \tag{5.35c}$$

Combining (5.35a) and (5.35b), we get

$$V_t(s(t))=\max_{c(t)}\{v_t(s(t),c(t))+V_{t+1}[h_t(s(t),c(t))]\}. \tag{5.36}$$

Equation (5.36) is *Bellman's functional recurrence equation.* It is called a functional equation because it implicitly defines the functional form of $V_t(s(t))$, which is the unknown entity. It provides the basis for an efficient method of solution called *backward induction,* which we explain below.

Example 5.2.1. We first illustrate how Bellman's equation works by using it to solve the control problem of Section 4.2. The problem is to find $c(t)$, $t=1,2,3$, to maximize

$$\sum_{t=1}^{3} \ln c(t)$$

subject to

$$s(t+1)=1.1s(t)-c(t),$$

$$s(1)=1, \quad s(4)=1.21.$$

Thus, in dynamic programming notation we have

$$v_t(s(t),c(t))=\ln c(t),$$

$$h_t(s(t),c(t))=1.1s(t)-c(t),$$

$$V_t(s(t))=\max_{c(\tau)} \sum_{\tau=t}^{3} \ln c(\tau),$$

with $s(\tau+1)=1.1s(\tau)-c(\tau)$, $s(t)$ given, and $s(4)=1.21$.

Bellman's equation states

$$V_t(s(t))=\max_{c(t)}[\ln c(t)+V_{t+1}(s(t+1))].$$

We begin with the last period, when we simply have

$$V_3(s_3)=\max_{c(3)} \ln c(3)$$

subject to $s(4)=1.1s(3)-c(3)$, $s(3)$ given, and $s(4)=1.21$. Since there is only one point in the choice set, we must have

$$c^*(3)=1.1s(3)-1.21.$$

Hence, $V_3(s(3))=\ln[1.1s(3)-1.21]$.

We now deal with the next period down the line:

$$V_2(s(2)) = \max_{c(2)}[\ln c(2) + V_3(s(3))]$$

subject to

$$s(3) = 1.1s(2) - c(2), \quad s(2) \text{ given.}$$

Substituting for V_3, we have

$$V_2(s(2)) = \max_{c(2)}[\ln c(2) + \ln(1.1s(3) - 1.21)]$$
$$= \max_{c(2)}[\ln c(2) + \ln(1.21s(2) - 1.1c(2) - 1.21)].$$

The first-order condition yields

$$c^*(2) = 0.55(s(2) - 1),$$

and substituting we have

$$V_2(s(2)) = \ln(0.33275) + 2\ln(s(2) - 1).$$

Next,

$$V_1(s(1)) = \max_{c(1)}[\ln c(1) + V_2(s(2))]$$

subject to

$$s(2) = 1.1s(1) - c(1), \quad s(1) \text{ given.}$$

Substituting for $V_2(s(2))$ and $s(2)$ yields

$$V_1(s_1) = \max_{c(1)}[\ln c(1) + \ln(0.33275) + 2\ln(1.1s(1) - c(1) - 1)].$$

The optimality condition yields

$$c^*(1) = (1.1s(1) - 1)/3.$$

Using $s(1) = 1$ and the expressions obtained for $c^*(1)$, $c^*(2)$, $c^*(3)$, and the transition equation, we work forward in time to obtain the optimal solution

$$c^*(1) = 0.0333,$$
$$s^*(2) = 1.1 - 0.0333 = 1.0667,$$
$$c^*(2) = 0.55(1.0667 - 1) = 0.0367,$$
$$s^*(3) = 1.1(1.0667) - 0.0367 = 1.1367,$$
$$c^*(3) = 1.1(1.1367) - 1.21 = 0.0403.$$

Thus, we see that the backward induction method consists of solving the last period first, taking as given the value of the state variable and working backward until the first period, when we actually know the value of the state variable. We then obtain the optimal solution by retracing our steps.

We now give a more formal account of the procedure. At time T, for given $s(T)$, we choose $c^*(T)$ that solves the problem facing the planner when there is only "one period to go":

$$\max_{c(T)} R_T = v_T(s(T), c(T)) \tag{5.37}$$

subject to

$$s(T+1) = h_T(s(T), c(T)),$$
$$s(T) \text{ given}, \quad s(T+1) = \bar{s} \text{ fixed}.$$

This problem yields $c^*(T)$ as a function of $s(T)$ (we suppress mention of its dependence on $\bar{s}$):

$$c^*(T) = g_T(s(T)).$$

By definition, $V_T(s(T))$ is the optimal value of R_T for given $s(T)$. Therefore,

$$V_T(s(T)) = v_T[s(T), g_T(s(T))]. \tag{5.38}$$

Working backward, at time $T-1$ we seek $c^*(T-1)$ that solves the problem facing the planner when there are "two periods to go":

$$\begin{aligned}\text{Maximize } R_{T-1} &= v_{T-1}(s(T-1), c(T-1)) + V_T(s(T)) \\ &= v_{T-1}(s(T-1), c(T-1)) + v_T[s(T), g_T(s(T))]\end{aligned} \tag{5.39}$$

subject to

$$s(T) = h_{T-1}(s(T-1), c(T-1)), \tag{5.40}$$
$$s(T-1) \text{ given}.$$

This gives $c^*(T-1)$ as a function of $s(T-1)$:

$$c^*(T-1) = g_{T-1}(s(T-1)). \tag{5.41}$$

The optimal value of R_{T-1} is obtained by substituting (5.40) and (5.41) into (5.39):

$$\begin{aligned}V_{T-1}(s(T-1)) = v_{T-1}&[s(T-1), g_{T-1}(s(T-1))] \\ &+ v_T\{h_{T-1}[s(T-1), g_{T-1}(s(T-1))], \\ &\quad g_T[h_{T-1}[s(T-1), g_{T-1}(s(T-1))]]\},\end{aligned} \tag{5.42}$$

which is a composite of known functions and has $s(T-1)$ for sole argument.

Proceeding in this way, we are faced, at each date $t > 1$, with the simple problem of finding $c^*(t)$ that maximizes

$$R_t = v_t(s(t), c(t)) + V_{t+1}(s(t+1)) \tag{5.43}$$

subject to

$$s(t+1) = h_t(s(t), c(t)), \tag{5.44}$$
$$s(t) \text{ given.}$$

The solution is expressed as

$$c^*(t) = g_t(s(t)). \tag{5.45}$$

The process is repeated until we reach $t = 1$, and the problem is then to find $c^*(1)$ that maximizes

$$R_1 = v_1(s(1), c(1)) + V_2(s(2))$$

subject to

$$s(2) = h_1(s(1), c(1)), \tag{5.46}$$
$$s(1) \text{ given } (= s_1).$$

This problem yields $c^*(1)$. The optimal value $s^*(2)$ is then computed using (5.46). Next (5.45) is used to obtain $c^*(2)$, which is in turn substituted into (5.44) to find $s^*(3)$. Again, (5.45) is applied to obtain $c^*(3)$, and so on.

We now apply the backward solution method to an economics problem.

Example 5.2.2: consumption policy. Let $s(t)$ denote an individual's stock of wealth at time t, and $c(t)$ the individual's consumption in period t. Let r be the rate of interest and let $\beta = 1 + r$. Then

$$s(t+1) = \beta[s(t) - c(t)]. \tag{5.47}$$

The individual wishes to find the time path $c(t)$, $t = 1, 2, \ldots, T$, that maximizes the sum of discounted utilities:

$$\text{Maximize } \sum_{t=1}^{T} \alpha^t u(c(t)) \tag{5.48}$$

subject to (5.47) and the boundary conditions

$$s(1) = s_1 \text{ fixed}, \quad s(T+1) = 0.$$

The term α^t is called the discount factor. We assume that $0 < \alpha < 1$. The function $u(c)$ is assumed to take the form

$$u(c) = [K/(1-\gamma)]c^{1-\gamma} \quad (K > 0,\ \gamma > 0,\ \gamma \neq 1). \tag{5.49}$$

Obviously, if we multiply the right-hand side of (5.49) by a positive constant, or add any constant to it, the solution $c^*(t)$, $t = 1, 2, \ldots, T$, will remain unchanged.

We proceed by applying the general approach described by (5.36)–(5.46) to the present example. For simplicity, we will write

$$A = K/(1-\gamma).$$

Solving backward, we first determine $c^*(T)$ that maximizes

$$R_T = \alpha^T A[c(T)]^{1-\gamma}$$

subject to

$$s(T+1) = \beta[s(T) - c(T)],$$
$$s(T+1) = 0, \quad s(T) \text{ given.}$$

This yields

$$c^*(T) = s(T). \tag{5.50}$$

Hence,

$$V_T(s(T)) = \alpha^T A[s(T)]^{1-\gamma}.$$

Next we find $c^*(T-1)$ that maximizes

$$R_{T-1} = \alpha^{T-1} A[c(T-1)]^{1-\gamma} + \alpha^T A[s(T)]^{1-\gamma} \tag{5.51}$$

subject to

$$s(T) = \beta[s(T-1) - c(T-1)], \tag{5.52}$$
$$s(T-1) \text{ given.}$$

Use (5.52) to substitute for $s(T)$ in (5.51), differentiate the resulting expression with respect to $c(T-1)$, and equate the derivative to zero:

$$c(T-1)^{-\gamma} = \alpha\beta^{1-\gamma}[s(T-1) - c(T-1)]^{-\gamma}.$$

Hence,

$$\alpha\beta^{1-\gamma} = \left[\frac{s(T-1) - c(T-1)}{c(T-1)}\right]^{\gamma},$$

$$\frac{s(T-1)}{c(T-1)} = 1 + (\alpha\beta^{1-\gamma})^{1/\gamma}.$$

Let

$$d = (\alpha\beta^{1-\gamma})^{1/\gamma}. \tag{5.53}$$

Then

$$c^*(T-1) = s(T-1)/(1+d). \tag{5.54}$$

Substitute this into (5.51) to obtain the optimal value of R_{T-1}:

$$V_{T-1}(s(T-1)) = \alpha^{T-1} A[s(T-1)]^{1-\gamma}(1+d)^{\gamma}. \tag{5.55}$$

From (5.55) we obtain the expression for R_{T-2},

$$R_{T-2} = \alpha^{T-2}A[c(T-2)]^{1-\gamma} + \alpha^{T-1}A[s(T-1)]^{1-\gamma}(1+d)^{\gamma},$$

which must be maximized subject to

$$s(T-1) = \beta[s(T-2) - c(T-2)],$$
$$s(T-2) \text{ given.}$$

This yields

$$c^*(T-2) = s(T-2)/(1+d+d^2). \tag{5.56}$$

Hence,

$$V_{T-2}(s(T-2)) = \alpha^{T-2}A[s(T-2)]^{1-\gamma}(1+d+d^2)^{\gamma}. \tag{5.57}$$

From (5.54)–(5.57) is it natural to guess that for any integer i ($i = 1, 2, \ldots, T-1$), the functions $c^*(T-i)$ and V_{T-i} are of the form

$$c^*(T-i) = s(T-i)/(1+d+d^2+\cdots+d^i), \tag{5.58}$$

$$V_{T-i}(s(T-i)) = \alpha^{T-i}A[s(T-i)]^{1-\gamma}(1+d+d^2+\cdots+d^i)^{\gamma}. \tag{5.59}$$

To show that our guess is correct, the method of proof by induction may be used. This method consists of showing that if (5.58) and (5.59) are correct for $i = m$, they are also correct for $i = m+1$. We leave this task to the reader.

It is appropriate at this stage to make some comments on the dynamic programming approach to control problems in discrete time. When we consider equations (5.50), (5.54), (5.56), or the general case (5.58), we see that our method of solution consists of devising a strategy or policy rule, given an arbitrary state of the system at each date. When we work back to initial time, for which we actually know the state of the system, we can begin to derive the particular optimal solution corresponding to this initial condition. The solution as described by (5.58) is known as a *closed-loop control* because the optimal value of the control variable during period $(T-i)$ is given as a function of the state variable at the beginning of that period. This is in contrast to an *open-loop control,* in which the solution is given as a function of time only. Typically dynamic programming yields a closed-loop solution, whereas the maximum principle yields an open-loop solution (e.g., see equation (4.45)). The reader must be warned that closed-loop solutions, such as (5.58), are not readily obtained for more complicated problems. The strength of dynamic programming in discrete time lies mainly in its applicability to numerical problems, and its usefulness in analytical economics is rather limited.

As we shall see in the next section, dynamic programming compares even less favorably with optimal control when time is represented as a continuous (i.e., real-valued) variable.

5.3 Dynamic programming in continuous time

Bellman's principle of optimality, stated in Section 5.2, is clearly applicable to the continuous-time version of problem (5.29). In this section we show that the continuous-time counterpart of the functional recurrence equation (5.36) yields a useful equation called the *Hamilton–Jacobi–Bellman equation,* which provides an alternative method to optimal control theory for solving continuous-time control. In addition, the Hamilton–Jacobi–Bellman equation can be used to derive heuristically the basic version of the maximum principle as presented in Chapter 4. We must hasten to add that for more complicated control problems the techniques provided by optimal control theory are more powerful; this is why we will concentrate on optimal control theory from Chapter 6 onward.

Since we will differentiate our equations with respect to time, it is useful to write our net benefit function as $v(s(t), c(t), t)$ and our return function as $V(s(t), t)$. By definition,

$$V(s(t), t) = \max \int_t^T v(s(\tau), c(\tau), \tau)\, d\tau \tag{5.60}$$

subject to

$$s(t) \text{ given}, \quad s(T) = s_T \text{ given},$$
$$\dot{s}(\tau) = f(s(\tau), c(\tau), \tau).$$

It follows from (5.60) and from the principle of optimality that

$$V(s(t), t) = \max_c \left[\int_t^{t+\Delta t} v(s(\tau), c(\tau), \tau)\, d\tau + V(s(t+\Delta t), t+\Delta t) \right]. \tag{5.61}$$

For a sufficiently small Δt, we can rewrite (5.61) as

$$V(s(t), t) = \max_{c(t)} [v(s(t), c(t), t)\Delta t + V(s(t+\Delta t), t+\Delta t)] + O(\Delta t), \tag{5.62}$$

where $O(\Delta t)$ is the sum of higher-order terms in Δt and has the property

$$\lim_{\Delta t \to 0} \frac{O(\Delta t)}{\Delta t} = 0.$$

Now, assuming that V is continuously differentiable, we can write

$$\begin{aligned} V(s(t+\Delta t), t+\Delta t) &= V(s(t), t) + V_s \Delta s + V_t \Delta t + O(\Delta t) \\ &= V(s(t), t) + [V_s \dot{s} + V_t]\Delta t + O(\Delta t). \end{aligned}$$

Substitute the above equation into (5.62):

$$V(s(t),t)=\max_{c(t)}[v(s(t),c(t),t)\Delta t+V(s(t),t)$$
$$+V_s f(s(t),c(t),t)\Delta t+V_t\Delta t]+O(\Delta t).$$

Cancel $V(s(t),t)$ on both sides, divide the resulting equation by Δt, and take the limit $\Delta t\to 0$:

$$0=\max_{c(t)}[v(s(t),c(t),t)+V_s f(s(t),c(t),t)+V_t]. \tag{5.63}$$

Equation (5.63) is called the Hamilton–Jacobi–Bellman equation. If we define the Hamiltonian

$$H(s,c,\pi,t)=v(s,c,t)+\pi f(s,c,t), \tag{5.64}$$

where

$$\pi=V_s(s,t), \tag{5.65}$$

then (5.63) can be written as

$$-V_t=\max_{c(t)} H. \tag{5.66}$$

Equation (5.63), or (5.66), is a partial differential equation because it involves the partial derivatives of V with respect to s and t. In general, this type of equation is difficult to solve, even for very simple v and f functions.

In order to give the reader a better appreciation of the Hamilton–Jacobi–Bellman equation, we illustrate it with a simplified version of problem (5.23), with $\delta=r=0$. Find $c(t)$ that maximizes

$$\int_0^T \ln(c(t))\,dt \tag{5.67}$$

subject to

$$\dot{s}(t)=-c(t),\quad s(0)=s_0,\quad s(T)=s_T.$$

V is defined by

$$V(s_t,t)=\max_{c(\tau)}\int_t^T \ln(c(\tau))\,d\tau \tag{5.68}$$

subject to

$$\dot{s}(\tau)=-c(\tau),\quad s(t)=s_t \text{ given},\quad s(T)=s_T.$$

The Hamilton–Jacobi–Bellman equation takes the form

$$0=\max_{c(t)}[\ln(c(t))+(V_s)(-c(t))+V_t]. \tag{5.69}$$

The first-order condition yields $c^{-1}=V_s$, and substituting in equation (5.69), we have

$$0=-\ln(V_s)-1+V_t. \tag{5.70}$$

This is a partial differential equation the solution of which would yield the function $V(s_t, t)$ and thus V_s and c. Unfortunately, the analytical solution of partial differential equations is a formidable problem, and we will not attempt to solve (5.70). We wish, however, to convince the reader that the solution obtained through dynamic programming is the same as that obtained from the maximum principle. To this end we derive $V(s_t, t)$ directly from its definition in (5.68) using optimal control techniques and verify that the solution satisfies (5.70).

Form the Hamiltonian $H(s, c, \pi) = \ln c - \pi c$ and apply the maximum principle to obtain $c^{-1} = \pi$, $\dot{\pi} = 0$, and $\dot{s} = -c$; hence, $c(\tau) = A$, $s(\tau) = B - A\tau$, and the boundary conditions $s_t = B - At$ and $s_T = B - AT$ determine the constants

$$A = (s_t - s_T)/(T - t),$$
$$B = (s_t T - s_T t)/(T - t).$$

Consequently, the return function is

$$V(s_t, t) = \int_t^T [\ln(s_t - s_T) - \ln(T - t)]\, d\tau$$
$$= (T - t)[\ln(s_t - s_T) - \ln(T - t)]. \tag{5.71}$$

It is easy to verify that

$$V_s = (T - t)/(s_t - s_T), \qquad V_t = 1 - \ln(s_t - s_T) + \ln(T - t)$$

do satisfy the partial differential equation (5.70).

This completes our exposition of dynamic programming. Although not well suited to the analytical approach we wish to follow in this book, it is a very useful technique. Interested readers may consult Bellman and Dreyfus (1962).

Exercises

1. Use Euler's equation to solve the following calculus of variations problems with boundary conditions:
   (a) Maximize $\int_0^1 -(\dot{s})^2 dt$, $s(0) = 0$, $s(1) = 5$,
   (b) Maximize $\int_0^1 -(bs - (\dot{s})^2)\, dt$, $s(0) = 0$, $s(1) = 2$,
   (c) Maximize $\int_0^1 (4s - (\dot{s})^2 - (s)^2)e^{-0.1t} dt$, $s(0) = 0$, $s(1) = 4$.
2. Let k be capital stock, $f(k)$ output, and c consumption. Let $U(c)$ be the utility function, where $c = f(k) - \dot{k}$. Use Euler's equation to find the necessary conditions for the following problem: maximize $\int_0^T U(f(k) - \dot{k})\, dt$ subject to $k(0) = k_0$, $k(T) = k_T$. Show that the resulting second-order differential equation in k can be transformed into a pair of first-order differential equations in c and k.

3. Show that if the integrand in equation (5.7) does not have t as a separate argument, then Euler's equation implies $F - \dot{s}F_{\dot{s}} = A$, where A is an arbitrary constant. (*Hint:* Take the time derivative of equation (5.7), divide by $\dot{s}$, and compare it with Euler's equation.)
4. Derive the Hamilton–Jacobi–Bellman equation for the following problem: Maximize $\int_0^T \ln(c(t))\,dt$ subject to $\dot{s}(t) = rs(t) - c(t)$, $s(0) = s_0$, $s(T) = s_T$ and show that the following return function satisfies that equation:

$$V(s_t, t) = (T-t)\ln\left(\frac{s_t e^{-rt} - s_T e^{-rT}}{T-t}\right) + \frac{r}{2}(T^2 - t^2).$$

Use optimal control theory to obtain this return function. (*Hint:* Use $c(\tau) = Be^{r\tau}$ in $\dot{s}(\tau) = rs(\tau) - c(\tau)$; hence, $s(\tau)e^{-r\tau} = A - B\tau$. Then solve for A and B using $s(t) = s_t$ and $s(T) = s_T$.)
5. Consider the problem of maximizing $\int_0^T \ln(c(t))e^{-\delta t}\,dt$ $(\delta > 0)$ subject to $\dot{s}(t) = rs(t) - c(t)$, $s(0) = s_0$, and $s(T) = s_T$. Show that the following return function satisfies the Hamilton–Jacobi–Bellman equation,

$$V(s_t, t) = \delta^{-1}(e^{-\delta t} - e^{-\delta T})\ln(\delta B)$$
$$+ \delta^{-1}(r-\delta)(te^{-\delta t} - Te^{-\delta T}) + \delta^2(r-\delta)(e^{-\delta t} - e^{-\delta T}),$$

where $B = e^{-r(t+T)}(e^{rt}s_T - e^{rT}s_t)/(e^{-\delta T} - e^{-\delta t})$.

CHAPTER 6

The general constrained control problem

In Chapter 4, we studied the problem of finding a function $c(t)$ that maximizes

$$V=\int_0^T v(s(t),c(t),t)\,dt$$

subject to

$$\dot{s}=f(s(t),c(t),t),$$

$$s(0)=s_0,\quad s(T)=s_T.$$

We now wish to consider a more general control problem involving many state variables and many control variables. We also wish to introduce constraints on the values that the control variables may take on at any point of time and also constraints on their overall paths from time 0 to time T. For ease of exposition, we retain the assumption that initial time and terminal time are exogenously specified, as are the initial and terminal values of the state variables; these conditions will be relaxed in Chapter 7.

6.1 The set of admissible controls

In many economic problems, the set of values that the control variables may take on is restricted. For example, consumption cannot be greater than output, and the rate of extraction of a natural resource may not exceed a certain upper bound specified by environmental control laws.

In order to be more specific let us consider a version of the optimal control problem studied in Chapter 4. Let $s(t)$, $c(t)$, and $I(t)$ denote respectively, the stock of capital, the flow of consumption, and the flow of gross investment at time t. Let $m>0$ be the rate of natural depreciation of capital. Then the rate of change in capital stock is

$$\dot{s}(t)=I(t)-ms(t). \tag{6.1}$$

We assume that output at time t is given by $e^{\gamma t}F(s(t))$, where γ is the rate of technical progress and F the production function. At any time t, the following constraints are imposed on the control variables $I(t)$ and $c(t)$:

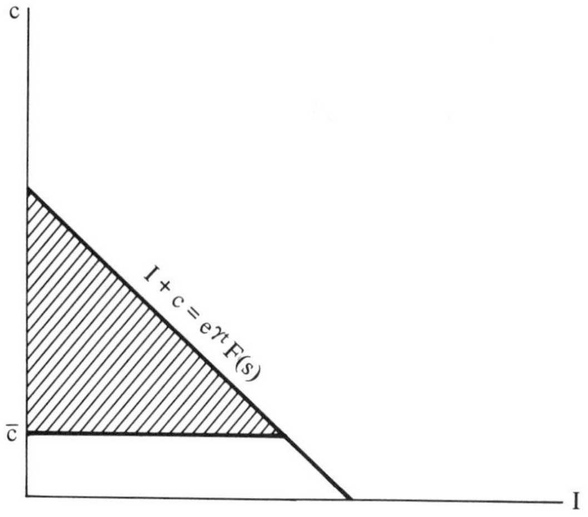

Figure 6.1

$$I(t)+c(t) \leq e^{\gamma t}F(s(t)), \tag{6.2}$$

$$I(t) \geq 0, \tag{6.3}$$

$$c(t) \geq \bar{c}. \tag{6.4}$$

Condition (6.2) states that the sum of consumption and gross investment cannot exceed total output at any instant. Condition (6.3) means that gross investment cannot be negative (negative gross investment, which was allowed in Chapter 4, means that capital stock can be consumed). Condition (6.4) stipulates a lower bound on the rate of consumption. The *set of admissible controls* is therefore the striped area in Figure 6.1. Note that the northeast frontier of the set moves with time and depends on the value of the state variable s.

Since the set of admissible controls depends on t and $s(t)$, we shall denote it by the symbol $W(s(t), t)$. Formally,

$$W(s(t), t) \equiv \{(I, c): I \geq 0,\ c \geq \bar{c},\ I+c \leq e^{\gamma t}F(s(t))\}.$$

It is important to note that (6.2) is a constraint on the values of $I(t)$ and $c(t)$ for given values of t and $s(t)$, and not a constraint on the value of the state variable $s(t)$. (It is, of course, possible to specify constraints on the values of the state variables alone, e.g., $s(t) \leq \bar{s}$ or $s(t) \geq \hat{s}$; however, these constraints give rise to much more complicated necessary conditions, since they also imply a constraint on $\dot{s}(t)$ when $s(t) = \bar{s}$ or $s(t) = \hat{s}$; for this reason we will not deal with constraints on state variables in this chapter and defer this topic until Chapter 10.)

We are now ready to state the constrained problem that is the subject of this chapter in its most general form.

The constrained control problem

Assume that there are r control variables, n state variables, and m constraints on the control variables. Suppose that the first m' constraints are inequality constraints and that the remaining $m-m'$ constraints are equality constraints. Then the set of admissible controls can be represented by m' inequalities and $m-m'$ equations.

For notational convenience, we now adopt the convention of writing $\mathbf{s}(t)$ and $\mathbf{c}(t)$ for the vectors of state variables and control variables, where $\mathbf{s}(t)=[s_1(t),s_2(t),\ldots,s_m(t)]'$ and $\mathbf{c}(t)=[c_1(t),c_2(t),\ldots,c_r(t)]'$. We can then write the constraints as

$$g^i(\mathbf{s}(t),\mathbf{c}(t),t)\geq 0, \quad i=1,2,\ldots,m'; \tag{6.5a}$$

$$g^j(\mathbf{s}(t),\mathbf{c}(t),t)=0, \quad j=m'+1,\ldots,m. \tag{6.5b}$$

The set of admissible controls is

$$W(\mathbf{s}(t),t)\equiv\{\mathbf{c}(t)\,|\,g^i(\mathbf{s}(t),\mathbf{c}(t),t)\geq 0,\ i=1,2,\ldots,m'; \\ g^j(\mathbf{s}(t),\mathbf{c}(t),t)=0,\ j=m'+1,\ldots,m\}. \tag{6.6}$$

Since we are dealing with constraints on control variables, it is important to note the requirement that each constraint contain at least one control variable. In addition, since we will have to find values of the control variables that maximize a Hamiltonian function subject to the constraints (6.5a) and (6.5b), we must require that these constraints satisfy one of the "constraint qualifications" discussed in Chapter 1 (see Lemma 1.4.1). The most convenient constraint qualification is the *rank condition,* which we restate here.

The rank condition. If the vector $\mathbf{c}(t)=(c_1(t),\ldots,c_r(t))'$ satisfies constraints (6.5a)–(6.5b) for given values of $s_1(t),\ldots,s_n(t)$ and t, and if p of these constraints are satisfied with equality ($p\geq m-m'$, because $m-m'$ is the number of equality constraints), then it is required that the matrix (of order $p\times r$) of partial derivatives of these p constraints with respect to the control variables be of rank p.

This constraint qualification implies in particular that while the number of constraints (m) may exceed the number of control variables (r), we require that the number of active constraints (p) not be greater than the number of control variables. In what follows we shall assume that the rank condition is satisfied.

There is an additional requirement, or rather a lack of one, on the control variables. Heretofore we have been implicitly assuming that the variables were differentiable functions of time. We know that if the variables were simply continuous, we would have to deal with left-hand-side and right-hand-side derivatives, and this would somewhat complicate the algebra. However, control theory actually encompasses far more general problems. It is capable, and this is one of its great achievements, of dealing with control variables that have a finite number of discontinuities of the "jump" kind. Formally we say that a variable is a *piecewise-continuous* function of time if and only if it is continuous almost everywhere, that is, anywhere but at a finite number of points, where it may exhibit jump discontinuities. It would seem rather queer to complicate things so much were it not for the fact that even in some simple problems, no solution would exist if jump discontinuities were disallowed. To prove our point it suffices to consider a trivial example.

Find $c(t)$ that maximizes

$$\int_0^2 -[s(t)-1]^2\,dt$$

with $\dot{s}(t)=c(t)$ and a double restriction on the control $0 \le c(t) \le 1$; $s(0)=0$, $s(2)=1$.

Clearly, to minimize the squared deviation of s from 1 (given that $s(2)$ must equal 1) we must increase s at the highest speed, $c=1$, until the goal is reached, $s=1$, and then leave s at 1 by setting $c=0$. Therefore, the optimal solution involves setting $c=1$ from time 0 to time 1 and henceforth setting $c=0$. The optimal control has a jump discontinuity at time 1.

Unless we wish to disregard such problems, we must impose only the requirement of piecewise continuity on control variables. All results in this chapter formally mention this qualification, but we will not study examples of discontinuous controls until Chapter 8. We will, however, encounter optimal controls that are continuous but not everywhere differentiable.

In the remainder of this chapter we shall see that the maximum principle for the constrained problem essentially involves working with a Lagrangean-type expression that incorporates the constraints (with multipliers) as well as the Hamiltonian, rather than the Hamiltonian alone. Before we proceed, we must deal separately with a seemingly different kind of constraint.

6.2 Integral constraints

In addition to constraints on control variables at any instant, we may wish to allow for another kind of constraint that imposes a restriction on the overall path of the variables. These will naturally take the form of integral constraints. For example, Robinson Crusoe may wish to choose the time paths of the control variables x_1, x_2, c_1, c_2, and I (respectively his own labor, Man Friday's labor, his consumption, Man Friday's consumption, and gross investment) so as to maximize his own utility,

$$V_1 = \int_0^T v_1(x_1(t), c_1(t), s(t), t)\,dt, \tag{6.7}$$

subject to the constraint that Man Friday's utility be equal to a given constant $\bar{V}$,

$$\int_0^T v_2(x_2(t), c_2(t), s(t), t)\,dt = \bar{V}, \tag{6.8}$$

and other constraints, such as

$$\dot{s}(t) = I(t) - ms(t),$$

$$c_1(t) + c_2(t) + I(t) \leq F(x_1(t), x_2(t), s(t)),$$

$$s(0) = s_0, \quad s(T) = s_T.$$

On reflection, an equality integral constraint such as (6.8) presents no new problem, for it can be replaced by a new differential equation, with two boundary conditions, for a newly defined state variable $k(t)$. Thus, (6.8) can be replaced by

$$\dot{k}(t) = v_2(x_2(t), c_2(t), s(t), t), \tag{6.8a}$$

$$k(0) = 0, \tag{6.8b}$$

$$k(T) = \bar{V}. \tag{6.8c}$$

To verify that (6.8a)–(6.8c) are equivalent to (6.8), we note that (6.8a) can be integrated to give

$$k(t) = \int_0^t v_2(x_2(\tau), c_2(\tau), s(\tau), \tau)\,d\tau,$$

and hence $k(T)$ is the left-hand side of (6.8).

Similarly, any inequality integral constraint (e.g., (6.8) with $= \bar{V}$ replaced by $\geq \bar{V}$) can be replaced by a differential equation together with two boundary conditions: an equality boundary condition and an inequality one ($k(0) = 0$, $k(T) \geq \bar{V}$). Although we have not dealt with inequality boundary conditions yet, they will be examined in the next chapter.

In what follows, we shall assume that all integral constraints have been transformed into differential equations with boundary conditions. This simplifies the exposition without loss of generality.

6.3 The maximum principle with equality constraints only

In this section we state the maximum principle for the case in which all constraints are equality constraints, and work out a simple example. We consider the following problem: find piecewise-continuous functions $\mathbf{c}(t)=[c_1(t), c_2(t), \ldots, c_r(t)]'$ that maximize

$$V=\int_0^T v(\mathbf{s}(t), \mathbf{c}(t), t)\, dt \tag{6.9}$$

subject to n differential equations,

$$\dot{s}_i(t)=f^i(\mathbf{s}(t), \mathbf{c}(t), t), \quad i=1,2,\ldots,n, \tag{6.10a}$$

m equality constraints,

$$g^j(\mathbf{s}(t), \mathbf{c}(t), t)=0, \quad j=1,2,\ldots,m, \tag{6.10b}$$

and $2n$ boundary conditions,

$$s_i(0)=s_{i0}, \; s_i(T)=s_{iT}, \quad i=1,2,\ldots,n. \tag{6.10c}$$

In this problem, both the initial time and the terminal time are fixed. The boundary values s_{i0} and s_{iT}, $i=1,\ldots,n$, are also exogenously specified. The functions v, f^i, g^j are assumed to possess continuous second-order partial derivatives.

As one would expect, the necessary conditions for this problem are quite similar to the ones presented in Chapter 4. There are, however, certain differences. First, since there are n state variables, we must have n costate variables: $\pi_1(t), \pi_2(t), \ldots, \pi_n(t)$. The Hamiltonian is therefore

$$H(\mathbf{s}(t), \mathbf{c}(t), \boldsymbol{\pi}(t), t)=v(\mathbf{s}(t), \mathbf{c}(t), t)+\sum_{i=1}^{n} \pi_i(t) f^i(\mathbf{s}(t), \mathbf{c}(t), t). \tag{6.11}$$

Second, at any given time t and for given values of $s_i(t)$ and $\pi_i(t)$, $i=1,\ldots,n$, the control variables $c_1(t), \ldots, c_r(t)$ must maximize the value of the Hamiltonian subject to the constraint that the values taken by the controls belong to the set of admissible controls $W(\mathbf{s}, t)$ as defined by (6.6), with $m'=0$. This means that one must introduce a multiplier $\lambda_j(t)$ for each constraint g^j, form the Lagrangean function,

$$\mathcal{L}=H(\mathbf{s}(t), \mathbf{c}(t), \boldsymbol{\pi}(t), t)+\sum_{j=1}^{m} \lambda_j(t) g^j(\mathbf{s}(t), \mathbf{c}(t), t), \tag{6.12}$$

find the derivatives $\partial\mathcal{L}/\partial c_i$, $i=1,\ldots,r$, and equate them to zero. Third, the differential equations for the costate variables are now

$$\dot{\pi}_i(t)=-\frac{\partial\mathcal{L}}{\partial s_i(t)}$$

(and not just $-\partial H/\partial s_i(t)$ as before), so as to take into account the effects of present changes in the state variables on the future set of admissible values of the control variables represented by the constraints $g^1, g^2, \ldots, g^m$ included in the Lagrangean function.

For future reference, we state the necessary conditions that constitute the maximum principle in a more formal and complete manner.

Theorem 6.3.1: necessary conditions for the equality-constrained problem. Let $\mathbf{c}^*(t) = [c_1^*(t), \ldots, c_r^*(t)]'$ be an optimal solution to the control problem (6.9)–(6.10) and $\mathbf{s}^*(t) = [s_1^*(t), \ldots, s_n^*(t)]'$ be the corresponding time path of the state variables. Then there exist costate variables $\boldsymbol{\pi}(t) = [\pi_1(t), \ldots, \pi_n(t)]'$ and Lagrange multipliers $\boldsymbol{\lambda}(t) = [\lambda_1(t), \ldots, \lambda_m(t)]'$ such that:

(i) At any time t, for given vectors $\mathbf{s}^*(t)$ and $\boldsymbol{\pi}(t)$, the control variable vector $\mathbf{c}^*(t)$ maximizes the Hamiltonian

$$H(\mathbf{s}^*(t), \mathbf{c}(t), \boldsymbol{\pi}(t), t) \equiv v(\mathbf{s}^*(t), \mathbf{c}(t), t) + \sum_{i=1}^{n} \pi_i(t) f^i(\mathbf{s}^*(t), \mathbf{c}(t), t) \tag{6.13}$$

subject to the condition that $\mathbf{c}(t)$ belongs to the set of admissible controls defined by (6.10b).

(ii) The costate variables $\pi_i(t)$, $i = 1, 2, \ldots, n$, are continuous functions of t and have piecewise-continuous derivatives (with respect to time) that satisfy the condition

$$\dot{\pi}_i(t) = -\frac{\partial \mathcal{L}^*}{\partial s_i(t)}; \quad i = 1, 2, \ldots, n, \tag{6.14}$$

where the asterisk indicates that the partial derivatives of $\mathcal{L}$ from (6.12) are evaluated at $(\mathbf{s}^*(t), \mathbf{c}^*(t))$;

(iii) Equation (6.10a) holds, or from the definition of the Lagrangean,

$$\dot{s}_i^*(t) = \frac{\partial \mathcal{L}^*}{\partial \pi_i(t)} = f^i(\mathbf{s}^*(t), \mathbf{c}^*(t), t). \tag{6.15a}$$

It is also required that $\mathbf{s}^*(t)$ satisfy the boundary conditions (6.10c):

$$s_i(0) = s_{i0}, \quad s_i(T) = s_{iT}, \quad i = 1, 2, \ldots, n. \tag{6.15b}$$

(iv) The Lagrange multipliers $\lambda_i(t)$, $i = 1, 2, \ldots, m$, are piecewise-continuous on $0 \le t \le T$ and are continuous whenever $\mathbf{c}^*(t)$ is continuous. (6.16)

(v) The Lagrangean $\mathcal{L}(\mathbf{s}^*(t), \mathbf{c}^*(t), \boldsymbol{\pi}(t), \boldsymbol{\lambda}(t), t)$ is a continuous function of time on $0 \le t \le T$. On each interval of continuity of $\mathbf{c}^*(t)$, the Lagrangean is differentiable totally with respect to t, and

$$\frac{d\mathcal{L}^*}{dt} = \frac{\partial \mathcal{L}^*}{\partial t}, \tag{6.17}$$

where $\partial \mathcal{L}^*/\partial t$ is the partial derivative of $\mathcal{L}(\mathbf{s}^*(t), \mathbf{c}^*(t), \boldsymbol{\pi}(t), \boldsymbol{\lambda}(t), t)$ with respect to the last argument. (Condition (6.17) can be derived from (6.13)–(6.17); see Section 6.5.)

Remark (a). Condition (6.13) can be stated formally as

$$H(\mathbf{s}^*(t), \mathbf{c}^*(t), \boldsymbol{\pi}(t), t) \geq H(\mathbf{s}^*(t), \mathbf{c}(t), \boldsymbol{\pi}(t), t), \tag{6.18}$$

for all $\mathbf{c}(t)$ satisfying $g^j(\mathbf{s}^*(t), \mathbf{c}(t), t) = 0$, $j = 1, 2, \ldots, m$, or since we assumed that the rank condition is satisfied,

$$\frac{\partial \mathcal{L}^*}{\partial c_i(t)} = 0, \quad i = 1, 2, \ldots, r, \tag{6.18$'$}$$

where the asterisk indicates that the partial derivatives of (6.12) are evaluated at $(\mathbf{s}^*(t), \mathbf{c}^*(t))$.

Remark (b). We should, strictly speaking, include an additional constant π_0 in our statement of the necessary conditions, where π_0 is the constant multiplier associated with the integrand of the objective function. In other words, we should have defined the Hamiltonian as

$$H = \pi_0 v(\mathbf{s}(t), \mathbf{c}(t), t) + \sum_{i=1}^{n} \pi_i(t) f^i(\mathbf{s}(t), \mathbf{c}(t), t).$$

If $\pi_0 \neq 0$, we may set $\pi_0 = 1$ because the equations for $\dot{\pi}_i(t)$ are linear in $(\pi_0, \pi_1, \ldots, \pi_n)$. Only in pathological cases does π_0 equal zero; we will not be concerned with these cases. The reader may consult Long and Vousden (1977, pp. 16–17) or Athans and Falb (1966, p. 291) for further discussions.

We now show how the necessary conditions can be used to solve a control problem involving one state variable, two control variables, and an equality constraint on the control variables.

Example 6.3.1: extraction of an exhaustible resource. Let $s(t)$ denote the stock of an exhaustible resource and $x(t)$ the rate of extraction from that stock, so that

$$\dot{s}(t) = -x(t). \tag{6.19}$$

The output of the finished good is a function of both the rate of extraction and the stock size $s(t)$. This output is denoted by $F(s(t), x(t))$. It is assumed that the function F is positive and increasing in each argument and that output is zero if $x(t)$ is zero. This may reflect the fact that higher-yielding ore is obtained when s is relatively large. Let $c(t)$ denote the flow of consumption of the finished good, and $u(c(t))$ the utility function with the usual properties ($u'(0)=\infty$, $u'(c)>0$, $u''(c)<0$). We also assume that the output of the finished good cannot be stored. Therefore, consumption must equal output:

$$F(s(t),x(t))-c(t)=0. \tag{6.20}$$

Assuming for simplicity that the rate of discount is zero, the problem is to find the time paths of the control variables $c(t)$ and $x(t)$ that maximize

$$V=\int_0^T u(c(t))\,dt \tag{6.21}$$

subject to (6.19), (6.20), and the boundary conditions

$$s(0)=s_0, \quad s(T)=s_T \ (<s_0).$$

(The assumptions that $u'(0)=\infty$ and $F(s,0)=0$ guarantee that both $c(t)$ and $x(t)$ are positive along an optimal path; therefore, we do not have to specify the constraints $c(t)\geq 0$ and $x(t)\geq 0$.)[1]

Let $\pi(t)$ be the costate variable. The Hamiltonian for this problem is

$$H(s(t),c(t),x(t),\pi(t))=u(c(t))-\pi(t)x(t),$$

and the Lagrangean is

$$\mathcal{L}=u(c(t))-\pi(t)x(t)+\lambda(t)[F(s(t),x(t))-c(t)].$$

The maximum principle yields the following conditions:

$$\frac{\partial\mathcal{L}}{\partial c(t)}=u'(c(t))-\lambda(t)=0, \tag{6.22}$$

$$\frac{\partial\mathcal{L}}{\partial x(t)}=-\pi(t)+\lambda(t)F_x=0, \tag{6.23}$$

$$\dot{\pi}(t)=-\frac{\partial\mathcal{L}}{\partial s(t)}=-\lambda(t)F_s, \tag{6.24}$$

$$\dot{s}(t)=\frac{\partial\mathcal{L}}{\partial\pi(t)}=-x(t). \tag{6.25}$$

We now attempt to interpret these necessary conditions. From (6.22) we see that λ is the imputed value of the finished good; it is positive since

[1] In this simple example we could obviously use (6.20) to eliminate c and reduce the problem to an unconstrained one with a single control. Our aim, however, is to introduce the reader to the use of the maximum principle in a constrained problem.

$u'(c) > 0$. It follows from this and (6.23)–(6.24) that the value of the resource, π, is positive and is decreasing over time (recall that F is increasing in both arguments). Equation (6.24) says that the rate of depreciation of the value of the resource, $-\dot{\pi}$, is equal to the marginal product of the resource in the production of the finished good multiplied by the value of the finished good. Thus, our interpretation in Section 4.5 carries over to problems with constraints on the control variables. In equation (6.23), π is to be interpreted as the cost of a marginal increase in x; the other term is the marginal value of the contribution of x to the finished good.

We now proceed to show how an explicit solution may be obtained when both the utility function and the production function are specified. To this end we assume that

$$u(c) = (1/\gamma)c^{\gamma} \quad \text{and} \quad F(s,x) = s^{\alpha}x^{\beta}, \tag{6.26}$$

where $\gamma < 1$, $\alpha > 0$, $\beta > 0$, and $\alpha + \beta < 1$. Using (6.26), conditions (6.20) and (6.22)–(6.25) take the special form

$$c = s^{\alpha}x^{\beta}, \tag{6.20$'$}$$

$$c^{\gamma-1} = \lambda, \tag{6.22$'$}$$

$$\pi = \lambda\beta s^{\alpha}x^{\beta-1}, \tag{6.23$'$}$$

$$\dot{\pi} = -\lambda\alpha s^{\alpha-1}x^{\beta}, \tag{6.24$'$}$$

$$\dot{s} = -x. \tag{6.25$'$}$$

Taking the ratio of (6.24′) and (6.23′) yields

$$\dot{\pi}/\pi = -(\alpha/\beta)(x/s) = (\alpha/\beta)(\dot{s}/s). \tag{6.27}$$

Substituting (6.20′) and (6.22′) into (6.23′) to eliminate λ, we obtain

$$\pi = \beta s^{\alpha\gamma}x^{\beta\gamma-1}. \tag{6.28}$$

Differentiating (6.28) with respect to time yields, after simplification (using (6.28) itself),

$$\dot{\pi}/\pi = \alpha\gamma s^{-1}\dot{s} + (\beta\gamma - 1)x^{-1}\dot{x}.$$

Since $x = -\dot{s}$ and $\dot{x} = -\ddot{s}$, we have

$$\dot{\pi}/\pi = \alpha\gamma(\dot{s}/s) + (\beta\gamma - 1)(\ddot{s}/\dot{s}). \tag{6.29}$$

Together (6.27) and (6.29) yield

$$(\dot{s}/s)(\alpha\beta^{-1} - \alpha\gamma) = (\beta\gamma - 1)(\ddot{s}/\dot{s}),$$

and finally, provided that $(\beta\gamma - 1) \neq 0$ so that we can divide both sides by it,

$$\ddot{s}/\dot{s} = -(\alpha/\beta)(\dot{s}/s). \tag{6.30}$$

This is a second-order differential equation, but it is so simple that it can be solved directly by repeated integration. Recall that $\int(\dot{u}/u)\,dt = \ln|u|+C$, and integrate both sides of (6.30) to obtain

$$\ln|\dot{s}| = -(\alpha/\beta)\ln|s|+A.$$

Since $\dot{s}<0$ and $s>0$, this is equivalent to

$$\ln(-\dot{s}) = -(\alpha/\beta)\ln s+A.$$

Applying the exponential operator to each side, we have

$$\dot{s} = -e^{A}s^{-\alpha/\beta} \quad \text{or} \quad \dot{s}(s^{\alpha/\beta}) = -e^{A},$$

which can be integrated again to yield

$$[\beta/(\alpha+\beta)]s^{(\alpha+\beta)/\beta} = -te^{A}+B.$$

Recalling that $e^{A}>0$ and $\beta/(\alpha+\beta)>0$, we can rewrite the preceding equation as

$$s^{(\alpha+\beta)/\beta} = -Et+G,$$

where E and G are arbitrary constants, $E>0$. Finally,

$$s(t) = (-Et+G)^{\beta/(\alpha+\beta)}. \tag{6.31}$$

The constants of integrations E and G can be determined using the boundary conditions

$$s(0) = G^{\beta/(\alpha+\beta)} = s_0, \quad \text{or} \quad G = s_0^{(\alpha+\beta)/\beta},$$

$$s(T) = (-ET+G)^{\beta/(\alpha+\beta)} = s_T,$$

$$-ET+G = s_T^{(\alpha+\beta)/\beta},$$

$$-ET = s_T^{(\alpha+\beta)/\beta} - s_0^{(\alpha+\beta)/\beta} < 0.$$

It remains to find the optimal paths $x^*(t)$ and $c^*(t)$. From (6.31),

$$x^*(t) = -\dot{s}^* = (E\beta/(\alpha+\beta))(-Et+G)^{-\alpha/(\alpha+\beta)} > 0, \tag{6.32}$$

$$c^*(t) = (s^*)^{\alpha}(x^*)^{\beta} = [E\beta/(\alpha+\beta)]^{\beta} = \text{const}. \tag{6.33}$$

This completes our solution of the problem, given that the utility function and the production function take the special forms described by (6.26).

In economic theory, it is desirable to find out the extent to which a particular result (e.g., constant consumption) depends on the special functional forms assumed. The remainder of this section is devoted to this task.

First note that the constants E and G are independent of γ; hence, none of the optimal paths of x, s, and c depend on this parameter. Therefore, the particular value of γ does not affect the "primal" part of the solution. (It does affect π and λ, however.)

Furthermore, suppose that for some utility function u we have found an optimal path with the property that c is constant. Then consider the same problem, with u replaced by another form $\tilde{u}$, where $\tilde{u}(c)$ still satisfies the property that marginal utility is positive and decreasing. Equation (6.22′) is altered. Let us look at the previous optimal (asterisked) solution as a candidate solution. Equations (6.20′) and (6.24′) remain satisfied. Use (6.22) to define the constant $\tilde{\lambda}=\tilde{u}'(c^*)$ and similarly scale π to obtain $\tilde{\pi}=\pi^*\tilde{\lambda}/\lambda^*$. Equations (6.23′) and (6.24′) are automatically satisfied because of their linear homogeneity and because of the constancy of both λ^* and $\tilde{\lambda}$. We have thus found the "new" optimal solution, with x^*, s^*, and c^* being the same as before.

Since we have already found an optimal consumption path for $u=(1/\gamma)c^\gamma$, this procedure always yields the solution. Sufficiency is guaranteed because $\tilde{u}$ is concave. We can state our result more formally: in Example 6.3.1, the consumption and extraction policies are independent of the utility function. The constancy of the consumption path relies heavily on the Cobb–Douglas form of the production function. Indeed, any homogeneous and concave production function with a constant elasticity of substitution different from unity (hence, not a Cobb–Douglas production function) will yield a nonconstant consumption path. In fact, a slightly stronger assertion can be proved: within the class of homothetic and concave production functions $F(s,x)$, the optimal consumption path is a constant c^* only if the production function exhibits a unitary elasticity of substitution in some neighborhood of the set $\{(s,x): F(s,x)=c^*, s_0\le s\le s_T\}$. The proof of these two assertions is the subject of exercises at the end of this chapter.

6.4 The maximum principle with inequality constraints

We now turn our attention to the case in which the set of admissible controls can be described by m inequality constraints:

$$g^j(s_1(t),s_2(t),\dots,s_n(t),c_1(t),\dots,c_r(t),t)\ge 0, \quad j=1,2,\dots,m,$$

where each g^j function is continuously differentiable in all arguments. Our control problem is to find the vector of controls $\mathbf{c}(t)$ that maximizes

$$V=\int_0^T v(\mathbf{s}(t),\mathbf{c}(t),t)\,dt \tag{6.34}$$

subject to

$$\dot{s}_i(t) = f^i(\mathbf{s}(t), \mathbf{c}(t), t), \quad i = 1, 2, \ldots, n, \tag{6.35}$$

$$g^j(\mathbf{s}(t), \mathbf{c}(t), t) \geq 0, \quad j = 1, 2, \ldots, m, \tag{6.36}$$

$$s_i(0) = s_{i0}, \; s_i(T) = s_{iT}, \quad i = 1, 2, \ldots, n. \tag{6.37}$$

The necessary conditions for this problem are identical to those stated in Section 6.3 (Theorem 6.3.1), except that the set of admissible controls is now

$$W(\mathbf{s}^*(t), t) \equiv \{\mathbf{c}(t) \mid g^j(\mathbf{s}^*(t), \mathbf{c}(t), t) \geq 0, \; j = 1, \ldots, m\} \tag{6.38}$$

and the Lagrange multipliers $\lambda_1, \ldots, \lambda_m$ are nonnegative. Thus, the following familiar complementary slackness conditions hold:

$$\begin{aligned} &\lambda_j(t) \geq 0, \;\; g^j(\mathbf{s}^*(t), \mathbf{c}^*(t), t) \geq 0, \;\; \text{and} \\ &\lambda_j(t)\, g^j(\mathbf{s}^*(t), \mathbf{c}^*(t), t) = 0, \quad j = 1, \ldots, m. \end{aligned} \tag{6.38$'$}$$

(The reader may refer to Section 1.4, where the complementary slackness conditions are explained in detail.)[2]

For problems with a sufficiently simple inequality constraint and one control, it is sometimes possible to obtain a solution by using both the method of Chapter 4 when the constraint is not binding and the method of Section 6.3 when the constraint binds. The first example of this section falls in that category; subsequently, we present a more complex example.

Example 6.4.1: optimal consumption with irreversible investment. In Sections 4.4 and 4.5, we studied a problem of optimal consumption. We did not impose the restriction that consumption be no greater than output. Thus, along paths such as (i) in Figures 4.2, 4.3, and 4.6, consumption may exceed output, implying that the capital stock can be consumed directly (or converted into the consumption good without cost). In this section we wish to investigate the effect of the constraint that consumption not exceed output:

$$c(t) \leq F(s(t)).$$

We also adopt the assumption made in Section 4.5, that capital depreciates at the rate m, so that

$$\dot{s}(t) = F(s(t)) - c(t) - ms(t).$$

[2] For simplicity of notation, we shall adopt the convention of excluding from the Lagrangean any nonnegativity constraints on the control variables. As a consequence if the constraints contain inequalities such as $c_j(t) \geq 0$, the first-order condition (6.18′) must be modified to

$$\frac{\partial \mathcal{L}^*}{\partial c_j(t)} \leq 0, \quad c_j^*(t) \geq 0, \quad c_j^*(t) \frac{\partial \mathcal{L}^*}{\partial c_j(t)} = 0,$$

as for a Kuhn–Tucker problem.

We retain the assumption that instantaneous utility is a function of consumption and that this utility flow is discounted at a rate $\delta > 0$, with $u'(c) > 0$, $u'(0) = \infty$, $u''(c) < 0$; we also assume $F(0) = 0$, $F'(s) > 0$, and $F''(s) < 0$.

In addition, we shall assume as in Section 4.5 that $F'(0) > \delta + m > F'(\infty)$ so as to ensure the existence of a steady-state equilibrium point. As before, the assumption $u'(0) = \infty$ ensures that $c^*(t)$ is positive along an optimal path. We could have replaced this assumption by the requirement that $c(t) \geq \bar{c}$, where $\bar{c}$ is some prespecified nonnegative lower bound on consumption. We refrained from doing so in order to keep our first example as simple as possible.

Our problem is to find $c^*(t)$ that maximizes

$$V = \int_0^T u(c(t)) e^{-\delta t} \, dt \tag{6.39}$$

subject to

$$\dot{s}(t) = F(s(t)) - c(t) - ms(t),$$
$$F(s(t)) - c(t) \geq 0,$$
$$s(0) = s_0, \quad s(T) = s_T.$$

The Hamiltonian is

$$H(s, c, \pi, t) = e^{-\delta t} u(c(t)) + \pi(t)[F(s(t)) - c(t) - ms(t)],$$

and the Lagrangean is

$$\mathcal{L}(s, c, \pi, \lambda, t) = H(s, c, \pi, t) + \lambda(t)[F(s(t)) - c(t)].$$

The optimal solution must satisfy the following conditions (the asterisk is omitted where no ambiguity arises):

(i) $c^*(t)$ maximizes $H(s, c, \pi, t)$ subject to the constraint that $F(s(t)) - c(t) \geq 0$. In terms of the Lagrangean, this means

$$\frac{\partial \mathcal{L}}{\partial c} = e^{-\delta t} u'(c(t)) - \pi(t) - \lambda(t) = 0 \tag{6.40}$$

with

$$\lambda(t) \geq 0, \quad F(s(t)) - c(t) \geq 0, \quad \lambda(t)[F(s(t)) - c(t)] = 0. \tag{6.41}$$

(ii) $$\dot{\pi} = -\frac{\partial \mathcal{L}}{\partial s} = -[\pi(t) + \lambda(t)] F'(s(t)) + m\pi(t). \tag{6.42}$$

(iii) $$\dot{s} = \frac{\partial \mathcal{L}}{\partial \pi} = F(s(t)) - c(t) - ms(t). \tag{6.43}$$

Recall that in Chapter 4 we constructed phase diagrams in the (s, c) space after deriving a differential equation for the control variable (equations (4.61) or (4.85)). Unfortunately, that method will not work here because it would introduce a $\dot{\lambda}$ term that cannot be eliminated with (6.41). The way to deal with such problems is in general to construct a diagram in the (state, costate) space; this is done in the next example for a more complicated case. Here, however, there is a single control variable and the inequality constraint is a simple bound on the control. This makes it possible to proceed directly to a phase diagram in the (s, c) space. If the bound is inactive, we can use the same method as in Chapter 4; if the control is on the boundary, we make use of that information to obtain the solution. In what follows we apply this simple idea to the present example.

Let us consider the two cases separately: case A, in which the bound is active, and case B, in which it is not. In case B, the results are the same as in Section 4.5. Using equations (4.84) and (4.85),

$$\dot{s} = F(s) - ms - c,$$
$$\dot{c} = -[u'(c)/u''(c)][F'(s) - m - \delta],$$

the phase diagram is easy to draw. This is done in Figure 6.2. (It is similar to Figure 4.6, "capped" by $c = F(s)$.) Note that these phase lines are valid only when $c < F(s)$. When $\lambda > 0$, the equality holds and we are in case A, to which we now turn.

In case A the bound is active. This means that $c = F(s)$; hence, $\dot{s} = -ms$ and both $\dot{c}$ and $\dot{s}$ are negative. Therefore, starting from any point on the graph of $c = F(s)$ the optimal trajectory moves in the southwest direction. Let $\tilde{s}$ denote the value of s at the equilibrium. If $s < \tilde{s}$, this trajectory moves down along the graph of $c = F(s)$ and cannot stray away from it, since in the proximity of this graph, c must increase; this is displayed in Figure 6.2. If $s > \tilde{s}$, the optimal path follows the $c = F(s)$ graph downward; it may, however, leave this graph to enter region B, in which the constraint ceases to be active.

The choice of the optimal policy depends on the values of s_0, s_T, and T. We now have a good opportunity to enhance our understanding of dynamic optimality by comparing optimal trajectories with and without the constraint, for various specified values of the parameters. This is done in Figure 6.3, which is to be compared with Figure 4.6. We take it that the two problems are identical save for the constraint $c \leq F(s)$; in particular, in each case the values of s_0, s_T, and T are common to the two problems.

The case in which trajectory (i) of Figure (4.6) was optimal may now have no feasible solution. Indeed, if T is small enough, there may not be

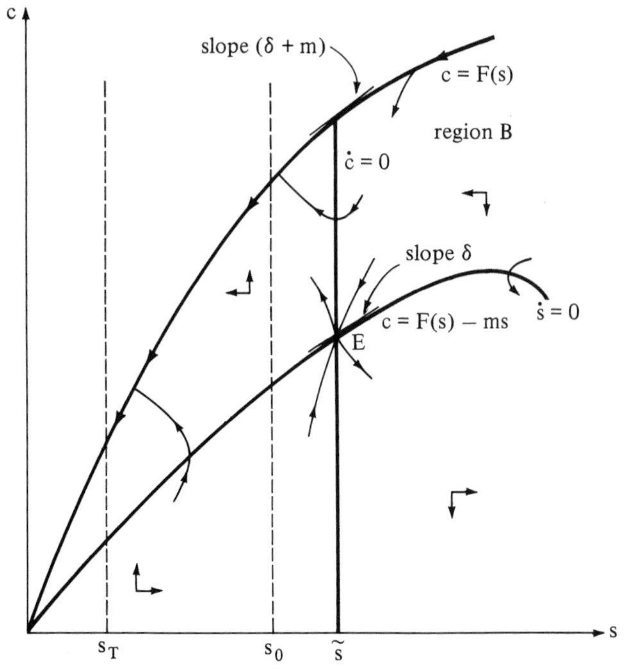

Figure 6.2

enough time to go from s_0 to s_T with the maximum consumption allowed $c = F(s)$.

When time is more plentiful but the optimal trajectory remains in region I with consumption rising all the while (path (ii)), we may now encounter the upper bound. If we were to follow the unconstrained path (ii) initially, we could not reduce the stock to s_T because when the constraint binds later on, the stock diminishes more slowly than along the unconstrained path. Therefore, we must start at a higher level of consumption initially. This results in the constrained optimal path labeled (ii'_c). These paths are plotted against time in Figure 6.4, where (ii_c) represents the (unfeasible) path that coincides with (ii) before reaching the bound. The kinks on consumption paths (ii_c) and (ii'_c) occur at the time the constraint begins to bind.

If the unconstrained path begins in region I and eventually crosses the curve $c = F(s)$, as (iii) does, we must now choose the slower trajectory (iii'_c). An interesting feature is that we now actually begin with a lower

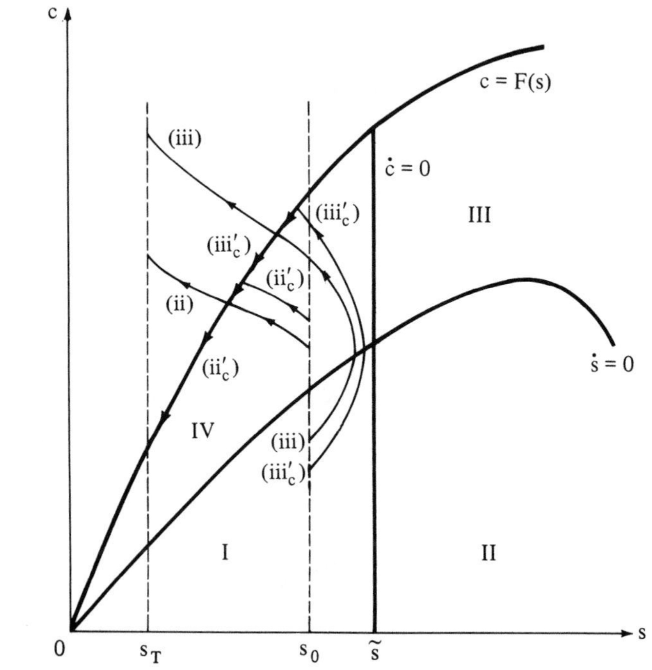

Figure 6.3

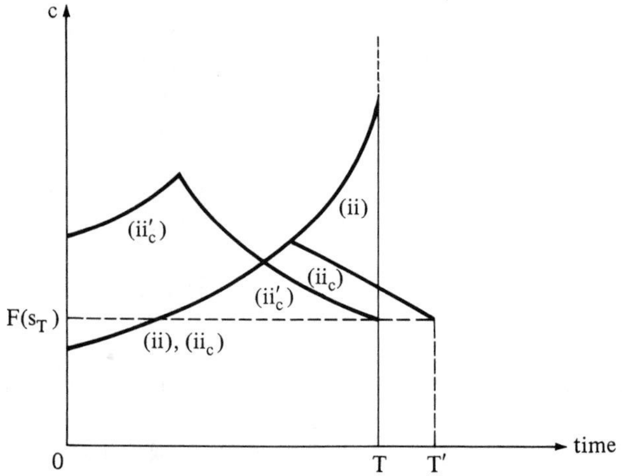

Figure 6.4

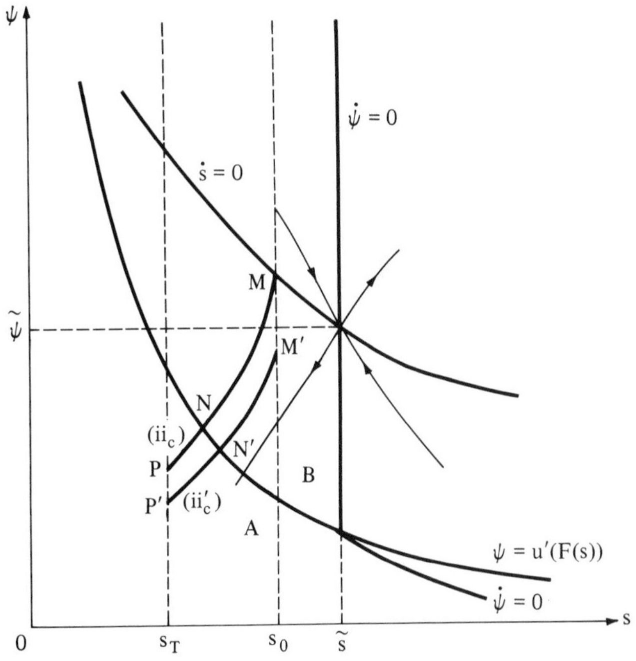

Figure 6.5

consumption. This serves to build up the stock of capital so that there is a much higher level of consumption in the intermediate period before consumption begins to decline. The imposition of the constraint thus has rather far-reaching effects on the optimal path. Whereas consumption was previously rising monotonically with time, it now may go through a peak. The planner compensates for lower consumption levels when the constraint is active with higher levels of consumption before that time; in order to reach these high levels at an earlier time, it must start with lower consumption at the beginning of the horizon in order to accumulate capital.

Before we leave this example let us mention that it is also possible to construct a phase diagram in the (s, ψ) space, where ψ is the current-value costate variable. The main obstacle is that it must be shown that the control and the multiplier can be expressed in terms of s and ψ only, so that the $(\dot{\psi}, \dot{s})$ system is autonomous. The reader will be guided through this task as an exercise at the end of this chapter. For now, let us simply display the result in Figure 6.5, which is comparable to Figure 4.3. There exists a unique equilibrium point $(\tilde{s}, \tilde{\psi})$, and it has the familiar saddle-point property. The differences between Figures 4.3 and 6.5 are now illustrated. Take, for example, the path *MNP* in Figure 6.5 and compare it with the path starting at the same point *M* in Figure 4.3. The two paths are denoted by (ii_c) and (ii), respectively, where the subscript c stands for "constrained." Clearly, both paths are identical along *MN*. However, once path (ii_c) enters region A the capital stock will fall at the rate ms, because consumption just equals output in that region. Along path (ii) consumption is increasing over time; hence, it exceeds output once the trajectory passes point *N*. Thus, (ii_c) takes a longer time (say, $T' > T$) to reach the prescribed final stock s_T. Therefore, for a common fixed time horizon T, if path (ii) is optimal for the unconstrained case, then path (ii_c) is not optimal for the constrained case, because it does not satisfy the boundary condition $s(T) = s_T$. The optimal path must therefore start with a higher rate of consumption; path (ii'_c) is one such path. These are the same consumption profiles illustrated in Figure 6.4.

Example 6.4.2: the mushroom grower's problem. In this example we model the husbanding of a resource, the growth of which can be controlled by both harvesting and cultivating. Let $s(t)$ denote the stock of mushrooms at time t. In the absence of harvesting for consumption or cultivation, this stock is assumed to grow at the rate $\dot{s} = \ln s(t)$. Let $c(t)$ denote consumption and $x(t)$ denote cultivation effort. We assume

$$\dot{s} = \ln s(t) - c(t) + F(x(t), s(t)),$$

where $F(x, s)$ takes the special form $F(x, s) = s[1 - e^{-x/s}]$. Note that this F function is nonnegatively valued, strictly increasing, homogeneous of degree 1, and concave in (x, s) when $x \geq 0$, $s \geq 0$. We assume that harvesting for consumption cannot exceed an upper bound, which depends on current stock s and the (fixed) quantity of harvesting equipment E; for simplicity we take the constraint to be $c \leq sE$, where $E = 1$. Consumption and cultivation effort must be nonnegative. The present value of instantaneous utility is assumed to be $v(c, x, t) = (\ln(c) - x)e^{-rt}$, where $r > 0$ is the discount rate. Recapitulating, the mushroom grower's problem is to find c and x that maximize

$$V = \int_0^T [\ln c - x] e^{-rt}\, dt \tag{6.44}$$

subject to

$$\dot{s} = \ln s - c + s[1 - e^{-x/s}], \tag{6.45}$$

$$c \leq s, \tag{6.46}$$

$$c \geq 0, \quad x \geq 0, \tag{6.47}$$

$$s(0)=s_0, \quad s(T)=s_T. \tag{6.48}$$

For concreteness we assume that $r<e^{-1}$; other cases will be discussed briefly. The current-value Hamiltonian and Lagrangean are, respectively,

$$H=\ln c-x+\psi[\ln s-c+s(1-e^{-x/s})], \tag{6.49}$$

$$\mathcal{L}=H+\lambda[s-c]. \tag{6.50}$$

Applying Theorem 6.3.1, modified as per equation (6.38′), we obtain the necessary conditions:

$$\frac{\partial \mathcal{L}}{\partial c}=c^{-1}-\psi-\lambda\leq 0, \quad c\geq 0, \; c(c^{-1}-\psi-\lambda)=0, \tag{6.51}$$

$$\frac{\partial \mathcal{L}}{\partial x}=-1+\psi e^{-x/s}\leq 0, \quad x\geq 0, \; x(-1+\psi e^{-x/s})=0, \tag{6.52}$$

$$\frac{\partial \mathcal{L}}{\partial \lambda}=s-c\geq 0, \quad \lambda\geq 0, \; \lambda(s-c)=0, \tag{6.53}$$

$$\dot{\psi}=r\psi-\frac{\partial \mathcal{L}}{\partial s}=\psi\left(r-s^{-1}-1+e^{-x/s}+\frac{x}{s}e^{-x/s}\right)-\lambda, \tag{6.54}$$

$$\dot{s}=\frac{\partial \mathcal{L}}{\partial \psi}=\ln s-c+s(1-e^{-x/s}). \tag{6.55}$$

Because there are two control variables constrained by inequalities, it is not possible to construct a two-dimensional phase diagram in the (state, control) space, and so we shall solve this problem in the (s, ψ) space.

First note that from (6.51), c is always positive as $\lim c^{-1}=+\infty$ when $c\to 0^+$. We can therefore use instead

$$c^{-1}=\psi+\lambda. \tag{6.51′}$$

We shall divide the nonnegative orthant in (s, ψ) into regions according to whether or not the constraints on the controls ($c\leq s$ and $x\geq 0$) bind. These and other features will be added one by one to make up Figure 6.6. The reader is invited to duplicate this building process on a separate graph:

(a) We define region A by $\lambda=0$. Therefore, by (6.53), $s\geq c$, and by (6.51′), $c=\psi^{-1}$; hence, $s\geq\psi^{-1}$, or $\psi\geq s^{-1}$ characterizes region A in Figure 6.6.
(b) Define region B by $\lambda>0$. Therefore, $s=c$ by (6.53), and (6.51′) yields $s^{-1}-\psi>0$, or $\psi<s^{-1}$, which characterizes region B.
(c) In region C, $x>0$, and by (6.52), $\psi=e^{x/s}>1$; thus, $\psi>1$ characterizes region C.

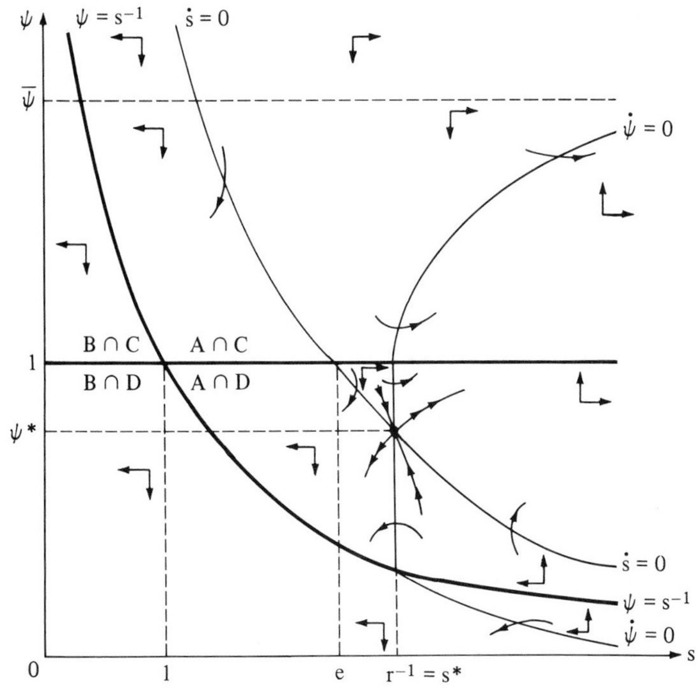

Figure 6.6

(d) In region D, $x = 0$, and by (6.52) $\psi \leq e^{-0/s} = 1$; thus, $\psi \leq 1$ characterizes region D.

The economic significance of these regions is of interest. When the shadow price ψ is high relative to the cost of effort, $(\psi > 1)$, it pays to cultivate the mushrooms $(x > 0)$; when ψ is low, no cultivation takes place. The extent to which mushrooms are harvested for consumption depends on two factors: the price ψ of the mushroom stock and the availability of mushrooms. If ψ is high and/or the available stock s is high, the inequality $\psi \geq s^{-1}$ will tend to hold; in this case mushrooms are not harvested at the maximum level $(c \leq s)$, although the size of the harvest may be very great if s is large. A low price and/or a meager stock of mushrooms will tend to imply $\psi < s^{-1}$ and $s = c$.

Clearly, the rectangular hyperbola $\psi = s^{-1}$ separating regions A and B, and the straight line $\psi = 1$ separating regions C and D, together delineate four sectors ($A \cap C$, $B \cap D$, etc.). The derivation of the $\dot{\psi} = 0$ locus and of the $\dot{s} = 0$ locus must proceed sector by sector.

Derivation of the $\dot{\psi}=0$ *locus*

(i) In sector $A \cap C$ the characteristics from (a) and (c) above are combined: $\lambda=0$, $e^{x/s}=\psi$. Substituting these in (6.54) we obtain

$$\dot{\psi}=\psi(r-s^{-1}-1+\psi^{-1}+\psi^{-1}\ln\psi). \tag{6.56}$$

Thus, $\dot{\psi}=0$ if and only if

$$g(s,\psi)\equiv r-s^{-1}-1+\psi^{-1}+\psi^{-1}\ln\psi=0, \tag{6.57}$$

since $\psi=0$ does not belong to this sector. Equation (6.57) yields $d\psi/ds=-g_s/g_\psi=\psi^2/(s^2\ln\psi)>0$; hence, $g(s,\psi)=0$ defines ψ as a strictly increasing function of s. Its graph passes through the point $(s=r^{-1},\ \psi=1)$; as $s\to+\infty$, $\psi\to(\bar{\psi})^-$, where $\bar{\psi}>1$ is obtained by writing (6.57) as $s=\psi/[(1+\ln\psi)-\psi(1-r)]$, and setting the denominator to zero. (A rough graph of $(1+\ln\psi)$ against $(1-r)\psi$ will convince the reader that $\bar{\psi}>1$ is unique.) Furthermore, $d\psi/ds=+\infty$ at $(s=r^{-1},\ \psi=1)$. Since $\partial g/\partial s=s^{-2}>0$, we have $\dot{\psi}>0$ to the right of the $g(s,\psi)=0$ curve and $\dot{\psi}<0$ to the left of it.

(ii) In sector $A\cap D$ we have, from (a) and (d) above, $\lambda=0$ and $x=0$; hence, (6.54) becomes

$$\dot{\psi}=\psi(r-s^{-1}). \tag{6.58}$$

Therefore, $\dot{\psi}=0$ along the vertical line $s=r^{-1}$, $\dot{\psi}>0$ to the right of it, and $\dot{\psi}<0$ to the left.

(iii) In sector $B\cap D$, gathering the information from (b) and (d), we have $x=0$, $c=s$, and $\lambda>0$; (6.51′) yields $\lambda=s^{-1}-\psi>0$ and (6.54) becomes

$$\dot{\psi}=\psi(r+1-s^{-1})-s^{-1}. \tag{6.59}$$

Therefore, $\dot{\psi}=0$ along the curve $\psi=((r+1)s-1)^{-1}$, which intersects the boundary curve $\psi=s^{-1}$ at $s=r^{-1}$ and goes to zero as s goes to infinity. It is easy to verify that the $\dot{\psi}=0$ locus remains below the boundary curve as s increases. Again $\dot{\psi}>0$ to the right of the curve as $d\dot{\psi}/ds>0$ in (6.59).

(iv) In sector $B\cap C$, $\psi=e^{x/s}$ and $\lambda>0$; hence,

$$\dot{\psi}=\psi(r-s^{-1}-1+\psi^{-1}+\psi^{-1}\ln\psi)-\lambda=\psi g(s,\psi)-\lambda. \tag{6.60}$$

We know from calculations in (i) that $g(s,\psi)<0$ to the left of the $g(s,\psi)=0$ graph. Therefore, $\dot{\psi}<0$ everywhere in sector $B\cap C$.

We now turn our attention to the locus of $\dot{s}=0$:

(i′) In sector $A \cap C$, $\lambda = 0$ and $x > 0$; hence, $c = \psi^{-1}$ and $\psi = e^{x/s} > 1$; these yield with (6.55)

$$\dot{s} = \ln s - \psi^{-1} + s(1 - \psi^{-1}) \equiv \phi(s, \psi). \tag{6.61}$$

The $\dot{s} = 0$ locus is described by $\psi = (1+s)/(s + \ln s)$. This curve passes through $(s = e, \psi = 1)$ and as $s \to \bar{s}^+$, $\psi \to +\infty$, where $\bar{s} \simeq 0.567$ is defined by $\bar{s} + \ln \bar{s} = 0$. Since ψ is positive, $s + \ln s$ is also positive; also, since $\psi > 1$, we have $\ln s < 1$; therefore, $d\psi/ds = (\ln s - 2 - s^{-1})/(s + \ln s)^2 < 0$, and the $\dot{s} = 0$ locus is negatively sloped. Finally, note that in this sector $\psi = (1+s)/(s + \ln s) > s^{-1}$ as $s^2 > \ln s$ and the $\dot{s} = 0$ locus lies above the $\psi = s^{-1}$ curve; it begins at $(s = e, \psi = 1)$ and goes up to infinity as s approaches $\bar{s}$ from above. In Figure 6.6 we have represented this locus under our assumption that $r < e^{-1}$. Since $\partial\phi/\partial s > 0$, as $\psi > 1$ in this sector, we see that $\dot{s} > 0$ to the right of the locus.

(ii′) In sector $A \cap D$, $x = 0$ and $\lambda = 0$; hence, $c = \psi^{-1}$ and (6.55) becomes

$$\dot{s} = \ln s - \psi^{-1}. \tag{6.62}$$

The $\dot{s} = 0$ locus is $\psi = (\ln s)^{-1}$, which passes through $(s = e, \psi = 1)$ and goes to zero as s goes to infinity. The curve $\psi = (\ln s)^{-1}$ is above the boundary curve $\psi = s^{-1}$, since $s > \ln s$. The signs of $\dot{s}$ are clear.

(iii′) In sector $B \cap D$, $x = 0$ and $c = s$; hence,

$$\dot{s} = \ln s - s < 0. \tag{6.63}$$

(iv′) In sector $B \cap C$, $c = s$ and $\psi = e^{x/s}$ and

$$\dot{s} = \ln s - s\psi^{-1} < 0, \tag{6.64}$$

since $\psi > 1$ and $s^{-1} > \psi$ imply $\ln s < 0$.

The diagram in Figure 6.6 is now complete, and we see that there exists a unique equilibrium in the $A \cap D$ sector. (Had we assumed that $e^{-1} < r < 1$ or $1 < r$, we would have obtained in either case an equilibrium in the $A \cap C$ sector. The reader is invited to rework the derivation of the $\dot{\psi} = 0$ locus in these cases; the $\dot{s} = 0$ locus is unaffected.) By inspection we see that the equilibrium is a saddle point, but we can confirm it locally by linearizing the system (6.54)–(6.55) around the equilibrium point $(s^* = r^{-1}, \psi^* = (\ln s^*)^{-1})$. We obtain from (6.62) and (6.58)

$$\dot{s} = \ln s - \psi^{-1},$$

$$\dot{\psi} = \psi(r - s^{-1}),$$

and

$$\begin{bmatrix} \dot{s} \\ \dot{\psi} \end{bmatrix} = \begin{bmatrix} (s^*)^{-1} & (\psi^*)^{-2} \\ \psi^*(s^*)^{-2} & 0 \end{bmatrix} \begin{bmatrix} s-s^* \\ \psi-\psi^* \end{bmatrix}. \tag{6.65}$$

The determinant of this matrix is negative, which characterizes a saddle point.

Note that at the equilibrium, the derivative of the *natural* growth rate of mushrooms with respect to s $((d/ds)\ln(s)=s^{-1})$ equals the rate of discount, r. This characterization has its parallel in the optimal growth model in which $F'(s)-m=\delta$ at equilibrium.

A variety of possible patterns emerges, depending on the values T, s_0, and s_T. We have not drawn these so as not to complicate the diagram, but the reader can easily do this. With $s_0<r^{-1}$, $s_T>r^{-1}$, and T large enough, the optimal path would go from sector $A\cap C$ into sector $A\cap D$, above the $\dot{s}$ locus, and perhaps back into sector $A\cap C$, below the $\dot{\psi}=0$ locus. Along this trajectory, x would go from positive, to zero, to positive again, while the upper bound on c would never be reached. In another instance, with s_0 and s_T both below r^{-1}, the optimal path may again enter sector $A\cap D$ from sector $A\cap C$ above the $\dot{s}=0$ locus, but then cross this locus to proceed to the $B\cap D$ sector, where the constraint $c\leq s$ binds. Many other scenarios may be envisaged, and it is important to keep track of the behavior of the control variables relative to the constraints on them.

It is obvious that we could not have (autonomous) phase diagrams in either the (s,c) or the (s,x) space. We can, of course, construct a phase diagram in (s,c) when $x=0$ and a diagram in (s,x) when $c=s$ (these correspond to regions D and B, respectively), but we cannot construct a two-dimensional diagram when $c<s$ and $x>0$; this is sector $A\cap C$. Therefore, in models such as the one treated in this example, with more than one control and inequality constraints on them that depend on the state variables, there is no alternative but to work in the (state, costate) space, although the phase diagrams we just mentioned may be of some interest.

6.5 Necessity and sufficiency theorems: the case with inequality and equality constraints

Many optimal control problems in economics involve both equality and inequality constraints. This type of problem creates no new difficulties. The necessary conditions are a straightforward generalization of those stated in the preceding sections. We shall state them here for easy reference. A sufficiency theorem is also stated and proved for this more general case.

The general constrained control problem with fixed endpoints consists of finding the optimal path $\mathbf{c}^*(t)$ for the control variables so as to maximize

$$V=\int_0^T v(\mathbf{s}(t),\mathbf{c}(t),t)\,dt \tag{6.66}$$

subject to n differential equations,

$$\dot{s}_i(t)=f^i(\mathbf{s}(t),\mathbf{c}(t),t), \quad i=1,2,\dots,n, \tag{6.67}$$

m' inequality constraints,

$$g^j(\mathbf{s}(t),\mathbf{c}(t),t)\geq 0, \quad j=1,2,\dots,m', \tag{6.68}$$

$m-m'$ equality constraints,

$$g^k(\mathbf{s}(t),\mathbf{c}(t),t)=0, \quad k=m'+1,\dots,m, \tag{6.69}$$

and $2n$ boundary conditions,

$$s_i(0)=s_{i0}, \ s_i(t)=s_{iT}, \quad i=1,2,\dots,n. \tag{6.70}$$

The time horizon T and the initial and terminal values of the state variables are exogenously specified.

As discussed in Section 6.1, we shall assume that constraints (6.68) and (6.69) satisfy the rank condition. We shall use the notation $W(\mathbf{s}(t),t)$ to denote the set of values of control variables that satisfy (6.68) and (6.69) simultaneously.

For problem (6.66)–(6.70) we define a Hamiltonian,

$$H(\mathbf{s}(t),\mathbf{c}(t),\boldsymbol{\pi}(t),t)\equiv v(\mathbf{s}(t),\mathbf{c}(t),t)+\sum_{i=1}^{n}\pi_i(t)f^i(\mathbf{s}(t),\mathbf{c}(t),t), \tag{6.71}$$

and a Lagrangean,

$$\mathcal{L}(\mathbf{s}(t),\mathbf{c}(t),\boldsymbol{\pi}(t),\boldsymbol{\lambda}(t),t)\equiv H(\mathbf{s}(t),\mathbf{c}(t),\boldsymbol{\pi}(t),t) + \sum_{j=1}^{m}\lambda_j(t)g^j(\mathbf{s}(t),\mathbf{c}(t),t), \tag{6.72}$$

where $\lambda_1(t),\dots,\lambda_m(t)$ are multipliers.

Theorem 6.5.1: necessity. Let $\mathbf{c}^*(t)$ be an optimal solution to the constrained problem (6.66)–(6.70) and $\mathbf{s}^*(t)$ be the corresponding time path of the state variables. Then there exist costate variables $\boldsymbol{\pi}(t)$ and (assuming the rank condition is satisfied) multipliers $\boldsymbol{\lambda}(t)$ such that:

(i) At any time t, for given vectors $\mathbf{s}^*(t)$ and $\boldsymbol{\pi}(t)$, the control variable vector $\mathbf{c}^*(t)$ maximizes the Hamiltonian (6.71) subject to the condition that $\mathbf{c}(t)$ belong to the set of admissible controls defined by (6.68) and (6.69). In view of the rank condition this implies that there exist multipliers $\boldsymbol{\lambda}(t)$ such that

$$\frac{\partial \mathcal{L}^*}{\partial c_i}=0, \quad i=1,2,\ldots,r, \tag{6.73}$$

$$\lambda_j(t)\geq 0,\ g^j(\mathbf{s}^*(t),\mathbf{c}^*(t),t)\geq 0,\ \lambda_j(t)g^j(\mathbf{s}^*(t),\mathbf{c}^*(t),t)=0, \quad j=1,2,\ldots,m', \tag{6.74a}$$

$$g^k(\mathbf{s}^*(t),\mathbf{c}^*(t),t)=0, \quad k=m'+1,\ldots,m, \tag{6.74b}$$

where the asterisk on $\mathcal{L}$ indicates that the derivatives are evaluated at $(\mathbf{s}^*(t),\mathbf{c}^*(t))$. The multipliers $\boldsymbol{\lambda}(t)$ are piecewise-continuous and continuous on each point of continuity of $\mathbf{c}^*(t)$.

(ii) The costate variables $\pi_i(t)$, $i=1,2,\ldots,n$, are continuous and have piecewise-continuous derivatives satisfying

$$\dot{\pi}_i(t)=-\frac{\partial \mathcal{L}^*}{\partial s_i(t)}, \quad i=1,2,\ldots,n. \tag{6.75}$$

(iii) $$\dot{s}_i^*(t)=\frac{\partial \mathcal{L}^*}{\partial \pi_i(t)}=f^i(\mathbf{s}^*(t),\mathbf{c}^*(t),t), \quad i=1,2,\ldots,n. \tag{6.76}$$

(iv) The Lagrangean $\mathcal{L}(\mathbf{s}^*(t),\mathbf{c}^*(t),\boldsymbol{\pi}(t),\boldsymbol{\lambda}(t),t)\equiv\phi(t)$ is a continuous function of t. On each interval of continuity of $\mathbf{c}^*(t)$, $\phi(t)$ is differentiable and

$$\phi'(t)\equiv\frac{d\mathcal{L}^*}{dt}=\frac{\partial \mathcal{L}^*}{\partial t}. \tag{6.77}$$

(v) The boundary conditions (6.70) must be satisfied.

A heuristic proof of Theorem 6.5.1 using the approach adopted in Section 4.5 is possible but slightly more involved. It is left to the interested reader. The necessity of (6.77) follows from (6.73)–(6.76). To see this, differentiate $\mathcal{L}$ totally with respect to t:

$$\frac{d\mathcal{L}^*}{dt}=\sum\frac{\partial \mathcal{L}^*}{\partial s_i}\frac{ds_i^*}{dt}+\sum\frac{\partial \mathcal{L}^*}{\partial c_i}\frac{dc_i^*}{dt}+\sum\frac{\partial \mathcal{L}^*}{\partial \pi_i}\frac{d\pi_i}{dt}+\sum\frac{\partial \mathcal{L}^*}{\partial \lambda_i}\frac{d\lambda_i}{dt}+\frac{\partial \mathcal{L}^*}{\partial t}.$$

But $\partial\mathcal{L}^*/\partial c_i=0$ by (6.73) and $(\partial\mathcal{L}^*/\partial s_i)\dot{s}_i^*=-(\partial\mathcal{L}^*/\partial\pi_i)\dot{\pi}_i$ by (6.75) and (6.76). It remains to show that $(\partial\mathcal{L}^*/\partial\lambda_i)(d\lambda_i/dt)=0$. If

$$g^i(\mathbf{s}^*(t),\mathbf{c}^*(t),\lambda)=0,$$

then $\partial\mathcal{L}^*/\partial\lambda_i=g^i(\mathbf{s}^*(t),\mathbf{c}^*(t),\lambda)=0$; if $g^i(\mathbf{s}^*(t),\mathbf{c}^*(t),t)>0$, then $\lambda_i(t)=0$ and hence $d\lambda_i/dt=0$. This completes our proof of the necessity of (6.77). □

We now show that if the Lagrangean is concave in the variables $(\mathbf{s}, \mathbf{c})$, then the necessary conditions just stated are also sufficient. The proof is much the same as that offered in Section 4.6. Let $(\mathbf{s}^*(t), \mathbf{c}^*(t))$ be a program that satisfies all the necessary conditions and let $(\boldsymbol{\pi}^*(t), \boldsymbol{\lambda}^*(t))$ be the associated costate variables and Lagrange multipliers. The asterisk for $\boldsymbol{\pi}$ and $\boldsymbol{\lambda}$ was suppressed in our statement of the necessary conditions so as to simplify the notation; it is reintroduced here, because we wish to emphasize that the function H, described for any feasible program $(\mathbf{s}(t), \mathbf{c}(t))$, is defined using the same values $(\boldsymbol{\pi}^*(t), \boldsymbol{\lambda}^*(t))$ that were found in the optimal program. Thus,

$$H^* \equiv v(\mathbf{s}^*(t), \mathbf{c}^*(t), t) + \sum \pi_i^*(t) f^i(\mathbf{s}^*(t), \mathbf{c}^*(t), t) \equiv v^* + \boldsymbol{\pi}^* \cdot \mathbf{f}^*, \tag{6.78}$$

$$\mathcal{L}^* \equiv H^* + \sum \lambda_j^*(t) g^j(\mathbf{s}^*(t), \mathbf{c}^*(t), t) \equiv v^* + \boldsymbol{\pi}^* \cdot \mathbf{f}^* + \boldsymbol{\lambda}^* \cdot \mathbf{g}^*, \tag{6.79}$$

$$H \equiv v(\mathbf{s}(t), \mathbf{c}(t), t) + \sum \pi_i^*(t) f^i(\mathbf{s}(t), \mathbf{c}(t), t) \equiv v + \boldsymbol{\pi}^* \cdot \mathbf{f}, \tag{6.80}$$

and

$$\mathcal{L} \equiv H + \sum \lambda_j^*(t) g^j(\mathbf{s}(t), \mathbf{c}(t), t) \equiv v + \boldsymbol{\pi}^* \cdot \mathbf{f} + \boldsymbol{\lambda}^* \cdot \mathbf{g}, \tag{6.81}$$

where $(\mathbf{s}, \mathbf{c})$ refer to any program satisfying (6.67)–(6.70) on $[0, T]$. Thus, $\mathcal{L}^*$ is the function $\mathcal{L}$ evaluated at $(\mathbf{s}, \mathbf{c}) = (\mathbf{s}^*, \mathbf{c}^*)$. For simplicity, the notation $\boldsymbol{\pi}^* \cdot \mathbf{f}$ denotes the inner product of the vector $\boldsymbol{\pi}^* = (\pi_1^*, \pi_2^*, \ldots, \pi_n^*)$ and the vector $\mathbf{f} = (f^1, f^2, \ldots, f^n)$, where f^i stands for $f^i(\mathbf{s}(t), \mathbf{c}(t), t)$. Similarly, the symbol $(\mathbf{c}^* - \mathbf{c}) \cdot \partial\mathcal{L}^*/\partial\mathbf{c}$ denotes the inner product of the vector

$$(\mathbf{c}^* - \mathbf{c}) \equiv (c_1^*(t) - c_1(t), \ldots, c_n^*(t) - c_n(t))$$

and the vector $(\partial\mathcal{L}/\partial c_1, \ldots, \partial\mathcal{L}/\partial c_n)$, where the derivatives are evaluated at $(\mathbf{s}^*(t), \mathbf{c}^*(t), \boldsymbol{\pi}^*(t), \boldsymbol{\lambda}^*(t), t)$.

Theorem 6.5.2: sufficiency. Let $(\mathbf{s}^*(t), \mathbf{c}^*(t))$ satisfy the conditions of Theorem 6.5.1 and assume that the Lagrangean (6.81) is concave in $(\mathbf{s}, \mathbf{c})$; then $(\mathbf{s}^*(t), \mathbf{c}^*(t))$ is an optimal path for the problem (6.66)–(6.70). If $\mathcal{L}$ is strictly concave, $(\mathbf{s}^*(t), \mathbf{c}^*(t))$ is the unique optimal solution.

Proof. This is a straightforward generalization of the proof offered in Section 4.6:

$$V^* - V = \int_0^T (v^* - v)\, dt$$

$$= \int_0^T [(H^* - \boldsymbol{\pi}^* \cdot \dot{\mathbf{s}}^*) - (H - \boldsymbol{\pi}^* \cdot \dot{\mathbf{s}})]\, dt$$

$$= \int_0^T [(H^* + \dot{\boldsymbol{\pi}}^* \cdot \mathbf{s}^*) - (H + \dot{\boldsymbol{\pi}}^* \cdot \mathbf{s})]\, dt + [\mathbf{s}^*(0) \cdot \boldsymbol{\pi}^*(0) - \mathbf{s}^*(T) \cdot \boldsymbol{\pi}^*(T)]$$

$$- [\mathbf{s}(0) \cdot \boldsymbol{\pi}^*(0) - \mathbf{s}(T) \cdot \boldsymbol{\pi}^*(T)] \quad \text{(by integration by parts)}$$

$$= \int_0^T [(H^* + \dot{\pi}^* \cdot \mathbf{s}^*) - (H + \dot{\pi}^* \cdot \mathbf{s})]\, dt$$

(because $\mathbf{s}(0) = \mathbf{s}^*(0) = \mathbf{s}_0$ and $\mathbf{s}(T) = \mathbf{s}^*(T) = \mathbf{s}_T$)

$$\geq \int_0^T [(H^* + \lambda^* \cdot \mathbf{g}^*) - (H + \lambda^* \cdot \mathbf{g}) + \dot{\pi}^* \cdot (\mathbf{s}^* - \mathbf{s})]\, dt$$

(since $\lambda_j^* g_j^* = 0$ and $\lambda_j^* g_j \geq 0$)

$$= \int_0^T [(\mathcal{L}^* - \mathcal{L}) + \dot{\pi}^* \cdot (\mathbf{s}^* - \mathbf{s})]\, dt$$

$$\geq \int_0^T \left[(\mathbf{c}^* - \mathbf{c}) \cdot \frac{\partial \mathcal{L}^*}{\partial \mathbf{c}} + (\mathbf{s}^* - \mathbf{s}) \cdot \frac{\partial \mathcal{L}^*}{\partial \mathbf{s}} + \dot{\pi}^* \cdot (\mathbf{s}^* - \mathbf{s}) \right] dt$$

(by the concavity of the Lagrangean)

$$= 0$$

(because the asterisked solution satisfies the maximum principle).

If the Lagrangean is strictly concave in $(\mathbf{s}, \mathbf{c})$, then $V^* > V$ and the optimal solution is unique. □

Corollary 6.5.1. The concavity of the Lagrangean is ensured if the following conditions are met:

(i) v is concave in $(\mathbf{s}, \mathbf{c})$;
(ii) each term $\pi_i^*(t) f^i(\mathbf{s}, \mathbf{c}, t)$ is concave in $(\mathbf{s}, \mathbf{c})$; thus, if $f^i(\mathbf{s}, \mathbf{c}, t)$ is concave (resp. convex), the condition is satisfied, provided that $\pi_i^*(t) \geq 0$ (resp. ≤ 0);[3]
(iii) each of the m' inequality constraints $g^j(\mathbf{s}, \mathbf{c}, t) \geq 0$ is concave in $(\mathbf{s}, \mathbf{c})$ (recall that $\lambda_j^*(t) \geq 0$ for these inequality constraints);
(iv) each of the $m - m'$ equality constraints $g^k(\mathbf{s}, \mathbf{c}, t) = 0$ has the property that $\lambda_k^*(t) g^k(\mathbf{s}, \mathbf{c}, t)$ is concave in $(\mathbf{s}, \mathbf{c})$; thus, if g^k is concave (resp. convex), the condition is satisfied provided that $\lambda_k^*(t) \geq 0$ (resp. ≤ 0).

Just as we did at the end of Chapter 4, we now use the maximized Hamiltonian to restate the maximum principle and present a more general sufficiency theorem that places restrictions on the maximized Hamiltonian.

Definition 6.5.1. For the problem (6.66)–(6.70) we define the *maximized Hamiltonian* as $H^0(\mathbf{s}(t), \pi(t), t) = \max_{\mathbf{c}(t)} H(\mathbf{s}(t), \mathbf{c}(t), \pi(t), t)$ subject to (6.68)–(6.69), where H is as in (6.71) and $\mathbf{c}(t)$ is piecewise-continuous.

This is a formal definition; some necessary conditions for $\mathbf{c}(t)$ to be optimally chosen are given in (6.73)–(6.74); presumably these conditions can

[3] For conditions ensuring that $\pi_i(t) \geq 0$, see Léonard (1981).

be used to obtain $\mathbf{c}(t)$ in terms of $\mathbf{s}(t)$, $\boldsymbol{\pi}(t)$, and t, and this can be substituted in the Hamiltonian.

The maximum principle can be restated.

Theorem 6.5.3: necessity. Let $\mathbf{c}^*(t)$ be an optimal solution to the problem (6.66)–(6.70) and $\mathbf{s}^*(t)$ be the corresponding path of the state variables. Then there exist costate variables $\boldsymbol{\pi}(t)$ such that

(i)
$$H^0(\mathbf{s}^*(t), \boldsymbol{\pi}(t), t) \equiv H(\mathbf{s}^*(t), \mathbf{c}^*(t), \boldsymbol{\pi}(t), t) = \max_{\mathbf{c}(t)} H(\mathbf{s}^*(t), \mathbf{c}(t), \boldsymbol{\pi}(t), t) \qquad (6.82a)$$

subject to (6.68)–(6.69), where H is as in (6.71) and $\mathbf{c}(t)$ is piecewise-continuous;

(ii) the costate variables $\pi_i(t)$, $i = 1, \ldots, n$, have piecewise-continuous derivatives satisfying

$$\dot{\pi}_i(t) = -\frac{\partial H^0(\mathbf{s}^*(t), \boldsymbol{\pi}(t), t)}{\partial s_i(t)}, \quad i = 1, \ldots, n; \qquad (6.82b)$$

(iii)
$$\dot{s}_i^*(t) = \frac{\partial H^0(\mathbf{s}^*(t), \boldsymbol{\pi}(t), t)}{\partial \pi_i(t)}, \quad i = 1, \ldots, n; \qquad (6.82c)$$

(iv) the maximized Hamiltonian $H^0(\mathbf{s}^*(t), \boldsymbol{\pi}(t), t) \equiv \phi(t)$ is a continuous function of t. On each interval of continuity of $\mathbf{c}^*(t)$, $\phi(t)$ is differentiable and

$$\phi'(t) \equiv \frac{dH^0}{dt} = \frac{\partial H^0}{\partial t}; \qquad (6.82d)$$

(v) the boundary conditions (6.70) are satisfied. (6.82e)

Remark. The main differences in the forms of Theorem 6.5.1 and Theorem 6.5.3 occur between conditions (6.75)–(6.76) and conditions (6.82b)–(6.82c). It is instructive to verify that these are indeed equivalent to one another. The alert reader will notice the strong similarities between this argument and the proof of Theorem 1.2.8. Indeed, this is but another instance of the envelope theorem, where the derivatives of the maximum value function H^0 are equal to those of the Lagrangean $\mathcal{L}$.

Let us define a function $\mathbf{c} = \boldsymbol{\theta}^0(\mathbf{s}, \boldsymbol{\pi}, t)$ by conditions (6.73) and (6.74); that is, $\boldsymbol{\theta}^0$ represents the optimal value of $\mathbf{c}$, given $\mathbf{s}$, $\boldsymbol{\pi}$, and t, and taking into account the constraints (6.68) and (6.69). Then $H^0(\mathbf{s}, \boldsymbol{\pi}, t) = H(\mathbf{s}, \boldsymbol{\theta}^0, \boldsymbol{\pi}, t)$ and (6.82b) can be expressed as

$$\dot{\boldsymbol{\pi}} = \frac{-\partial H^0}{\partial \mathbf{s}} = \frac{-\partial H(\mathbf{s}, \boldsymbol{\theta}^0, \boldsymbol{\pi}, t)}{\partial \mathbf{s}} + \frac{\partial (\boldsymbol{\theta}^0)'}{\partial \mathbf{s}} \cdot \left(\frac{-\partial H(\mathbf{s}, \boldsymbol{\theta}^0, \boldsymbol{\pi}, t)}{\partial \mathbf{c}} \right)$$

$$= -\frac{\partial H}{\partial \mathbf{s}} + \frac{\partial (\boldsymbol{\theta}^0)'}{\partial \mathbf{s}} \cdot \sum_{j=1}^{m} \lambda_j \frac{\partial g^j(\mathbf{s}, \boldsymbol{\theta}^0, t)}{\partial \mathbf{c}} \quad \text{by (6.73).}$$

All the constraints for which $g^j > 0$ have $\lambda_j = 0$ and can be ignored. For those that hold as equalities, the controls have been chosen, for given $\mathbf{s}$ and $\boldsymbol{\pi}$, such that g^j remains zero. Hence, for given $\boldsymbol{\pi}$, $dg^j = 0$ implies

$$\frac{\partial g^j(\mathbf{s}, \theta^0, t)}{\partial \mathbf{s}} + \frac{\partial (\theta^0)'}{\partial \mathbf{s}} \bullet \left(\frac{\partial g^j(\mathbf{s}, \theta^0, t)}{\partial \mathbf{c}} \right) = \mathbf{0}$$

and we have

$$\dot{\boldsymbol{\pi}} = \frac{-\partial H(\mathbf{s}, \theta^0, \boldsymbol{\pi}, t)}{\partial \mathbf{s}} - \sum_{j=1}^{m} \lambda_j \frac{\partial g^j(\mathbf{s}, \theta^0, t)}{\partial \mathbf{s}},$$

which is (6.75). Briefly, we do the same for (6.76),

$$\dot{\mathbf{s}} = \frac{\partial H^0}{\partial \boldsymbol{\pi}} = \frac{\partial H}{\partial \boldsymbol{\pi}} + \frac{\partial (\theta^0)'}{\partial \boldsymbol{\pi}} \bullet \frac{\partial H}{\partial \mathbf{c}} = \frac{\partial H}{\partial \boldsymbol{\pi}} - \sum_{j=1}^{m} \lambda_j \frac{\partial (\theta^0)'}{\partial \boldsymbol{\pi}} \bullet \frac{\partial g^j}{\partial \mathbf{c}} = \frac{\partial H}{\partial \boldsymbol{\pi}},$$

since

$$\frac{\partial (\theta^0)'}{\partial \boldsymbol{\pi}} \bullet \frac{\partial g^j(\mathbf{s}, \theta^0, t)}{\partial \mathbf{c}} = \mathbf{0} \quad (\text{when } g^j = 0).$$

Example 6.5.1. We now return to the mushroom grower's problem (Example 6.4.2) and derive equations (6.56) and (6.58)–(6.64) using the maximized Hamiltonian. Note that the controls are (c, x) and the current-value costate is ψ. The multipliers will be (λ, μ) because we need to attach a multiplier $\mu \geq 0$ to the constraint $x \geq 0$ and the Lagrangean of (6.50) is replaced by $\mathcal{L} = H + \lambda(s - c) + \mu x$; this is done so that (6.73) represents the first-order conditions accurately. We know $c > 0$ because $\ln c$ is in the maximand; hence, (6.51) and (6.52) become $c^{-1} - \psi + \lambda = 0$ and $-1 + \psi e^{-x/s} + \mu = 0$. From this we obtain

$$c = (\psi + \lambda)^{-1} \quad \text{and} \quad x = s \ln(\psi/(1 - \mu)). \tag{6.83}$$

These equations do not correspond to the θ^0 that we introduced on our remark on Theorem 6.5.3 because they still contain the multipliers λ and μ. (It is not possible to eliminate the multipliers until we know which constraints bind.) We say that equation (6.83) defines $(c, x) = \tilde{\theta}$, say. Substituting $\tilde{\theta}$ into H of (6.49) we obtain

$$\tilde{H} = \psi(s + \ln s) - s\left(1 - \mu + \ln\left(\frac{\psi}{1 - \mu}\right)\right) - \frac{\psi}{\psi + \lambda} - \ln(\psi + \lambda).$$

In this problem we cannot obtain H^0 as a single expression, but we must distinguish cases A, B, C, and D as in Section 6.4. We briefly summarize their relevant properties (in terms of the multipliers):

$$\begin{aligned}
&\text{A:} \quad \lambda = 0,\\
&\text{B:} \quad \lambda > 0, \text{ and } s = c \text{ or } s = (\psi + \lambda)^{-1},\\
&\text{C:} \quad \mu = 0,\\
&\text{D:} \quad \mu > 0 \text{ and } x = 0 \text{ or } \psi = 1 - \mu.
\end{aligned} \tag{6.84}$$

In each case, AC, AD, BD, and BC, when we know which constraints bind, we can calculate H^0 and then use (6.82b) and (6.82c) to obtain $\dot{\pi}$ and $\dot{s}$. We now do this.

AC: $H^0 = -(1+s)(1+\ln\psi)+\psi(s+\ln s)$; hence,
$\dot{\psi} = r\psi - \partial H^0/\partial s = \psi[r-1-s^{-1}+\psi^{-1}+\psi^{-1}\ln\psi]$, which is (6.56);
$\dot{s} = \partial H^0/\partial\psi = -(1+s)\psi^{-1}+s+\ln s$, which is (6.61);

AD: $H^0 = -\ln\psi+\psi\ln s - 1$;
$\dot{\psi} = \psi(r-s^{-1})$ and $\dot{s} = \ln s - \psi^{-1}$, which are (6.58) and (6.62);

BD: $H^0 = (1+\psi)\ln s - s\psi$;
$\dot{\psi} = \psi(r+1-s^{-1})$ and $\dot{s} = \ln s - s$, which are (6.59) and (6.63);

BC: $H^0 = (1+\psi)\ln s - s(1+\ln\psi)$;
$\dot{s} = \ln s - s\psi^{-1}$, which is (6.66);
$\dot{\psi} = \psi(r-s^{-1}-1+\psi^{-1}+\psi^{-1}\ln\psi)-s^{-1}+\psi$, which is (6.60) with $\lambda = s^{-1}-\psi$ from case B.

It is important to realize that obtaining $\dot{\pi}$ and $\dot{s}$ directly from $\tilde{H}$ (or H) and then eliminating λ and μ with (6.84) according to each case would *not* yield the correct results. In each case the correct expression for H^0 must be obtained before $\dot{\pi}$ and $\dot{s}$ are derived. The reader is invited to verify this.

Remark. It is always possible, and sometimes convenient, to use an approach intermediate between the one illustrated here and the method used in Section 6.4. For instance, we can use (6.83) to eliminate c and x from $\mathcal{L} = H+\lambda(s-c)+\mu x$; we obtain an "optimized Lagrangean":

$$\tilde{\mathcal{L}} = \psi(s+\ln s)+s(\mu-1)\left(1+\ln\frac{\psi}{1-\mu}\right)+s\lambda-\ln(\psi+\lambda)-1.$$

Thus,

$$\dot{\psi} = r\psi - \frac{\partial\tilde{\mathcal{L}}}{\partial s} = \psi\left(r-1-s^{-1}+(1-\mu)\psi^{-1}\ln\frac{\psi}{1-\mu}+1-\mu+\lambda\right),$$

and

$$\dot{s} = \frac{\partial\tilde{\mathcal{L}}}{\partial\psi} = s+\ln s+(\mu-1)\frac{s}{\psi}-\frac{1}{\psi+\lambda},$$

which are valid everywhere and do not explicitly contain the controls. Using (6.84) it is then easy to specialize the two preceding differential equations to obtain (6.56) and (6.58)–(6.64).

Theorem 6.5.4: sufficiency. Let $(\mathbf{s}^*(t), \mathbf{c}^*(t))$ satisfy the conditions of Theorem 6.5.3 and assume that the maximized Hamiltonian H^0 of (6.84) is concave in $\mathbf{s}$. Then $(\mathbf{s}^*(t), \mathbf{c}^*(t))$ is an optimal path for the problem (6.66)–(6.70); if H^0 is strictly concave, it is the unique optimal solution.

Note once again that attention must be paid to the properties of H and g^j, $j=1,\dots,m$, in terms of $\mathbf{c}$ so that condition (6.82a) is satisfied. Concavity of H^0, by itself, does not guarantee optimality.

6.6 Concluding notes

In this chapter we have explored the techniques of solving control problems involving equality and inequality constraints. The reader will have noted that sometimes it requires a certain ingenuity to obtain the optimal solution from the necessary conditions, which are themselves easy to derive.

We have been assuming that the terminal values of the state variables are fixed. We have done this to keep the exposition as similar as possible to the introductory account given in Chapter 4. The time has now come to relax this assumption. This is the main purpose of the next chapter.

Exercises

1. Consider the following modified version of the mushroom grower's problem (Example 6.4.2). Let the utility function and the natural growth function be $2\ln(c+1)-x$ and $\ln(s+1)$, respectively; all other specifications are unchanged. Derive the necessary conditions and solve the problem using the phase diagram method. Note that there will be a region in which the optimal consumption is zero.
2. Modify the mushroom grower's problem (Example 6.4.2) by assuming that the utility function depends on both c and s, and there is no possibility of cultivation $(x=0)$. We choose $u(c,s)=\ln(cs)$ and $\dot{s}=s(1-s)-c$. The constraint $s-c\geq 0$ still applies. Construct the phase diagram in the (state, costate) space and show that the equilibrium is a saddle point.
3. Generalize the mushroom grower's problem (Example 6.4.2) by not assuming specific functional forms. Restrict instead the functions as follows: assume for the utility function $u'(c)>0$, $u''(c)<0$, $u'(0)=+\infty$, and for the natural growth function $G(s)$ $\exists\bar{\bar{s}}>\bar{s}>0$ such that $G(0)=G(\bar{\bar{s}})=0$, $G'(\bar{s})=0$, $G'(0)>r$, and $G''(s)<0$. The constraint $s-c\geq 0$ still applies, but we omit the possibility of cultivation $(x=0)$. Derive and interpret the necessary conditions. Construct the state–costate phase diagram and verify that the equilibrium is a saddle point.
4. At the end of Example 6.3.1, it was claimed that if the production function is homogeneous and has a constant elasticity of substitution, then the optimal consumption path is constant only if the elasticity of substitution is equal to 1. Now prove this result, using the following steps:
   (a) Write the production function in the form $F(s,x)=z^h$, where $z=f(s,x)$ is homogeneous of degree 1. Substitute this in (6.23), take the time derivative, and use (6.24) to establish that $f_{xx}\dot{x}+f_{xs}\dot{s}=-f_s$. (First note that the constancy of c implies that of F, z, and λ.)
   (b) Differentiate the identity $c=z^h$ with respect to time and deduce that c is constant if and only if $f_x\dot{x}+f_s\dot{s}=0$.

(c) Apply Euler's theorem to f and f_x to show that

$$f_x x + f_s s = f \quad \text{and} \quad f_{xx} x + f_{xs} s = 0.$$

Use this and the results of (a) and (b) to show that along the optimal path with constant consumption, $1 = (f_{xs} f)/(f_x f_s) = \sigma$. (*Hint:* Recall that $\dot{s} = -x$; use the results (a) and (b) to solve for this x term; use Euler's theorem results to solve for x also.)

5. For Example 6.4.1, we asserted without proof that Figure 6.5 is the phase diagram in the (ψ, s) space. Construct this diagram using the necessary conditions derived from the current-value Lagrangean.
6. There is a resource that is freely available in unlimited amounts; however, some equipment is required to harvest it, and the equipment is costly to build. The problem is to find the optimal building and harvesting policy that maximizes the total present value of profit over the horizon $[0, T]$. The following notation is used: $x(t)$ is the flow of resource harvested at date t; $b(t)$ the flow of equipment built at date t; $E(t)$ the stock of equipment available at date t; $R(x(t))$ the revenue, at date t, from selling $x(t)$ units of resource at date t; $C(b(t))$ the cost, at date t, of building $b(t)$ units of equipment at date t; and δ the positive rate of discount.

   We assume that $R' > 0$, $R'' < 0$, $C' > 0$, $C'' > 0$, and there exists a positive value $\bar{E}$ such that $R'(\bar{E}) = \delta C'(0)$. The constraints are that $x(t) \geq 0$, $b(t) \geq 0$ at all times, and if the stock of equipment is $E(t)$ at date t, no more than $E(t)$ units of the resource may be harvested (i.e., $x(t) \leq E(t)$); T, $E(0)$, and $E(T)$ are specified positive constants. Set up the problem in optimal control format, apply the maximum principle, and show that $E(t) = x(t) > 0$ at all times. Draw a phase diagram in the (φ, E) space, where φ is the current-value costate. Identify the region where $b = 0$. Describe in words the optimal policy for selected values of T, $E(0)$, and $E(T)$.
7. Consider the operation of a commercial fishing fleet. The amount of fish caught depends only on the size of the fleet. The fleet deteriorates in constant proportion to its size, but this can be counteracted by boat building, which is costly. Profit is simply revenue from the amount of fish caught minus the cost of boat building. At instant t, $s(t)$ is the size of the fleet with which a quantity $F(s(t))$ of fish is caught. The fleet deteriorates at the proportional rate m; $x(t)$ is the flow of boat building and $C(x(t))$ is its cost; fish sell at price p per unit and the discount rate on profit is δ. The planning horizon is $[0, T]$ and the size of the fleet at time zero is s_0. We assume that all of T, δ, m, p, and s_0 are specified positive constants, F is positively valued, strictly increasing, and concave with $F'(\infty) = 0$, while C is strictly positively valued, strictly increasing, and convex with $C'(0) > 0$; x is restricted to nonnegative values (boats cannot be taken apart and sold). We also assume that $C'(0) < pF'(0)/(m + \delta)$.

   (a) Suppose that you have leased this fleet for T periods and that the contract specifies the size of the fleet when it must be returned, say $s_T > 0$. Formulate the problem of maximizing the total present value of profit over the horizon $[0, T]$ subject to the above restrictions. Apply the maximum principle and give an economic interpretation of all variables and conditions.

(b) Draw a phase diagram in the (x,s) space. (It is advisable to deal separately with $x>0$ and $x=0$). Identify the intercepts of the $\dot{x}=0$ locus with the axes ($\bar{x}$ and $\bar{s}$, say). As a first step you might want to use the following numerical example: $p=1$, $\delta=0.1$, $m=0.4$, $F(s)=2(s+1)^{1/2}-2$, and $C(x)=0.5(x+1)^2-0.5$. Give a complete account of the optimal policy for selected values of T and s_T (some above $\bar{s}$, some below). Can you tell whether x is ever zero for any length of time?

(c) Draw a phase diagram in the (φ,s) space where φ is the current-value costate. Identify the region where $x=0$ and describe the optimal policy for selected values of T and s_T.

8. Yabbies are an Australian variety of small freshwater crayfish, prized as a delicacy by connoisseurs of crustaceans. Consider a pond containing a yabby population of size $s(t)$; its natural growth rate is $f(s(t))$, but this can be reduced by $c(t)$, the catch of yabbies at time t (they are very easy to catch); this catch can be sold and provides a revenue $R(c(t))$ at time t. The exponential rate of discount for revenue is δ. Both the revenue function and the growth rate function are strictly concave and have a global maximum; more precisely,

$$\exists \bar{c}, \bar{\bar{c}}, \text{ with } \bar{\bar{c}}>\bar{c}>0,\ R'(\bar{c})=0,\ R(0)=R(\bar{\bar{c}})=0, \text{ and } R''(c)<0, \text{ all } c.$$
$$\exists \bar{s}, \bar{\bar{s}}, \text{ with } \bar{\bar{s}}>\bar{s}>0,\ f'(\bar{s})=0,\ f(0)=f(\bar{\bar{s}})=0, \text{ and } f''(s)<0, \text{ all } s;$$
$$f'(0)>\delta.$$

Our aim is to find the catching policy $c(t)\geq 0$ over some specified horizon $[0,T]$ that maximizes the present value of revenue flows subject to $\dot{s}(t)=f(s(t))-c(t)$ and $s(0)=s_0$. For the time being we ignore constraints on $s(T)$. First draw a rough graph of R and f according to the above assumptions. Show that there exists $\hat{s}$, with $0<\hat{s}<\bar{s}$ and $f'(\hat{s})=\delta$.

(a) Carefully derive the conditions necessarily obeyed by an optimal policy and interpret them.

(b) Draw a phase diagram in the (c,s) space under the special assumption $f(\bar{s})<\bar{c}$. How many equilibrium points are there? Which one is preferred? Draw a phase diagram in the (φ,s) space under the above assumption – φ is the current-value costate; identify the region where $c=0$. Choose arbitrary positive values for T, s_0, and s_T and show that they determine the exact optimal path.

(c) Redo part (b) with $R(c)=0.5c(1-c)$, $f(s)=s(1-s)$, and $r=0.2$.

9. Reconsider exercise 8 without the special assumption $f(\bar{s})<\bar{c}$.

(a) Redo part (b) of exercise 8 with the special assumption $f(\hat{s})<\bar{c}<f(\bar{s})$. How many equilibrium points are there?

(b) Redo part (b) with the special assumption $\bar{c}<f(\hat{s})$.

(c) Examine the stability of the various equilibria. If T were very large, can you guess which one would be the optimal path? Can you identify a crucial value of s_0 that determines whether $c(t)=\bar{c}$ will be the optimal solution $\forall t$? Why is this solution desirable? While $c(t)=\bar{c}$, how much would one pay for more yabbies to stock the pond?

CHAPTER 7

Endpoint constraints and transversality conditions

In Chapters 4 and 6 we assumed that the time horizon T, the initial values of state variables, s_{i0}, and their terminal values s_{iT} were exogenously specified. Obviously these are very restrictive assumptions. In many economic problems we want to allow some of these values to be determined endogenously (subject to constraints). For example, the optimal consumption problem (Example 6.4.1) may be modified to allow the planner to select the value of the terminal stock $s(T)$, subject only to some constraint such that $s(T)$ may not be less than a certain lower bound $\bar{s}$, or even to select the economy's doomsday T after which all activities cease. Obviously, when $s(T)$ or T is not fixed, we need additional necessary conditions to determine the new unknown ($s^*(T)$ or T^*); these conditions are called the *transversality conditions.*

We shall look at various cases, beginning with the simplest. Section 7.8 contains a general statement synthesizing the various transversality conditions. Because there are many kinds of boundary conditions, there are also many kinds of transversality conditions. This array of special cases sometimes appears formidable to students of optimal control theory. For this reason a summary table is provided in Section 7.10. The table lists various features of control problems, and for each one gives the associated transversality condition. If a problem has several of these features, all corresponding transversality conditions apply.

Each of the following sections presents one type of problem and derives the associated transversality condition. The bulk of each section is devoted to the solution of an example of that type so as to illustrate the importance of the transversality condition for the determination of the optimal path.

We shall deal with various modified versions of the problem of Section 6.5, which we reproduce here for convenience.

$$\text{Maximize } V = \int_0^T v(\mathbf{s}(t), \mathbf{c}(t), t)\, dt \tag{7.1a}$$

subject to

$$\dot{s}_i = f^i(\mathbf{s}(t), \mathbf{c}(t), t), \quad i = 1, 2, \ldots, n, \tag{7.1b}$$

$$g^j(\mathbf{s}(t), \mathbf{c}(t), t) \geq 0, \quad j = 1, 2, \ldots, m', \tag{7.1c}$$

$$g^k(\mathbf{s}(t), \mathbf{c}(t), t) = 0, \quad k = m'+1, \ldots, m, \tag{7.1d}$$

$$s_i(0) = s_{i0}, \quad i = 1, 2, \ldots, n, \tag{7.1e}$$

$$s_i(T) = s_{iT}, \quad i = 1, 2, \ldots, n, \tag{7.1f}$$

where s_{i0}, s_{iT}, and T are exogenously given. Because the s_{iT} values are fixed at the outset, such problems are called *fixed-endpoint problems.*

7.1 Free-endpoint problems

In this section we modify (7.1f) by allowing some $s_i(T)$ to be free. For concreteness, let

$$s_i(T) \text{ free}, \quad i = 1, 2, \ldots, n', \tag{7.2a}$$

$$s_j(T) = s_{jT} \text{ given}, \quad j = n'+1, \ldots, n. \tag{7.2b}$$

Intuitively, since there are now n' more objects of choice, as reflected in (7.2a), we must obtain n' additional necessary conditions to help determine the optimal value of each $s_i(T)$, $i = 1, 2, \ldots, n'$. These additional necessary conditions are the transversality conditions for this problem.

Transversality conditions for free endpoint

Theorem 7.1.1. If problem (7.1) is modified as in (7.2), the additional necessary conditions, called the transversality conditions, are

$$\pi_i^*(T) = 0, \quad i = 1, 2, \ldots, n'. \tag{7.3}$$

Proof. For simplicity we shall offer only a heuristic proof of the necessity of (7.3). For any given $\mathbf{s}_T$, let $V^*(\mathbf{s}_T)$ denote the optimal value of the integral (7.1a). The optimal choice of the free s_{iT} values must maximize the function $V^*(\mathbf{s}_T)$. In other words, s_{iT}^* must be chosen such that

$$\partial V^*(\mathbf{s}_T)/\partial s_{iT} = 0, \quad i = 1, 2, \ldots, n'. \tag{7.4}$$

But in Chapter 4 (see equation (4.80)) we have shown that

$$\partial V^*(\mathbf{s}_T)/\partial s_{iT} = -\pi_i^*(T). \tag{7.5}$$

This equation and (7.4) imply (7.3). □

The transversality condition (7.3) has an economic meaning: since the planner attaches no value to the terminal stock and is not constrained to meet a certain target s_{iT}, the stock should be used until its marginal contribution is zero at the end of the planning horizon. We now illustrate the use of the transversality condition by considering a variation of Example 6.4.1.

Example 7.1.1. Our problem is to find $c(t)$ and s_T that maximize

$$V=\int_0^T u(c(t))e^{-\delta t}\,dt \tag{7.6}$$

subject to

$$\dot{s}(t)=F(s(t))-c(t)-ms(t), \tag{7.7}$$

$$F(s(t))-c(t)\geq 0, \tag{7.8}$$

$$s(0)=s_0 \text{ fixed}, \tag{7.9}$$

$$s(T)=s_T \text{ free}. \tag{7.10}$$

We note that s_T is free in the sense that it is not exogenously specified, but the planner's freedom to choose s_T is in fact limited because (7.7)–(7.9) imply a certain lower bound on s_T: since s_T must obey the law of motion (7.7) and $F(s(t))-c(t)\geq 0$, the state variable cannot fall at a rate exceeding $ms(t)$. The smallest feasible value of s_T is therefore $s(0)e^{-mT}$. However, this restriction on the choice of s_T need not be stated as a separate constraint, because it has been reflected in conditions (7.7)–(7.9).

The necessary conditions for problem (7.6) are (6.40)–(6.43) and the transversality condition

$$\pi(T)=0. \tag{7.11}$$

Recall that the current-value shadow price was defined as

$$\psi(t)=\pi(t)e^{\delta t};$$

the transversality condition in terms of the current-value shadow price is

$$\psi(T)=0. \tag{7.11$'$}$$

We shall use (7.11′) and Figure 6.5 to characterize the solution of the free-endpoint problem (7.6).

Figure 7.1 is a reproduction of Figure 6.5. Suppose path PP' is optimal for the fixed-endpoint problem, with $s(T)=Q$. For the free-endpoint problem, path PP' remains feasible but no longer optimal, because at time T it reaches the point P'; hence, $\psi(T)$ is positive, violating the transversality condition (7.11′). For $\psi(T)$ to be zero, the optimal path must end at a point on the line $\psi=0$, that is, on the horizontal axis. Paths such as GG' and EE' satisfy this condition. The time t_1 at which path EE' cuts the vertical line QP' is less than T because we know that it takes exactly T units of time for path PP' to travel from P to P' and that the capital stock decumulates more quickly the farther down the trajectory is in the diagram (recall that to lower values of ψ correspond higher values of c). Similarly, the time t_2 at which path GG' cuts the line QP' is less than T but greater than t_1. Which one of the paths EE' and GG' is optimal

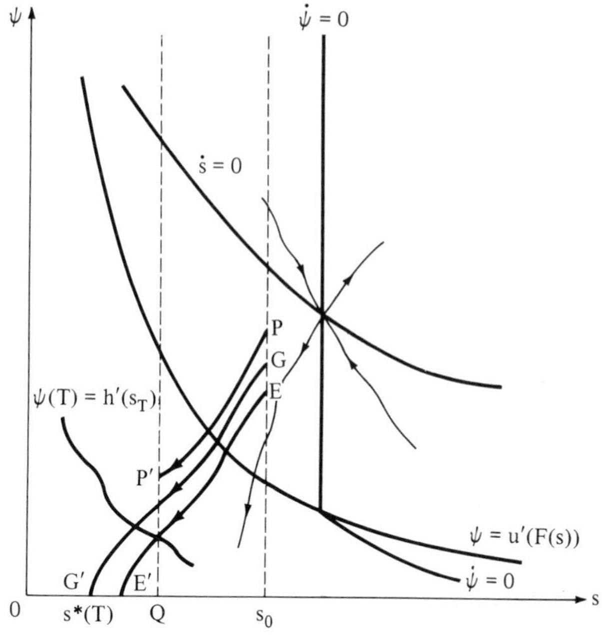

Figure 7.1

depends on the length of the time horizon, T. Given T, there is exactly one optimal path. Say that GG' is the optimal path; then the optimal value of the stock at terminal time is read at $s^*(T)$. Note also that since the optimal path lies below path PP', the consumption $c^*(t)$ along the optimal path is higher than the consumption along path PP'.

In order to reinforce the reader's grasp of the solution of control problems with a transversality condition, we now solve another simple free-endpoint problem, this time without the use of a phase diagram.

Example 7.1.2. An economy produces a consumption good. Output of the consumption good, $c(t)$, is a function of effort, $n(t)$:

$$c(t) = \alpha n(t), \quad \alpha > 0. \tag{7.12}$$

Consumption adds to the stock of pollution, $p(t)$:

$$\dot{p}(t) = (c(t))^2 - mp(t), \tag{7.13}$$

where $m > 0$ is the natural rate of decay of the pollution stock.

Utility is a function of consumption, effort, and pollution:

$$U = \ln c - \beta n^2 - bp, \tag{7.14}$$

where $\beta > 0$ and $b > 0$. The planner's problem is to find $c(t)$, $n(t)$, and p_T that maximize

$$V = \int_0^T (\ln c - \beta n^2 - bp) e^{-\delta t}\, dt \tag{7.15}$$

subject to (7.12), (7.13), and the boundary conditions

$$p(0) = p_0 \text{ given,} \tag{7.16}$$

$$p(T) = p_T \text{ free.} \tag{7.17}$$

Substitute (7.12) into (7.15) to eliminate n; the Hamiltonian is

$$H = (\ln c - \lambda c^2 - bp) e^{-\delta t} + \pi (c^2 - mp),$$

where $\lambda = \beta/\alpha^2 > 0$ and π is the costate variable. The necessary conditions are

$$\frac{\partial H}{\partial c} = \left(\frac{1}{c} - 2\lambda c\right) e^{-\delta t} + 2\pi c = 0, \tag{7.18}$$

$$\dot{\pi} = -\frac{\partial H}{\partial p} = b e^{-\delta t} + m\pi, \tag{7.19}$$

$$\dot{p} = \frac{\partial H}{\partial \pi} = c^2 - mp, \tag{7.20}$$

and the transversality condition

$$\pi(T) = 0. \tag{7.21}$$

(The boundary condition $p(0) = p_0$ must, of course, be satisfied.)

From (7.18), the optimal consumption at any time is given by

$$c = (2\lambda - 2\pi e^{\delta t})^{-1/2}. \tag{7.22}$$

(We shall show that π is nonpositive along an optimal path.) Substitute (7.22) into (7.20):

$$\dot{p} = (2\lambda - 2\pi e^{\delta t})^{-1} - mp. \tag{7.23}$$

The pair of first-order differential equations (7.19) and (7.23) together with the two boundary conditions (7.16) and (7.21) yield unique time paths for $\pi(t)$ and $p(t)$. This can be done by first integrating (7.19):

$$\pi(t) = A e^{mt} - (b/(\delta + m)) e^{-\delta t}, \tag{7.24}$$

where A is a constant of integration that can be determined by using the transversality condition (7.21):

$$0 = \pi(T) = Ae^{mT} - (b/(\delta + m))e^{-\delta T}.$$

Hence,

$$A = (b/(\delta + m))e^{-(\delta + m)T}.$$

Therefore,

$$\pi(t)e^{\delta t} = (b/(\delta + m))(e^{-(m+\delta)(T-t)} - 1), \tag{7.25}$$

which is negative for all $t < T$. (This negativity could have been obtained using the result of Léonard, 1981.) Thus, the transversality condition enables us to get the exact form of $\pi(t)$. The rest of the solution follows.

The optimal consumption path is obtained by substituting (7.25) into (7.22):

$$c(t) = [2\lambda + (2b/(\delta + m))(1 - e^{(\delta + m)(t - T)})]^{-1/2}.$$

It is easy to verify that $\dot{c}(t) > 0$.

The optimal time path of pollution can be obtained from (7.23). First, rearrange (7.23):

$$(2\lambda - 2\pi e^{\delta t})^{-1} = \dot{p} + mp.$$

Multiply both sides by e^{mt}:

$$e^{mt}(2\lambda - 2\pi e^{\delta t})^{-1} = e^{mt}(\dot{p} + mp) = \frac{d}{dt}(p(t)e^{mt}). \tag{7.26}$$

Integration from 0 to t and using τ as the variable time yield

$$p(t)e^{mt} - p(0) = \int_0^t (2\lambda - 2\pi(\tau)e^{\delta\tau})^{-1} e^{m\tau}\, d\tau.$$

It is clear that $p(t)$ is not necessarily a monotone function of time. As an exercise, the reader may construct the phase diagram in the (p, ψ) space where ψ is the current-value shadow price:

$$\psi(t) = \pi(t)e^{\delta t}.$$

It should be noted that $\psi(t)$ is negative along an optimal path; see (7.25). This makes economic sense: since the stock of pollution is a "bad," its marginal value is negative for all $t < T$.

7.2 Problems with free endpoint and a scrap value function

In the preceding section as well as in all problems heretofore considered, it was assumed that the planner derives no benefit from the stock of capital left over at the terminal time T. We now consider a more general problem in which the planner attaches a value to what is left over at T. Thus, we must deal with problem (7.1) modified by adding a "scrap value function" $\phi(\mathbf{s}_T, T)$ to the integral of (7.1a). For convenience it is restated in full here.

Our problem is to find the vector of control $\mathbf{c}(t)$ and the terminal stock $\mathbf{s}_T$ that maximize

$$W=\int_0^T v(\mathbf{s}(t),\mathbf{c}(t),t)\,dt+\phi(\mathbf{s}_T,T) \tag{7.27}$$

subject to

$$\dot{s}_i(t)=f^i(\mathbf{s}(t),\mathbf{c}(t),t),\quad i=1,2,\dots,n, \tag{7.28a}$$

$$g^j(\mathbf{s}(t),\mathbf{c}(t),t)\geq 0,\quad j=1,2,\dots,m', \tag{7.28b}$$

$$g^k(\mathbf{s}(t),\mathbf{c}(t),t)=0,\quad k=m'+1,\dots,m, \tag{7.28b'}$$

$$s_i(0)=s_{i0}\text{ given},\quad i=1,2,\dots,n, \tag{7.28c}$$

$$s_i(T)\text{ free},\quad i=1,2,\dots,n', \tag{7.28d}$$

$$s_i(T)=s_{iT}\text{ given},\quad i=n'+1,\dots,n. \tag{7.28e}$$

Notice that in (7.27) the function $\phi(\mathbf{s}_T,T)$ is added to the integral. A possible economic interpretation is that this function represents the maximum value of an integral of future utility flow starting from time T with an initial capital stock $\mathbf{s}_T$, in just the same way as $V^*(\mathbf{s}_0,\mathbf{s}_T)$ represents the value of the integral expression in (7.27) for given initial stock $\mathbf{s}_0$ and terminal stock $\mathbf{s}_T$.

The fact that the planner attaches some value to the terminal stock has a bearing on the transversality conditions.

Transversality conditions for free endpoint with scrap value

Theorem 7.2.1. For problem (7.27)–(7.28) the following conditions are necessary:

$$\pi_i(T)=\frac{\partial\phi(\mathbf{s}_T,T)}{\partial s_{iT}},\quad i=1,2,\dots,n'. \tag{7.29}$$

Condition (7.29) makes economic sense: it equates the marginal benefit of an increase in s_{iT} (through its contribution to the scrap value) to the marginal cost of such an increase measured over the whole horizon and represented by $\pi_i(T)$ (see equation (4.80)).

Proof. We now provide a heuristic proof that relies on solving problem (7.27)–(7.28) using a two-step procedure. For simplicity we assume that $V^*(\mathbf{s}_0,\mathbf{s}_T)$ is differentiable, although this assumption is not needed for (7.29).

In the first step, we arbitrarily fix the terminal stock at some value $\mathbf{s}_T$ and find $\mathbf{c}(t)$ to maximize

$$V=\int_0^T v(\mathbf{s}(t),\mathbf{c}(t),t)\,dt \tag{7.30}$$

subject to (7.28a)–(7.28c) and the fixed-endpoint condition

$$\mathbf{s}(T) = \mathbf{s}_T \text{ fixed.}$$

Let V^* be the maximized value of V; then V^* depends on $\mathbf{s}_0$ and $\mathbf{s}_T$. The second step consists of finding the optimal $\mathbf{s}_T^*$ that maximizes

$$W(\mathbf{s}_0, \mathbf{s}_T) \equiv V^*(\mathbf{s}_0, \mathbf{s}_T) + \phi(\mathbf{s}_T, T). \tag{7.31}$$

The necessary conditions for $\mathbf{s}_T^*$ to be the solution of problem (7.31) are

$$\frac{\partial W}{\partial s_{iT}} = 0, \quad i = 1, 2, \ldots, n',$$

which are equivalent to

$$-\frac{\partial V^*}{\partial s_{iT}} = \frac{\partial \phi}{\partial s_{iT}}. \tag{7.32}$$

Equations (4.80) and (7.32) imply the transversality conditions (7.29). □

Example 7.2.1. In order to illustrate the use of condition (7.29), let us return to the pollution control problem of Example 7.1.2 and modify it by adding the scrap value function

$$\phi(p_T, T) = -\gamma p_T e^{-\delta T},$$

where $\gamma > 0$ is a constant. The scrap value function has a negative derivative with respect to p_T because pollution is a "bad."

Our problem is to choose $c(t)$ and p_T that maximize

$$W = \int_0^T (\ln c - \lambda c^2 - bp) e^{-\delta t}\, dt - \gamma p_T e^{-\delta T}$$

subject to

$$\dot{p} = c^2 - mp, \tag{7.33}$$

$$p(0) = p_0 \text{ given}, \tag{7.34}$$

$$p(T) = p_T \text{ free}. \tag{7.35}$$

The necesary conditions are (7.18)–(7.20), (7.34), and the transversality condition

$$\pi(T) = -\gamma e^{-\delta T}. \tag{7.36}$$

Condition (7.36) is used to determine the constant of integration A in (7.24):

$$-\gamma e^{-\delta T} = A e^{mT} - (b/(\delta + m)) e^{-\delta T}.$$

This yields

$$A = \left(-\gamma + \frac{b}{\delta + m}\right) e^{-(\delta + m)T},$$

and with (7.24)

$$e^{\delta t}\pi(t) = \left(-\gamma + \frac{b}{\delta + m}\right) e^{-(\delta + m)(T-t)} - \left(\frac{b}{\delta + m}\right).$$

Hence, from (7.22),

$$c(t) = \left[2\lambda - 2\left(-\gamma + \frac{b}{\delta + m}\right) e^{-(\delta + m)(T-t)} + \left(\frac{2b}{\delta + m}\right)\right]^{-1/2}.$$

It follows that the larger is γ, the smaller is the consumption flow. This makes economic sense: if it is more costly to dispose of the stock of "bad" at the terminal time T, then the consumption flow, which adds to pollution, must be reduced.

Remark. If the control problem is autonomous (in the sense that the only independent time term is $e^{-\delta t}$ appearing as a multiplicative factor in the integrand) and if the scrap value function takes the form $\phi(s_T, T) = e^{-\delta T} h(s_T)$, then the transversality condition can be represented in a phase diagram, thus providing useful qualitative information about the optimal path. For example, if problem (7.6) is modified so that the objective function is

$$W = \int_0^T u(c(t)) e^{-\delta t}\, dt + e^{-\delta T} h(s_T),$$

where $h'(s_T) > 0$, $h''(s_T) < 0$, then the transversality condition can be represented by the curve

$$\psi(T) = h'(s_T).$$

This curve is depicted in Figure 7.1. Any optimal path starting from $s(0) = s_0$ must end at a point on this curve, and not on the horizontal axis as when there was no scrap value.

7.3 Lower bound constraints on endpoint

Quite often the terminal value of a state variable may be constrained to be not less than a prespecified constant. For example, a sand-mining firm may be required to leave a stock of sand, $s(T)$, not less than some lower bound s_L. Clearly, in many economic problems we wish to impose nonnegativity constraints on terminal capital stocks.

In this section we consider the problem of finding s_T and $c(t)$ that solve problem (7.1) with (7.1f) modified as

$$s_i(T) \geq s_{iL} \text{ given}, \quad i=1,2,\ldots,n'. \tag{7.37}$$

The rest of (7.1f) remains unchanged.

Transversality conditions for lower bound constraints on endpoint

Theorem 7.3.1. For problem (7.1) modified as by (7.37), the following conditions are necessary:

$$\pi_i(T) \geq 0, \quad s_i(T) - s_{iL} \geq 0, \quad \pi_i(T)[s_i(T) - s_{iL}] = 0, \qquad i=1,2,\ldots,n'. \tag{7.38}$$

Condition (7.38) is essentially the same as the complementary slackness condition in static optimization and has the same economic interpretation: if a stock is not used up to its maximum extent (i.e., if the lower bound constraint is not binding), then its price must be zero.

To provide a heuristic proof of the necessity of (7.38), we use a by now familiar argument that relies on the two-step procedure. In the first step, we consider the solution of problem (7.1) with $s_i(T) = s_{iT}$ (fixed) for all i. In the second step we find the "best" s_T, that is, the value s_T^* that maximizes $V^*(s_T)$ subject to the constraint

$$s_{iT} - s_{iL} \geq 0, \quad i=1,2,\ldots,n'.$$

The necessary condition characterizing the best s_T is

$$\frac{\partial V^*}{\partial s_{iT}} \leq 0, \quad s_{iT}^* - s_{iL} \geq 0, \quad \left(\frac{\partial V^*}{\partial s_{iT}}\right)(s_{iT}^* - s_{iL}) = 0, \quad i=1,2,\ldots,n'. \tag{7.39}$$

This condition and (4.80) imply the transversality condition (7.38). □

Example 7.3.1. Let us illustrate the use of condition (7.38) on a simple example. A firm has a fixed time interval $[0, T]$ over which it can extract an exhaustible resource. Let $s(0)$ denote the initial stock of resource and $c(t)$ the rate of extraction, so that

$$\dot{s}(t) = -c(t). \tag{7.40}$$

The profit flow derived from the extraction and sale of the resource good is $v(c(t))$. The firm's objective is to find the time path $c(t)$ that maximizes

$$\int_0^T v(c(t))e^{-rt}\,dt \tag{7.41}$$

subject to (7.40) and

$$c(t) \geq 0, \tag{7.42}$$

$$s(0) = s_0 \text{ fixed}, \tag{7.43}$$

$$s(T) \geq 0. \tag{7.44}$$

We note that in this case the lower bound s_L is zero. We do not need the constraint $s(t) \geq 0$ for all t in $[0, T]$, because conditions (7.42) and (7.44), together with (7.40), ensure that $s(t)$ is always nonnegative.

The Hamiltonian for this problem is

$$H = v(c)e^{-rt} - \pi c, \tag{7.45}$$

$c(t)$ is required to be nonnegative, and the necessary conditions are

$$\frac{\partial H}{\partial c} = v'(c^*)e^{-rt} - \pi \leq 0, \tag{7.45a}$$

$$(v'(c^*)e^{-rt} - \pi)c^* = 0, \quad c^* \geq 0, \tag{7.45b}$$

$$\dot{\pi} = -\frac{\partial H}{\partial s} = 0, \tag{7.45c}$$

$$\dot{s} = \frac{\partial H}{\partial \pi} = -c^*, \tag{7.45d}$$

$$s(0) = s_0, \tag{7.45e}$$

$$\pi(T) \geq 0, \quad s_T^* \geq 0, \quad \pi(T)s_T^* = 0. \tag{7.45f}$$

To sharpen our results, we assume that the profit function $v(c)$ has the following properties:

$$v'(0) > 0, \tag{7.46a}$$

$$v''(c) < 0 \quad \text{for all } c \geq 0, \tag{7.46b}$$

$$v'(c_M) = 0 \quad \text{for some } c_M > 0. \tag{7.46c}$$

Condition (7.46a) ensures that extraction is profitable at some positive rate of extraction, condition (7.46b) ensures that for any given $\pi(t)$ there exists a unique $c^*(t)$ that maximizes the Hamiltonian, and condition (7.46c) implies that it never pays to extract at a rate $c(t)$ exceeding c_M because marginal profit is negative beyond c_M.

From (7.45c) and (7.45f), $\pi(t)$ is a nonnegative constant, say K:

$$\pi(t) = \pi(0) = K \geq 0.$$

The constancy of π and condition (7.45a) imply that the discounted marginal profit must be a constant during any time interval of positive extraction. (This result is known as *Hotelling's rule,* in honor of Hotelling's contribution to the theory of the mine.)

There are two possible cases: $K>0$ and $K=0$. We shall show that $K=0$ only if the time horizon is so short that the total cumulative extraction that yields the highest profit cannot exceed the initial stock. In this case the firm can extract the resource at the myopic profit-maximizing level c_{M} for all t in $[0,T]$ and still have $s(T)>0$, thereby justifying the use of the myopic rule. Under these circumstances, any addition to the stock will be of no value to the firm. (Recall that the firm has a fixed time horizon T; after T, it no longer has the right to extract.) We now seek to confirm our intuitive reasoning by a simple manipulation of the necessary conditions (including the transversality condition). Let us define

$$\tilde{T}=s_0/c_{\mathrm{M}}. \tag{7.47}$$

If the fixed time horizon T is equal to $\tilde{T}$, then extraction at the constant rate c_{M} will just exhaust the resource stock at time $\tilde{T}$, because

$$s_T-s_0=\int_0^T \dot{s}(t)\,dt=-c_{\mathrm{M}}T, \tag{7.48}$$

so that if $T=\tilde{T}$, then $s_T=0$. If T is less than $\tilde{T}$, then extraction at the constant rate c_{M} will leave some positive stock at time T, as can be seen from (7.47) and (7.48). So if $T\le\tilde{T}$, the optimal extraction policy is $c^*(t)=c_{\mathrm{M}}$ for all t. The necessary conditions (7.45a)–(7.45f) will be satisfied with $\pi(t)=0$. Sufficiency is satisfied because of the concavity of the Hamiltonian in the control and state variables.

We now show that if $T>\tilde{T}$, then K must be positive. Suppose $K=0$ when $T>\tilde{T}$. Then from (7.45a) and (7.45b), $c^*(t)=c_{\mathrm{M}}$ for all t in $[0,T]$, but this is not possible, because at the rate of extraction c_{M}, the resource stock is exhausted at $\tilde{T}<T$. Thus, K must be positive when $T>\tilde{T}$ and the transversality condition (7.45f) implies $s^*(T)=0$.

In the remainder of this section we examine the time profile of the optimal extraction path when $T>\tilde{T}$. Recall that the transversality condition and $K>0$ imply $s^*(T)=0$. We derive some properties of the optimal path in this case.

We first show that $c(t)$ must be monotone decreasing. Consider any pair (t_a,t_b) with $t_a<t_b$ and $c(t_a)>0$, $c(t_b)>0$; then from (7.45a)

$$v'(c(t_a))e^{-rt_a}=v'(c(t_b))e^{-rt_b}=K. \tag{7.49}$$

Since $t_a<t_b$, this implies $v'(c(t_a))<v'(c(t_b))$ and hence $c(t_a)>c(t_b)$.

Next we show that if c ever becomes zero, it will remain zero until the end of the horizon. Again consider $t_a<t_b$ with $c(t_a)=0$. We will establish that if $c(t_b)>0$, we would have a violation of the necessary conditions. Suppose $c(t_b)>0$; then by (7.45a)

$$K=v'(c(t_b))e^{-rt_b}\ge v'(0)e^{-rt_a}. \tag{7.50}$$

Hence,

$$v'(c(t_b)) \geq v'(0)e^{r(t_b-t_a)} > v'(0), \tag{7.51}$$

but this is impossible because $v'(0) > v'(c)$ for all $c > 0$.

Finally, we show that the optimal path is continuous. This is done in several stages. We can rule out jumps from zero to positive with the preceding result. We use the strict concavity of v to show that it is never optimal to have any other jump discontinuities in $c(t)$. Suppose there is one at $t_1 < T$. Let $c(t_1^-)$ and $c(t_1^+)$ denote the left-hand and right-hand limits of $c(t)$. (Recall that $c(t)$ is only required to be piecewise-continuous.) If both limits are positive, then from (7.45a) and the piecewise continuity of $c(t)$, there exists $\epsilon > 0$ such that for all positive $\delta < \epsilon$,

$$v'(c(t_1+\delta))e^{-r(t_1+\delta)} = v'(c(t_1-\delta))e^{-r(t_1-\delta)} = K. \tag{7.52}$$

Taking the limit as δ tends to zero,

$$v'(c(t_1^+)) = v'(c(t_1^-)).$$

The strict concavity of v implies that the preceding equation is satisfied if and only if $c(t_1^+) = c(t_1^-)$. Alternatively, suppose that $c(t_1^+) = 0$, so that (7.52) no longer holds. Then for $t_1 < T$, we have

$$v'(c(t_1+\delta))e^{-r(t_1+\delta)} \leq K = v'(c(t_1-\delta))e^{-r(t_1-\delta)}. \tag{7.53}$$

Again taking the limit as $\delta \to 0$,

$$v'(c(t_1^+)) \leq v'(c(t_1^-)),$$

but this inequality cannot be satisfied for $c(t_1^+) = 0$ and $c(t_1^-) > 0$, because $v'(c)$ is strictly decreasing.

To sum up, $c(t)$ is continuous everywhere, decreasing while positive, and if it reaches zero it will remain at that level.

Let us see how the optimal path varies as T takes on values larger than $\tilde{T}$. The path is determined by

$$v'(c(0)) = K = v'(c(t))e^{-rt} \tag{7.54}$$

as long as $c(t)$ remains positive. Thus, $c(t)$ remains below c_M for all t and any small increase in T above $\tilde{T}$ results in a small decrease in c for all t. The exact path is determined once the value $c(0)$ is chosen and it is done so as to exhaust s_0 since the transversality condition requires $s^*(T) = 0$. This does not mean that $s(t)$ necessarily remains strictly positive for all $t < T$: it may be optimal to exhaust the resource stock at some time $t_E < T$, provided that T is large enough (in relation to r and s_0) and provided that $v'(0)$ is finite, as we now demonstrate.

Let us try to understand why this may occur. Given a finite supply of the resource, we tend to accumulate more benefit if we reduce the flow of the resource to a trickle but for a longer period; this is due to the strict concavity of the net benefit function $v(c)$. This phenomenon, however, must be balanced by the preference for earlier benefit due to the positive discount rate. For this reason there may exist a time $t_E < T$ such that c becomes nil at that date and according to (7.54)

$$v'(c(0)) = v'(0)e^{-rt_E}, \tag{7.55}$$

$$t_E = r^{-1}[\ln v'(0) - \ln v'(c(0))], \tag{7.56}$$

where $c(0)$ is to be determined. If $v'(0) = \infty$, there is no such time, for T is finite. But if $v'(0)$ is finite, there is a time t_E at which c becomes zero. To determine t_E and $c(0)$, note that the choice of $c(0)$ determines $c(t)$ through (7.56) and that we must have

$$\int_0^{t_E} c(t)\,dt = s_0. \tag{7.57}$$

We must choose $c(0)$ such that the path $c(t)$ determined by (7.54) satisfies (7.57) subject to (7.56).

Let us apply our results to a special case, where $v(c)$ is quadratic:

$$v(c) = ac - (bc^2/2), \quad a > 0, \ b > 0,$$

$$v'(c) = a - bc.$$

The function $v(c)$ attains its maximum at c_M:

$$c_M = a/b.$$

The time horizon $\tilde{T}$, defined by (7.47), is in this case

$$\tilde{T} = bs_0/a.$$

The time t_E can also be determined. Note that (7.55) yields

$$a - bc(0) = ae^{-rt_E} \tag{7.55$'$}$$

and (7.54) yields

$$c(t) = (a/b) - b^{-1}[a - bc(0)]e^{rt}. \tag{7.54$'$}$$

Thus, the integral (7.57) gives

$$(a/b)t_E - [a - bc(0)](e^{rt_E} - 1)/br = s_0. \tag{7.58}$$

Substitute (7.55′) into (7.58) to obtain

$$(a/br)[e^{-rt_E} - 1 + rt_E] = s_0, \tag{7.58$'$}$$

and for each s_0 a unique t_E can be determined, because the expression inside the square brackets is a strictly increasing function of t_E. Having determined t_E, we can use (7.55′) to solve for the optimal $c(0)$, if the time horizon T exceeds or equals t_E. It is clear that both $c(0)$ and t_E are increasing functions of the initial stock size.

Finally, if the time horizon T is shorter than t_E but longer than $\tilde{T}$, then using the results that $c(t)$ is positive for all $t \leq T$ and the stock is exhausted in this case, we can compute $c(0)$ from (7.54′) and the exhaustion condition:

$$\int_0^T c(t)\,dt = \frac{a}{b}T - (e^{rT}-1)[a-bc(0)]\frac{1}{br} = s_0.$$

Therefore,

$$c(0) = (a/b) + [(a/b)T - s_0][r/(e^{rT}-1)],$$

which is positive because $(a/b)T > (a/b)\tilde{T} = s_0$.

The reader should be able to draw a phase diagram in the space (s, ψ), where ψ is the current-value shadow price,

$$\psi(t) = \pi(t)e^{rt}.$$

We also invite the reader to verify that all the qualitative results obtained so far for the resource extraction problem, with $v(c)$ satisfying (7.49a)–(7.49c), remain valid when the discount rate is not a constant, that is, when we replace the discount factor e^{-rt} by a general function $\alpha(t) > 0$, where $\alpha'(t) < 0$.

7.4 Problems with lower bound constraints on endpoint and a scrap value function

In this section we consider a slightly more general class of problems that combines the features of problems in Sections 7.2 and 7.3: we retain the lower bound constraints on endpoint as in (7.37) of Section 7.3 and modify the objective function to allow for a scrap value function; thus, the objective function is as in (7.27) of Section 7.2. Both these equations are reproduced below and renumbered for clarity. It will come as no surprise that the transversality condition is a combination of those of the preceding two sections.

Transversality conditions for problems with lower bound constraints on endpoint and scrap value function

Theorem 7.4.1. For problem (7.1) modified by (7.59a) and (7.59b), the transversality conditions (7.60) are necessary:

$$s_{iT} \geq s_{iL} \text{ given}, \quad s_{jT} \text{ free}, \quad i = 1, 2, \ldots, n', \; j = n'+1, \ldots, n, \tag{7.59a}$$

$$W=\int_0^T v(\mathbf{s}(t),\mathbf{c}(t),t)\,dt+\phi(\mathbf{s}_T,T). \tag{7.59b}$$

Equation (7.59b) replaces (7.1a) as the objective function. Then, for $i=1,2,\ldots,n'$,

$$\pi_i(T)-\frac{\partial\phi}{\partial s_{iT}}\geq 0,\quad s_{iT}-s_{i\mathrm{L}}\geq 0,\quad \left(\pi_i(T)-\frac{\partial\phi}{\partial s_{iT}}\right)(s_{iT}-s_{i\mathrm{L}})=0 \tag{7.60a}$$

and for $j=n'+1,\ldots,n$,

$$\pi_j(T)=\frac{\partial\phi}{\partial s_{jT}}. \tag{7.60b}$$

Equations (7.60) are the transversality conditions for this problem.

The reader is invited to give a heuristic proof of the necessity of this condition by using the familiar two-step procedure, the second step being the maximization of

$$W(\mathbf{s}_T)\equiv V^*(\mathbf{s}_T)+\phi(\mathbf{s}_T,T)$$

subject to

$$s_{iT}\geq s_{i\mathrm{L}},\quad s_{jT}\ \text{free},\quad i=1,\ldots,n',\ j=n'+1,\ldots,n.$$

Example 7.4.1. We now illustrate the use of (7.60) on a model of optimal saving.

An individual has the utility function

$$u(c)=(1-\gamma)^{-1}c^{1-\gamma}+A,\quad \gamma>0,\ A=\text{const},$$

where $c(t)$ denotes consumption at time t. Her stock of financial assets is $s(t)$, which yields the interest income $\beta s(t)$, where $\beta>0$ is the interest rate. Her wage income is an exogenous flow and is denoted by $w(t)$. The difference between total income and consumption is her net saving, which is the net addition to her stock of financial assets:

$$\dot{s}(t)=\beta s(t)+w(t)-c(t). \tag{7.61}$$

The initial stock of financial assets is $s(0)$, exogenously given. The terminal stock, s_{T}, is bequeathed to her son on her retirement. The individual's valuation of this bequest is

$$\phi(s_{\mathrm{T}},T)=e^{-\delta T}ms_{\mathrm{T}},\quad m\geq 0, \tag{7.62}$$

where $\delta>0$ is the rate of discount.

The individual is free to choose s_{T}, but there is a lower bound constraint

$$s_T\geq s_{\mathrm{L}}.$$

(s_L is exogenously given; if $s_L = 0$, the constraint means that the individual cannot bequeath a negative stock of financial assets, i.e., debts.)

The time horizon T is fixed. The individual's problem is to find $c(t)$ and s_T that maximize

$$\int_0^T \{A+(1-\gamma)^{-1}[c(t)]^{1-\gamma}\}e^{-\delta t}\,dt + e^{-\delta T} m s_T \tag{7.63}$$

subject to (7.61) and

$$s(0) = s_0 \text{ given,}$$
$$s(T) \geq s_L, \quad s_L \text{ given; we assume } s_L < s_0.$$

Notice that we did not write down the constraint $c(t) \geq 0$ because the utility function we have chosen has the property $u'(0) = \infty$ and this ensures that $c(t)$ is strictly positive along an optimal path.

The Hamiltonian for problem (7.63) is

$$H = [A+(1-\gamma)^{-1}c^{1-\gamma}]e^{-\delta t} + \pi(\beta s + w - c),$$

and the necessary conditions are

$$\frac{\partial H}{\partial c} = c^{-\gamma}e^{-\delta t} - \pi = 0, \tag{7.64}$$

$$\dot{\pi} = -\frac{\partial H}{\partial s} = -\beta\pi, \tag{7.65}$$

$$\dot{s} = \frac{\partial H}{\partial \pi} = \beta s + w - c, \tag{7.66}$$

$$s(0) = s_0. \tag{7.67}$$

From (7.60) we obtain the transversality condition

$$\pi(T) - me^{-\delta T} \geq 0, \quad s_T - s_L \geq 0, \quad (\pi(T) - me^{-\delta T})(s_T - s_L) = 0. \tag{7.68}$$

Condition (7.64) says that the discounted marginal utility of consumption is to be equated with the discounted shadow price of the stock of financial assets. This condition and (7.68) imply that if the constraint $s_T \geq s_L$ is not binding, then for an optimum the discounted marginal utility of consumption at T must be equated with the discounted marginal valuation of the bequest. However, if m is "very small," then this equality cannot be satisfied and the individual values consumption more than bequest and would have wished to erode further her financial assets but the constraint $s_T \geq s_L$ prevents her from doing so. We now proceed to verify our intuitive reasoning.

From (7.65), we have

$$\pi(t)=\pi(0)e^{-\beta t}. \tag{7.69}$$

Conditions (7.64) and (7.69) yield

$$[c(t)]^{-\gamma}e^{-\delta t}=\pi(t)=\pi(0)e^{-\beta t}=[c(0)]^{-\gamma}e^{-\beta t}. \tag{7.70}$$

Hence,

$$c(t)=c(0)\exp[(\beta-\delta)t/\gamma], \tag{7.71}$$

where $\exp[y]$ denotes e^y for any y. Substitute (7.71) into (7.66):

$$\dot{s}-\beta s=w(t)-c(0)\exp[(\beta-\delta)t/\gamma].$$

Multiply both sides by $\exp[-\beta t]$ and integrate from 0 to T:

$$\exp[-\beta T]s(T)=s(0)+Y$$
$$-c(0)\frac{\gamma}{\delta+\beta\gamma-\beta}\left\{1-\exp\left[\frac{(\beta-\delta-\beta\gamma)T}{\gamma}\right]\right\}, \tag{7.72}$$

where Y is the present value of the stream of wage income

$$Y=\int_0^T w(t)\exp[-\beta t]\,dt.$$

The right-hand side of (7.72) is a decreasing function of $c(0)$, because if $\delta+\beta\gamma-\beta$ is positive (negative) then the last exponential is smaller (larger) than unity. Therefore, the terminal stock $s(T)$ is smaller the larger $c(0)$ is. Furthermore, if both sides of (7.72) are multiplied by $\exp[\beta T]$ we have

$$s(T)=(s(0)+Y)e^{\beta T}-c(0)\alpha e^{\beta T}, \tag{7.72$'$}$$

where

$$\alpha=[\gamma/(\delta+\beta\gamma-\beta)]\{1-\exp[(\beta-\delta-\beta\gamma)T/\gamma]\}>0. \tag{7.73}$$

Since we have assumed $s(0)>s_{\rm L}$, the first term on the right-hand side of (7.72′) exceeds $s_{\rm L}$ and if $c(0)$ is sufficiently large, then $s(T)=s_{\rm L}$; we denote this particular value of $c(0)$ by $c_{\rm L}(0)$:

$$c_{\rm L}(0)=(s(0)+Y-e^{-\beta T}s_{\rm L})/\alpha. \tag{7.74}$$

Using (7.70) we obtain the value of $\pi(T)$ that corresponds to $c_{\rm L}(0)$:

$$\pi_{\rm L}(T)=[c_{\rm L}(0)]^{-\gamma}e^{-\beta T}.$$

If $\pi_{\rm L}(T)\geq me^{-\delta T}$, then the transversality condition (7.68) is satisfied, with $s_{\rm T}=s_{\rm L}$, for the consumption path starting with $c(0)=c_{\rm L}(0)$, and therefore this path is the optimal path. (Recall that by construction the path (7.71) with the initial condition given by (7.74) satisfies all the necessary conditions; sufficiency follows from the usual concavity property.) In contrast, if $\pi_L(T) < me^{-\delta T}$, then the consumption path starting with $c_L(0)$ is not optimal because the transversality condition is violated. In economic terms, this is the case in which the marginal value of bequest, m, is so high that if the stock of financial assets were driven to s_L at time T, the marginal value of consumption would be lower than m. What is the optimal consumption path in this case? Since the optimal terminal stock must be strictly greater than s_L, the transversality condition (7.68) implies

$$\pi(T) = me^{-\delta T}.$$

The optimal value $c^*(0)$ can then be calculated from (7.70):

$$[c^*(0)]^{-\gamma} e^{-\beta T} = me^{-\delta T}$$

and

$$c^*(t) = m^{-1/\gamma} \exp[(\beta - \delta)\gamma^{-1}(t - T)].$$

This completes our analysis of the determination of the optimal path. We now draw the reader's attention to some interesting features of the solution. First, the lower bound constraint on s_T is binding if and only if

$$[(s(0) + Y - s_L e^{-\beta T})/\alpha]^{-\gamma} \geq me^{(\beta - \delta)T}.$$

Therefore, if $s(0)$ or Y is sufficiently large, given s_L, the constraint will not bind. In retrospect, this result is intuitively obvious.

Second, from (7.71) $\dot{c}(t) > 0$ if and only if the rate of interest exceeds the rate of discount, that is, $\beta > \delta$. This result is plausible: if the rate of interest is sufficiently high, the individual will have an incentive to save more during the earlier part of the planning horizon. This result relies on the assumption that marginal utility of consumption at any time depends only on $c(t)$.

Third, using (7.71) we can calculate *for any given* s_T the present value of the utility flow:

$$\begin{aligned} V(s_0, s_T) &= \int_0^T \{A + (1-\gamma)^{-1}[c(t)]^{1-\gamma}\} e^{-\delta t}\, dt \\ &= (1-\gamma)^{-1}[c(0)]^{1-\gamma} \frac{\gamma}{\delta + \beta\gamma - \beta} \left\{1 - \exp\left[\frac{(\beta - \beta\gamma - \delta)T}{\gamma}\right]\right\} \\ &\quad + \frac{A}{\delta}\{1 - \exp[-\delta T]\}, \end{aligned} \tag{7.75}$$

where from (7.72)

$$c(0) = (s(0) + Y - e^{-\beta T} s_T)/\alpha$$

and α is given by (7.73). The derivative of V with respect to s_T is

$$\frac{\partial V}{\partial s_{\mathrm{T}}} = [c(0)]^{-\gamma} \frac{\partial c(0)}{\partial s_{\mathrm{T}}} \frac{\gamma}{\delta+\beta\gamma-\beta} \left\{1 - \exp\left[\frac{(\beta-\beta\gamma-\delta)T}{\gamma}\right]\right\}$$
$$= -[c(0)]^{-\gamma} e^{-\beta T}. \tag{7.76}$$

This result and (7.70) imply that

$$\frac{\partial V}{\partial s_{\mathrm{T}}} = -\pi(T),$$

which serves to verify in this instance our general proposition about the meaning of the shadow price. Similarly,

$$\frac{\partial V}{\partial s_0} = [c(0)]^{-\gamma} = \pi(0).$$

7.5 Free-terminal-time problems without a scrap value function

Thus far we have always assumed that the terminal time T is fixed. In many economic problems it makes sense to allow the planner to choose T. For example, mining firms are often free to discontinue exploiting a site before every grain of ore is extracted, purely for economic reasons. The optimal terminal time will be denoted by T^*. To determine the additional unknown T^*, we need another necessary condition.

Additional transversality condition for free-terminal-time problems without a scrap value function

Theorem 7.5.1. For problems without a scrap value function such as problem (7.1) or modified as in Sections 7.1 or 7.3, the additional transversality condition when terminal time is free is

$$H(\mathbf{s}^*(T^*), \mathbf{c}^*(T^*), \boldsymbol{\pi}(T^*), T^*) = 0, \tag{7.77}$$

provided that T^* is finite.

We now proceed to offer a heuristic proof of Theorem 7.5.1. Let us first consider the case in which the terminal stocks must take on fixed values as in problem (7.1):

$$s_i^*(T^*) = b_i, \;\; b_i \text{ fixed}, \quad i = 1, 2, \ldots, n. \tag{7.78}$$

We then have the following free-terminal-time, fixed-endpoint problem: find $\mathbf{c}(t)$ and T that maximize

$$V_{\mathrm{F}} = \int_0^T v(\mathbf{s}(t), \mathbf{c}(t), t)\, dt \tag{7.79}$$

subject to

$$\dot{s}_i(t) = f^i(\mathbf{s}(t), \mathbf{c}(t), t), \tag{7.80}$$

$$s_i(0) = s_{i0} \text{ fixed}, \tag{7.81a}$$

$$s_i(T) = b_i \text{ fixed}. \tag{7.81b}$$

Notice that we use b_i rather than s_{iT} in order to emphasize that when we vary T, the terminal stocks must remain fixed at the value b_i, $i = 1, 2, \ldots, n$. The subscript F in (7.79) indicates that this is a free-terminal-time problem.

For any fixed T, define

$$V(\mathbf{b}, T) \equiv V(b_1, b_2, \ldots, b_n, T) \equiv \max_{\mathbf{c}(t)} \int_0^T v(\mathbf{s}(t), \mathbf{c}(t), t)\, dt \tag{7.82}$$

subject to (7.80) and (7.81). Using the argument that we developed in Section 4.5 we have for an arbitrary function $\boldsymbol{\pi}(t)$

$$V(\mathbf{b}, T) = \int_0^T [H(\mathbf{s}^*, \mathbf{c}^*, \boldsymbol{\pi}, t) + \dot{\boldsymbol{\pi}} \cdot \mathbf{s}^*]\, dt - \boldsymbol{\pi}(T) \cdot \mathbf{b} + \boldsymbol{\pi}(0) \cdot \mathbf{s}_0. \tag{7.83}$$

Differentiate (7.83) with respect to T:

$$\begin{aligned} \frac{\partial V}{\partial T} &= H(\mathbf{s}^*(T), \mathbf{c}^*(T), \boldsymbol{\pi}(T), T) + \dot{\boldsymbol{\pi}}(T) \cdot \mathbf{s}^*(T) \\ &\quad + \int_0^T \left[(H_{\mathbf{s}} + \dot{\boldsymbol{\pi}}) \cdot \frac{d\mathbf{s}^*}{dT} + (H_{\mathbf{c}}) \cdot \frac{d\mathbf{c}^*}{dT} \right] dt - \frac{d}{dT}(\boldsymbol{\pi}(T)\mathbf{b}). \end{aligned} \tag{7.84}$$

Since the integral term is zero when $\boldsymbol{\pi}(t)$ is the optimal path of the costate variables and since $\mathbf{s}^*(T) = \mathbf{b}$ by (7.81), we have

$$\frac{\partial V(\mathbf{b}, T)}{\partial T} = H(\mathbf{s}^*(T), \mathbf{c}^*(T), \boldsymbol{\pi}(T), T). \tag{7.85}$$

Equation (7.85) says that when we lengthen the time horizon of problem (7.82) by a small increment ΔT, the value of the integral of the fixed-time problem (7.82) will increase or decrease according to whether the Hamiltonian, evaluated at T, is positive or negative. The best time horizon T^* is that value of T that maximizes $V(\mathbf{b}, T)$; thus, the derivative (7.85), evaluated at T^*, is necessarily zero provided that T^* is finite. This establishes the necessity of (7.77). If (7.85) were positive for all T, the optimal time horizon would be infinite.

A moment's reflection will convince the reader that (7.77) is also necessary if the b_i's are not fixed but are to be chosen optimally (possibly subject to some constraints). This is because for any optimally chosen set of values $(b_1^*, b_2^*, \ldots, b_n^*)$ condition (7.77) is necessary with respect to the choice of T^*, given that

$$s_i^*(T^*) = b_i^*.$$

In order to offer an economic interpretation of the transversality condition (7.77), we begin with a concrete example. Consider a resource extraction problem (similar to Example 7.3.1) with fixed endpoint $s(T)=b$ (fixed). For given T, the firm finds $c(t)$ that maximizes

$$\int_0^T v(c(t))e^{-rt}\,dt$$

subject to

$$\dot{s}(t)=-c(t),$$
$$s(0)=s_0,$$
$$s(T)=b.$$

Now consider an increase in T by an amount ΔT so that the terminal constraint would be

$$s(T+\Delta T)=b.$$

The gain to the firm would be the flow of profit from T to ΔT. This is approximately

$$v(c(T))e^{-rT}\Delta T.$$

But the stock at T would have to be larger:

$$s(T)=b-\int_T^{T+\Delta T}\dot{s}(t)\,dt\simeq b+(\Delta T)c(T).$$

In other words, the cumulative extraction from 0 to T would have fallen by the amount $(\Delta T)c(T)$. The shadow price of the resource being $\pi(T)$, the opportunity cost of this amount of stock is therefore $\pi(T)(\Delta T)c(T)$. If T is optimally chosen, the marginal gain must equal the marginal loss:

$$v(c(T))e^{-rT}\Delta T=\pi(T)(\Delta T)c(T). \tag{7.85'}$$

Dividing both sides of (7.85′) by ΔT, and taking the limit $\Delta T\to 0$, we have

$$v(c(T))e^{-rT}-\pi(T)c(T)=0.$$

But this is precisely the Hamiltonian at time T.

The marginal gain of increasing T is $v(s(T),c(T),T)\Delta T$ for more general models, and the marginal loss is $-\pi(T)\Delta Tf(s(T),c(T),T)$. The optimal choice of T must equate marginal gain with marginal loss.

Example 7.5.1. We now illustrate the use of the transversality condition (7.77) in the determination of the optimal terminal time T^*. Let us return to problem (7.63) and make some modifications: instead of a fixed time horizon T we now allow T to be optimally chosen; we also assume that

$m=0$, $w(t)=0$, $\beta=0$ for simplicity. We may interpret our problem as that of a resource-extracting firm, which derives a profit flow

$$u(c(t))=A+(1-\gamma)^{-1}[c(t)]^{1-\gamma}, \quad 1-\gamma>0,$$

where $c(t)$ is the rate of extraction and $A<0$ is a fixed rent per unit of time that must be paid as long as the firm remains in business. The assumption $1-\gamma>0$ is made so that we can identify the other term as a positive revenue. The firm must determine the closing-down date T^* and the extraction path $c^*(t)$ for t in $[0, T^*]$. The initial stock is given and the terminal stock must be at least as large as $s_L \geq 0$.

The Hamiltonian for this problem is

$$H=[A+(1-\gamma)^{-1}c^{1-\gamma}]e^{-\delta t}-\pi c. \tag{7.86}$$

The necessary conditions are

$$\frac{\partial H}{\partial c}=c^{-\gamma}e^{-\delta t}-\pi=0, \tag{7.87a}$$

$$\dot{\pi}=-\frac{\partial H}{\partial s}=0, \tag{7.87b}$$

$$\dot{s}=\frac{\partial H}{\partial \pi}=-c, \tag{7.87c}$$

and the transversality conditions are

$$\pi(T)\geq 0, \quad s_T-s_L\geq 0, \quad \pi(T)(s_T-s_L)=0, \tag{7.87d}$$

$$H(T)=[A+(1-\gamma)^{-1}c(T)^{1-\gamma}]e^{-\delta T}-\pi(T)c(T)=0. \tag{7.87e}$$

We have pointed out in the preceding section that $c^*(t)$ is strictly positive because $u'(0)=\infty$. Hence $\pi(t)$ is positive. Condition (7.87d) then implies that $s_T^*=s_L$.

Turning to condition (7.87e), we can see that this condition and (7.87a) yield a precise characterization of the optimal terminal rate of extraction:

$$[A+(1-\gamma)^{-1}c(T)^{1-\gamma}]/c(T)=[c(T)]^{-\gamma}; \tag{7.88}$$

that is, the choice of $c(T)$ must equate average profit with marginal profit. Since $A<0$ and $(1-\gamma)>0$, this equation yields a unique value $c^*(T)>0$:

$$c^*(T)=[-A(1-\gamma)/\gamma]^{1/(1-\gamma)}. \tag{7.89}$$

This is illustrated in Figure 7.2, which shows that the average profit is maximized at $c^*(T)$. This determines the optimal terminal rate of extraction, but by itself does not determine the terminal time. To determine T^* we must combine (7.89) with other necessary conditions. From (7.87)

$$c^*(T)=(s(0)-s_L)(\delta/\gamma)[e^{\delta T/\gamma}-1]^{-1}. \tag{7.90}$$

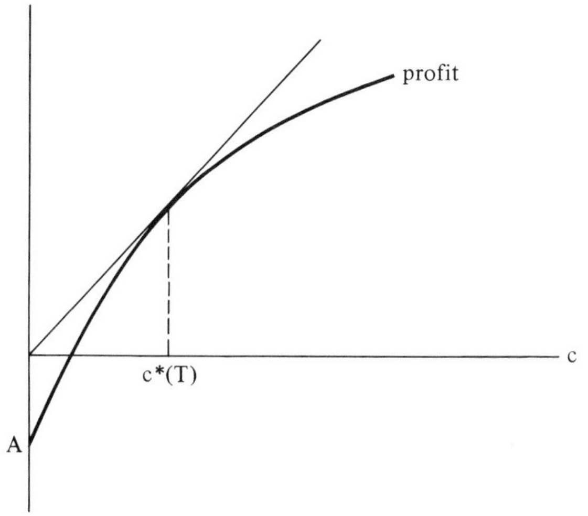

Figure 7.2

We obtained this by noting that $[c^*(t)]^{-\gamma}e^{-\delta t}=[c^*(T)]^{-\gamma}e^{-\delta T}$ and

$$\int_0^T c^*(t)\,dt = s(0)-s_{\mathrm{L}}.$$

We obtain T^* by equating the right-hand side of (7.89) with that of (7.90):

$$e^{\delta T/\gamma}-1=(s(0)-s_{\mathrm{L}})(\delta/\gamma)[(-A)(1-\gamma)/\gamma]^{-1/(1-\gamma)}. \tag{7.91}$$

Thus, T^* is an increasing function of $s(0)$ and a decreasing function of the fixed cost B (where $B\equiv -A>0$).

The alert reader will have noticed that the assumption $A<0$ is essential, given that $1-\gamma>0$; for if both A and $1-\gamma$ are positive, then for any fixed value of T the derivative $\partial V/\partial T$ $(=H(T))$ is positive, indicating that the value of the integral is monotone increasing in T and that the optimal time horizon is infinite. This can be verified directly from (7.75) by differentiating its right-hand side with respect to T.

Thus, problems with free terminal time have the special feature that the optimal path itself depends on the sign of the integrand. For instance, negatively valued utility functions are acceptable for fixed T but yield $T^*=0$ if T is free.

7.6 Free-terminal-time problems with a scrap value function

In free-terminal-time problems with a scrap value function, equation (7.77) is no longer the correct condition.

Additional transversality condition for free-terminal-time problems with a scrap value function

Theorem 7.6.1. For free-terminal-time problems with a scrap value function such as problem (7.27), the following transversality condition is also necessary:

$$H(\mathbf{s}^*(T^*), \mathbf{c}^*(T^*), \boldsymbol{\pi}(T^*), T^*) + \frac{\partial\phi(\mathbf{b}, T^*)}{\partial T} = 0, \tag{7.92}$$

where $\phi(\mathbf{b}, T)$ is the scrap value function. This applies whether or not endpoint values of the state variables are constrained.

To see the necessity of (7.92) recall that the choice of T must maximize

$$V(\mathbf{b}, T) + \phi(\mathbf{b}, T),$$

where $V(\mathbf{b}, T)$ is defined by (7.82). The optimal T^* must satisfy

$$\frac{\partial V}{\partial T} + \frac{\partial \phi}{\partial T} = 0,$$

provided that T^* is finite. But from (7.85) $\partial V/\partial T$ is $H(T)$ and (7.92) follows.

Example 7.6.1. Let us apply (7.92) to a version of the resource extraction problem considered in Section 7.5. Find the extraction path $c^*(t)$ and the closing-down time T^* that maximize

$$\int_0^T [A + (1-\gamma)^{-1} c^{1-\gamma}] e^{-\delta t}\, dt + mbe^{-\delta T}, \tag{7.93}$$

where b is terminal stock, $m > 0$, $A < 0$, $(1-\gamma) > 0$. The maximization is subject to

$$\dot{s}(t) = -c(t), \tag{7.94a}$$

$$s(0) = s_0 \text{ fixed}, \tag{7.94b}$$

$$s(T) = b, \; b \text{ fixed}. \tag{7.94c}$$

The Hamiltonian is (7.86), the necessary conditions are (7.87a)–(7.87c), and because $s(T) = b$ (b fixed) the only transversality condition is, by Theorem 7.6.1,

$$[A + (1-\gamma)^{-1} c(T)^{1-\gamma}] e^{-\delta T} - \pi(T) c(T) - \delta mbe^{-\delta T} = 0. \tag{7.95}$$

Use (7.87a) to simplify (7.95):

$$[-\delta mb + A + (1-\gamma)^{-1} c(T)^{1-\gamma}]/c(T) = [c(T)]^{-\gamma}. \tag{7.96}$$

This solves for $c^*(T)$ uniquely:

$$c^*(T) = [(\delta mb - A)(1-\gamma)\gamma^{-1}]^{1/(1-\gamma)}. \tag{7.97}$$

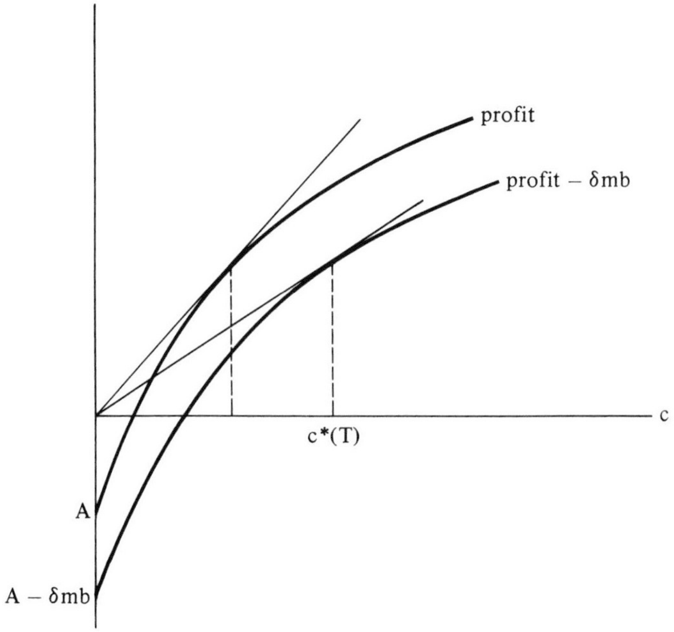

Figure 7.3

This condition and (7.90), with $b = s_L$,

$$c^*(T) = (s(0) - b)(\delta/\gamma)[e^{\delta T/\gamma} - 1]^{-1}, \tag{7.98}$$

determine the optimal time T^* uniquely:

$$T^* = (1/\delta)\ln\{1 + (s_0 - b)(\delta/\gamma)[(\delta mb - A)(1-\gamma)/\gamma]^{-1/(1-\gamma)}\}. \tag{7.99}$$

Equation (7.96) resembles (7.88) but for the presence of the scrap value term. It is illustrated in Figure 7.3, which shows that the value of $c^*(T)$ has been increased by the introduction of the scrap value. The terminal time T^* is smaller, the larger is the scrap value parameter m; see (7.99). Equation (7.95) gives us a concrete example that is helpful for interpreting the transversality condition. The first term consists of the additional discounted profit flow per unit of time that would be obtained if T^* were increased. The third term is the discounted interest cost: if the firm were to delay the closing time, it would have to delay the receipt of the scrap value, and the opportunity cost of this delay is the interest income foregone. The second term is the opportunity cost of the extra amount of stock needed, as we argued earlier (see (7.85′)).

Finally, let us consider the case in which the terminal stock b is free, subject only to a lower bound constraint:

$$s(T)=b\geq s_{\mathrm{L}},\quad s_{\mathrm{L}} \text{ fixed.} \tag{7.100}$$

In this case, in addition to the transversality condition (7.95) we have a second transversality condition:

$$\pi(T)-me^{-\delta T}\geq 0,\quad b-s_{\mathrm{L}}\geq 0,\quad (\pi(T)-me^{-\delta T})(b-s_{\mathrm{L}})=0. \tag{7.101}$$

(This condition is identical to (7.68).)

There are two possibilities: (i) $b^*=s_{\mathrm{L}}$ and (ii) $b^*>s_{\mathrm{L}}$. Let us try the first solution. Substitute s_{L} for b in (7.97) to obtain a specific value of $c^*(T)$, and denote this by $c_{\mathrm{L}}^*(T)$. Equation (7.87a) implies that $\pi(T)$ must take the corresponding value

$$\pi_{\mathrm{L}}(T)=[c_{\mathrm{L}}^*(T)]^{-\gamma}e^{-\delta T}. \tag{7.102}$$

If $[c_{\mathrm{L}}^*(T)]^{-\gamma}\geq m$, then the transversality condition (7.101) is satisfied and therefore $b^*=s_{\mathrm{L}}$ is the optimal solution. If $[c_{\mathrm{L}}^*(T)]^{-\gamma}<m$, then (7.101) is violated, indicating that $b^*>s_{\mathrm{L}}$ is optimal. Now since the lower bound constraint is not binding, we have $\pi(T)=me^{-\delta T}$, implying $[c(T)]^{-\gamma}=m$. This and (7.97) yield an equation that determines b^* uniquely:

$$m^{-1/\gamma}=[(\delta mb^*-A)(1-\gamma)\gamma^{-1}]^{1/(1-\gamma)}.$$

The optimal time T^* can in either case be determined using (7.99).

7.7 Other transversality conditions

In Sections 7.1–7.6 we derived transversality conditions corresponding to various endpoint or terminal-time specifications that are frequently encountered in economic problems. There are, of course, many other, less common specifications. For example, the initial condition $s_i(0)=s_{i0}$ (fixed) may be modified to $s_i(0)=a_i$ (a_i free). Similarly, the initial time condition $t_0=0$ has so far been taken for granted, but there is no reason why t_0 itself cannot be an object of choice. More generally, we may allow equality and inequality constraints involving the initial time t_0, the terminal time T, the initial values of the state variables s_{i0}, and their terminal values s_{iT}.

The reader is invited to use the approach adopted in Sections 7.1–7.6 to derive the following transversality conditions for the various cases.

(a) *Free initial time:*

$$H(t_0)=0. \tag{7.103}$$

(b) *Free initial conditions:* With $s_i(0)=a_i$ free, the transversality condition is

$$\pi_i(0)=0. \tag{7.104}$$

(c) *Free initial conditions, with initial purchase cost:* If the initial capital stock $s_i(t_0)$ can be chosen (i.e., it is "free" in the sense that it is not fixed) but at a cost that increases with its size, the objective function is

$$\int_{t_0}^{T} v(\mathbf{s},\mathbf{c},t)\,dt+\phi(\mathbf{b},T)-\theta(\mathbf{a},t_0), \tag{7.105}$$

where $\theta(\mathbf{a},t_0)$ is the cost of purchasing the stocks $\mathbf{s}(t_0)=\mathbf{a}$. If t_0 is fixed and a_i is free, the transversality condition is

$$\pi_i(t_0)=\frac{\partial\theta}{\partial a_i}. \tag{7.106}$$

If t_0 is also free, we have an additional transversality condition:

$$H(t_0)=-\frac{\partial\theta}{\partial t_0}. \tag{7.107}$$

It becomes apparent that it would be useful to have a general formula from which the various transversality conditions can be derived. Fortunately, a theorem by Hestenes (1966) provides us with exactly what we need. The following section is devoted to this theorem.

7.8 A general formula for transversality conditions

In order to obtain a general formula for transversality conditions, we assume (without loss of generality) that the following variables are potentially free, subject to some constraints:

the initial time t_0,
the terminal time T,
the initial values of the state variables $s_i(t_0)$, $i=1,2,\ldots,n$,
the terminal values of the state variables $s_i(T)$, $i=1,2,\ldots,n$.

For notational convenience, we write

$$t_0=a_0, \tag{7.108a}$$

$$s_i(t_0)=a_i,\quad i=1,2,\ldots,n, \tag{7.108b}$$

$$T=b_0, \tag{7.108c}$$

$$s_i(T)=b_i,\quad i=1,2,\ldots,n. \tag{7.108d}$$

Let $(\mathbf{a},\mathbf{b})=(a_0,a_1,\ldots,a_n,b_0,b_1,\ldots,b_n)$. We assume that the constraints on $(\mathbf{a},\mathbf{b})$ can be expressed in the form of E equality constraints plus I inequality constraints:

$$h^j(\mathbf{a}, \mathbf{b}) = 0, \quad j = 1, 2, \ldots, E,$$
$$h^i(\mathbf{a}, \mathbf{b}) \geq 0, \quad i = E+1, E+2, \ldots, E+I.$$

All the functions h^j and h^i are assumed to be differentiable. In most practical cases they take very simple forms. For example, if t_0 is fixed at 0, then one of the h^j, say h^1, takes the form

$$a_0 - 0 = 0. \tag{7.109}$$

If t_0 is free, but is constrained to lie in the interval [2, 8], say, then we can incorporate this constraint in the form of two h^i-type inequalities:

$$a_0 - 2 \geq 0, \tag{7.110a}$$
$$8 - a_0 \geq 0. \tag{7.110b}$$

Similarly, if $s_1(T)$ is fixed at $\bar{s}_1$, then one of the h^j's, say h^3, takes the form

$$b_1 - \bar{s}_1 = 0. \tag{7.111}$$

If $s_2(T)$ must be at least as large as a certain lower bound s_{2L}, we write as one of the h^i's,

$$b_2 - s_{2L} \geq 0. \tag{7.112}$$

If $s_4(T)$ must be at least twice as great as $s_3(T)$, we write

$$b_4 - 2b_3 \geq 0. \tag{7.113}$$

Our general control problem can now be formulated as follows: find the time path of the control variables $\mathbf{c}(t)$ and the values $(\mathbf{a}, \mathbf{b})$ to maximize

$$W \equiv \int_{t_0}^{T} v(\mathbf{s}(t), \mathbf{c}(t), t)\, dt + G(\mathbf{a}, \mathbf{b}) \tag{7.114}$$

subject to

$$\dot{s}_i(t) = f^i(\mathbf{s}(t), \mathbf{c}(t), t), \quad i = 1, 2, \ldots, n, \tag{7.115a}$$
$$g^j(\mathbf{s}(t), \mathbf{c}(t), t) \geq 0, \quad j = 1, 2, \ldots, m', \tag{7.115b}$$
$$g^k(\mathbf{s}(t), \mathbf{c}(t), t) = 0, \quad k = m'+1, \ldots, m, \tag{7.115c}$$
$$s_i(t_0) = a_i, \quad i = 1, 2, \ldots, n, \tag{7.115d}$$
$$s_i(T) = b_i, \quad i = 1, 2, \ldots, n, \tag{7.115e}$$
$$t_0 = a_0, \tag{7.115f}$$
$$T = b_0, \tag{7.115g}$$
$$h^j(\mathbf{a}, \mathbf{b}) = 0, \quad j = 1, 2, \ldots, E, \tag{7.115h}$$
$$h^i(\mathbf{a}, \mathbf{b}) \geq 0, \quad i = E+1, \ldots, E+I. \tag{7.115i}$$

Notice that the function $G(\mathbf{a}, \mathbf{b})$ in (7.114) generalizes the concepts of scrap value function and initial cost function. It might take a simple form such as

$$mb_1e^{-\delta b_0} - qa_1e^{-\delta a_0},$$

which has the interpretation of a scrap value function minus an initial purchase cost function, where δ is the rate of discount, m the disposal value per unit, and q the purchase price per unit.

We are now ready to state the theorem on transversality conditions.

Theorem 7.8.1. If problem (7.114) has an optimal solution, then this solution must satisfy the necessary conditions (6.73)–(6.77) and the following transversality conditions (where all derivatives and all functions are evaluated at $\mathbf{s}^*$, $\mathbf{c}^*$, $\mathbf{a}^*$, $\mathbf{b}^*$):

$$H(t_0) - \frac{\partial G}{\partial a_0} - \sum_{k=1}^{I+E} \mu_k \frac{\partial h^k}{\partial a_0} = 0, \tag{7.116a}$$

$$\pi_i(t_0) + \frac{\partial G}{\partial a_i} + \sum_{k=1}^{I+E} \mu_k \frac{\partial h^k}{\partial a_i} = 0, \tag{7.116b}$$

where $i = 1, 2, \ldots, n$;

$$H(T) + \frac{\partial G}{\partial b_0} + \sum_{k=1}^{I+E} \mu_k \frac{\partial h^k}{\partial b_0} = 0, \tag{7.116c}$$

$$\pi_i(T) - \frac{\partial G}{\partial b_i} - \sum_{k=1}^{I+E} \mu_k \frac{\partial h^k}{\partial b_i} = 0, \tag{7.116d}$$

where $i = 1, 2, \ldots, n$, and where μ_k are multipliers with the following properties:

(i) for $k = 1, 2, \ldots, E$, μ_k are constants and

$$h^k(\mathbf{a}^*, \mathbf{b}^*) = 0; \tag{7.116e}$$

(ii) for $k = E+1, \ldots, E+I$, μ_k are constants and

$$\mu_k \geq 0, \quad h^k(\mathbf{a}^*, \mathbf{b}^*) \geq 0, \quad \mu_k h^k(\mathbf{a}^*, \mathbf{b}^*) = 0. \tag{7.116f}$$

Remark. One should distinguish $\boldsymbol{\lambda}(t)$, piecewise-continuous multipliers associated with (7.115b)–(7.115c) and included in the Hamiltonian, from $\boldsymbol{\mu}$, constant multipliers associated with (7.115h)–(7.115i) and not included in the Hamiltonian.

We will not offer a proof of the theorem. The reader is invited to construct a heuristic proof along the lines we used to prove the various transversality conditions in Sections 7.1–7.6. The reader should also check that the conditions stated in those sections are special cases of (7.116a)–(7.116f).

We should warn the reader that Theorem 7.8.1 has not been stated in the most rigorous language. For example, we should have included a shadow price $\pi_0 \geq 0$, but we have chosen to set $\pi_0 = 1$ even though in some abnormal cases $\pi_0 = 0$. The reader should recall a similar remark in Section 6.3, where we referred to other sources for further discussion. For a precise statement and a formal proof of Theorem 7.8.1, the reader may consult Hestenes (1966, theorem 11.1); see also Long and Vousden (1977, pp. 14–19) for discussion.

7.9 Sufficiency theorems

In Chapter 6, we offered a sufficiency theorem for the case of fixed endpoint and fixed time horizon. That theorem is not applicable to the various cases considered in the present chapter. We now offer a number of sufficiency theorems which ensure that, under certain conditions, any solution that satisfies the necessary conditions (including the transversality conditions) is an optimal solution.

It turns out that if both T and t_0 are fixed and finite, then the required sufficiency theorem is only a simple adaptation of the one offered in Chapter 6. Matters are not so simple when either t_0 or T (or both) are free. For the first case, we offer the following theorem.

Theorem 7.9.1: sufficiency. Let both t_0 and T be fixed and finite. Without loss, we set $t_0 = 0$. Let $(\mathbf{s}^*(t), \mathbf{c}^*(t))$ with initial and terminal endpoints $(a_1^*, a_2^*, \ldots, a_n^*, b_1^*, b_2^*, \ldots, b_n^*)$ be a path satisfying (7.115a)–(7.115i), let $\boldsymbol{\pi}^*(t)$, $\boldsymbol{\lambda}^*(t)$ be costate variables and Lagrange multipliers associated with (7.115a)–(7.115c), and let $\boldsymbol{\mu}^*$ be constant multipliers associated with endpoint conditions (7.115h)–(7.115i). Then the necessary conditions stated in Theorem 7.8.1, that is, conditions (6.73)–(6.77) and (7.116a)–(7.116f), are also sufficient for a global maximum provided that the following conditions are satisfied:

(i) the Lagrangean $\mathcal{L} = v + \boldsymbol{\pi}^* \cdot \mathbf{f} + \boldsymbol{\lambda}^* \cdot \mathbf{g}$ is concave in the variables $(\mathbf{s}, \mathbf{c})$;

(ii) the function $G(\mathbf{a}, \mathbf{b})$ is concave in the variables $(a_1, a_2, \ldots, a_n, b_1, b_2, \ldots, b_n)$;

(iii) the functions $\mu_k^* h^k(\mathbf{a}, \mathbf{b})$, $k = 1, 2, \ldots, E+I$, are concave in $(a_1, a_2, \ldots, a_n, b_1, b_2, \ldots, b_n)$.

Proof. Let $\underline{\mathbf{a}} = (a_1, a_2, \ldots, a_n)$ and $\underline{\mathbf{b}} = (b_1, b_2, \ldots, b_n)$ so that G and $\mathbf{h}$ become $G(a_0, \underline{\mathbf{a}}, b_0, \underline{\mathbf{b}})$ and $\mathbf{h}(a_0, \underline{\mathbf{a}}, b_0, \underline{\mathbf{b}})$. Note that both a_0 and b_0 are fixed (t_0 and T being fixed). Let G^* and $\mathbf{h}^*$ denote the functions G and $\mathbf{h}$ evaluated at $(\underline{\mathbf{a}}^*, \underline{\mathbf{b}}^*)$. The notation $\boldsymbol{\pi}^* \cdot \mathbf{f}$, $\boldsymbol{\lambda}^* \cdot \mathbf{g}$, and $\boldsymbol{\mu}^* \cdot \mathbf{h}$ denotes the inner products as explained in Section 6.5. Then, recalling the proof of Theorem 6.5.2,

$$W^*-W=V^*-V+G^*-G$$
$$\geq \int_0^T [(H^*+\dot{\pi}^*\cdot s^*)-(H+\dot{\pi}^*\cdot s)]\,dt+\pi^*(0)\cdot(\underline{a}^*-\underline{a})$$
$$-\pi^*(T)\cdot(\underline{b}^*-\underline{b})+(\underline{a}^*-\underline{a})\cdot\frac{\partial G^*}{\partial \underline{a}}+(\underline{b}^*-\underline{b})\cdot\frac{\partial G^*}{\partial \underline{b}}$$
(by the concavity of G)
$$=\int_0^T [(H^*+\dot{\pi}^*\cdot s^*)-(H+\dot{\pi}^*\cdot s)]\,dt-\frac{\mu^*\cdot\partial h^*}{\partial \underline{a}}\cdot(\underline{a}^*-\underline{a})$$
$$-\frac{\mu^*\cdot\partial h^*}{\partial \underline{b}}\cdot(\underline{b}^*-\underline{b}) \quad \text{(by (7.116b) and (7.116d))}$$
$$\geq \int_0^T [(H^*+\dot{\pi}^*\cdot s^*)-(H+\dot{\pi}^*\cdot s)]\,dt-\mu^*\cdot(h^*-h)$$
(by the concavity of $\mu^*\cdot h$)
$$\geq \int_0^T [(H^*+\dot{\pi}^*\cdot s^*)-(H+\dot{\pi}^*\cdot s)]\,dt$$
(because $\mu^*\cdot h^*=0$ and $\mu^*\cdot h\geq 0$)
$$\geq \int_0^T [(\mathcal{L}^*-\mathcal{L})+\dot{\pi}^*\cdot(s^*-s)]\,dt. \tag{7.117}$$

The last integral is nonnegative by the concavity of $\mathcal{L}$; see the proof of Theorem 6.5.2. □

Remark. Condition (i), that the Lagrangean is concave in $(\mathbf{s},\mathbf{c})$, can be replaced by a weaker condition. Let us define the function

$$H^0(\mathbf{s},\pi,t)=\max_{\mathbf{c}} H(\mathbf{s},\mathbf{c},\pi,t),$$

where $\mathbf{c}$ belongs to the set of admissible controls defined by equation (6.6). We can replace (i) by

(i′) $H^0(\mathbf{s},\pi^*,t)$ is concave in $\mathbf{s}$.

To see that (i′) implies that (7.117) is nonnegative, note that

$$\mathcal{L}^*=H^0(\mathbf{s}^*,\pi^*,t)\equiv H^{0*},$$
$$\mathcal{L}\leq H^0(\mathbf{s},\pi^*,t).$$

Hence,

$$\mathcal{L}^*-\mathcal{L}\geq H^0(\mathbf{s}^*,\pi^*,t)-H^0(\mathbf{s},\pi^*,t)$$
$$\geq (\mathbf{s}^*-\mathbf{s})\cdot(\partial H^{0*}/\partial \mathbf{s}) \quad \text{(by the concavity of } H^0)$$
$$=-(\mathbf{s}^*-\mathbf{s})\cdot\dot{\pi}^*$$

(because $\partial H^{0*}/\partial \mathbf{s}=\partial\mathcal{L}^*/\partial \mathbf{s}$, as a result of the envelope theorem).

Theorem 7.9.1 applies to the case in which both t_0 and T are fixed and finite. Consider now the case in which T is free. Then to ensure that T^*

is a maximizing choice (and not a minimizing one) we must add to the list (i), (ii), (iii) of Theorem 7.9.1 a fourth condition:

(iv) The function $W = V + G(\mathbf{a}, \mathbf{b})$ is concave in b_0 (recall that $b_0 = T$).

Condition (iv) is rather stringent because V itself is a function of T. Take the simplest case, in which b_0 does not appear in $\mathbf{h}$ or in G. Then (iv) is satisfied if and only if $H(T)$ $(= \partial V/\partial T)$ is a decreasing function of T. This means that $H(T)$ must be positive for all $T < T^*$ and negative for all $T > T^*$, where T^* is the optimal terminal time. It is very difficult to check that these conditions are satisfied. (The reader may consult Seierstad, 1984, for further discussion.)

7.10 A summary table of common transversality conditions

Although Theorem 7.8.1 provides a very useful and general formula for obtaining transversality conditions for a variety of cases, for some purposes it may be more convenient to have a simple table summarizing these conditions for common cases.

For convenience, we restate problem (7.114) here, in a somewhat less general form. Find $\mathbf{c}(t)$ and, depending on the cases, t_0, T, $s_i(T)$, $s_j(T)$, $s_k(t_0)$, $s_h(t_0)$ for some or all i, j, k, or h, so as to maximize

$$W = \int_{t_0}^{T} v(\mathbf{s}(t), \mathbf{c}(t), t)\, dt + \phi(\mathbf{s}(T), T) - \theta(\mathbf{s}(t_0), t_0), \tag{7.118}$$

where ϕ is the scrap value function and θ the initial cost function. The maximization is subject to (7.115a)–(7.115c) and other conditions as stated in the table.

Table 7.1 lists six main features, (A1)–(A6). If a problem has several of these main features, all of the corresponding transversality conditions apply. Categories (B1)–(B6) generalize (A1)–(A6) to the case with a scrap value function and an initial value function.

7.11 Control parameters

In some economic problems the restrictions imposed on the actions of the central planner take on the peculiar form of requiring some controls to remain constant over the whole horizon; we label these *control parameters*. In Section 1.5 we discussed optimal peakload policies whereby the plant capacity was to be selected by the firm but had to remain fixed during the whole program; we will shortly reformulate this problem in an optimal control format with the capacity as a control parameter. Another example would be to include in an exhaustible resource problem the purchase of the mine itself with the quality of the ore treated as a control

Table 7.1. *Common transversality conditions*

	$s_i(T)$ free (1)	$s_j(T) \geq s_{j\mathrm{L}}$ (2)	Free terminal time T (3)	$s_k(t_0)$ free (4)	$s_h(t_0) \leq \bar{s}_h$ (5)	Free initial time t_0 (6)
(A) No scrap value, no initial cost	$\pi_i(T)=0$	$\pi_j(T) \geq 0$ $\pi_j(T)[s_j(T)-s_{j\mathrm{L}}]=0$	$H(T^*)=0$	$\pi_k(t_0)=0$	$\pi_h(t_0) \geq 0$ $\pi_h(t_0)[\bar{s}_h - s_h(t_0)]=0$	$H(t_0^*)=0$
(B) With scrap value[a] and initial cost[b]	$\pi_i(T)=\dfrac{\partial\phi}{\partial s_i(T)}$	$\pi_j(T)-\dfrac{\partial\phi}{\partial s_j(T)} \geq 0$ $\left[\pi_j(T)-\dfrac{\partial\phi}{\partial s_j(T)}\right]$ $\times[s_j(T)-s_{j\mathrm{L}}]=0$	$H(T^*)+\dfrac{\partial\phi}{\partial T}=0$	$\pi_k(t_0)=\dfrac{\partial\theta}{\partial s_k(t_0)}$	$\pi_h(t_0)-\dfrac{\partial\theta}{\partial s_h(t_0)} \geq 0$ $\left[\pi_h(t_0)-\dfrac{\partial\theta}{\partial s_h(t_0)}\right]$ $\times[s_h(t_0)-\bar{s}_h]=0$	$H(t_0^*)+\dfrac{\partial\theta}{\partial t_0}=0$

[a] $\phi(s(T),T)$. [b] $\theta(s(t_0),t_0)$.

parameter: once chosen, it would remain constant throughout exploitation and influence the efficiency of extraction; it would also figure in the purchase price of the mine. We shall see that some of the transversality conditions encountered earlier in this chapter can be seen as special cases of the conditions characterizing the choice of control parameters when these parameters are the terminal value of state variables or terminal time, for instance. We shall also see that control parameters can sometimes be treated as an additional state variable, say $\beta(t)$, with the equation of motion $\dot{\beta}(t)=0$ and free initial and terminal values.

We first state the main result. Consider again problem (7.1) with the addition of a vector of (constant) control parameters $\boldsymbol{\beta}=[\beta_1,\ldots,\beta_R]'$. Choose $\mathbf{c}(t)$ and $\boldsymbol{\beta}$ to maximize

$$V=\int_{t_0}^{T} v(\mathbf{s}(t),\mathbf{c}(t),\boldsymbol{\beta},t)\,dt-K(\boldsymbol{\beta}) \tag{7.119a}$$

subject to

$$\dot{s}_i(t)=f^i(\mathbf{s}(t),\mathbf{c}(t),\boldsymbol{\beta},t),\quad i=1,\ldots,n, \tag{7.119b}$$

$$g^j(\mathbf{s}(t),\mathbf{c}(t),\boldsymbol{\beta},t)\geq 0,\quad j=1,\ldots,m', \tag{7.119c}$$

$$g^j(\mathbf{s}(t),\mathbf{c}(t),\boldsymbol{\beta},t)=0,\quad j=m'+1,\ldots,m, \tag{7.119d}$$

$$s_i(t_0)=A_i^0(\boldsymbol{\beta}),\ s_i(T)=A_i^T(\boldsymbol{\beta}),\quad i=1,\ldots,n, \tag{7.119e}$$

$$t_0=a^0(\boldsymbol{\beta}),\quad T=a^T(\boldsymbol{\beta}). \tag{7.119f}$$

Theorem 7.11.1. In problem (7.119) the optimal choice of the vector of control parameters $\boldsymbol{\beta}$ implies the following necessary conditions – in addition to those of Theorem 6.5.1:

$$\mathcal{L}(t_0)\frac{\partial a^0}{\partial\beta_r}-\sum_{i=1}^{n}\pi_i(t_0)\frac{\partial A_i^0}{\partial\beta_r}-\left[\mathcal{L}(T)\frac{\partial a^T}{\partial\beta_r}-\sum_{i=1}^{n}\pi_i(T)\frac{\partial A_i^T}{\partial\beta_r}\right]$$
$$=-\frac{\partial K}{\partial\beta_r}+\int_{t_0}^{T}\frac{\partial\mathcal{L}(t)}{\partial\beta_r}\,dt,\quad r=1,\ldots,R, \tag{7.120}$$

where

$$\mathcal{L}(t)=v(\mathbf{s}(t),\mathbf{c}(t),\boldsymbol{\beta},t)+\sum_{i=1}^{n}\pi_i(t)f^i(\mathbf{s}(t),\mathbf{c}(t),\boldsymbol{\beta},t)$$
$$+\sum_{j=1}^{m}\lambda_j(t)g^j(\mathbf{s}(t),\mathbf{c}(t),\boldsymbol{\beta},t).$$

A heuristic derivation of (7.120) is most easily obtained by transforming V in the manner of Section 7.5, where for arbitrary $\boldsymbol{\beta}$ and $\boldsymbol{\pi}(t)$ we have

$$V(\boldsymbol{\beta})=\int_{a^0(\beta)}^{a^T(\beta)}\left[\mathcal{L}(t)+\sum_{i=1}^{n}\dot{\pi}_i(t)s_i(t)\right]dt$$
$$-\sum_{i=1}^{n}\pi_i(T)A_i^T(\boldsymbol{\beta})+\sum_{i=1}^{n}\pi_i(t_0)A_i^0(\boldsymbol{\beta})-K(\boldsymbol{\beta}).$$

Using Leibniz's rule (Appendix to Chapter 2) we can calculate $\partial V/\partial\beta_r$ and set it to zero:

$$\begin{aligned}\frac{\partial V}{\partial\beta_r} &= \mathcal{L}(T)\frac{\partial a^T}{\partial\beta_r} - \mathcal{L}(t_0)\frac{\partial a^0}{\partial\beta_r} + \left[\sum_{i=1}^n \dot{\pi}_i(T)s_i(T)\right]\frac{\partial a^T}{\partial\beta_r} \\ &\quad - \left[\sum_{i=1}^n \dot{\pi}_i(t_0)s_i(t_0)\right]\frac{\partial a^0}{\partial\beta_r} \\ &\quad + \int_{t_0}^T\left[\frac{\partial\mathcal{L}(t)}{\partial\beta_r} + \sum_{i=1}^n\left(\frac{\partial\mathcal{L}(t)}{\partial s_i} + \dot{\pi}_i(t)\right)\frac{\partial s_i(t)}{\partial\beta_r} + \sum_{j=1}^N \frac{\partial\mathcal{L}(t)}{\partial c_j}\frac{\partial c_j(t)}{\partial\beta_r}\right]dt \\ &\quad - \sum_{i=1}^n\left[\frac{d\pi_i(T)}{dT}A_i^T(\beta)\frac{\partial a_i^T}{\partial\beta_r}\right] + \sum_{i=1}^n\left[\frac{d\pi_i(t_0)}{dt_0}A_i^0(\beta)\frac{\partial a_i^0}{\partial\beta_r}\right] \\ &\quad - \sum_{i=1}^n \pi_i(T)\frac{\partial A_i^T}{\partial\beta_r} + \sum_{i=1}^n \pi_i(t_0)\frac{\partial A_i^0}{\partial\beta_r} - \frac{\partial K}{\partial\beta_r} = 0,\end{aligned}$$

where the first three lines come from Leibniz's rule. When we use the optimal $\pi_i(t)$ trajectories, this expression reduces to (7.120) with the usual simplifications.

Remark. We have proceeded under the implicit assumption that the choice of each β_r is unrestricted. If restrictions are imposed, the usual modifications are required. For instance, if we impose $\beta_r \geq 0$, the above equality must be changed to $\partial V/\partial\beta_r \leq 0$ with $\beta_r \geq 0$ and $\beta_r(\partial V/\partial\beta_r) = 0$. This would result in a "larger than or equal to" sign ($\geq$) in (7.120).

Although these calculations appear complicated, they take on a much more familiar aspect in some special cases.

Special case 1. The control parameter vector does not appear as a distinct argument in any of the functions v, f^i, or g^j. Then we have to maximize

$$\int_{t_0}^T v(\mathbf{s},\mathbf{c},t)\,dt - K(\boldsymbol{\beta})$$

subject to

$$\begin{aligned}&\dot{s}_i = f^i(\mathbf{s},\mathbf{c},t), \quad i=1,\ldots,n,\\ &g^j(\mathbf{s},\mathbf{c},t) \geq 0, \quad j=1,\ldots,m',\\ &g^j(\mathbf{s},\mathbf{c},t) = 0, \quad j=m'+1,\ldots,m,\\ &s_i(t_0) = A_i^0(\boldsymbol{\beta}),\ s_i(T) = A_i^T(\boldsymbol{\beta}), \quad i=1,\ldots,n,\\ &t_0 = a^0(\boldsymbol{\beta}), \quad T = a^T(\boldsymbol{\beta}).\end{aligned}$$

This is easily recognized as a problem similar to the one of Section 7.8, where a general formula for transversality conditions was derived. Indeed, if we take $\boldsymbol{\beta}=(\mathbf{a},\mathbf{b})$ of (7.115) with $K(\boldsymbol{\beta})=-G(\mathbf{a},\mathbf{b})$, $A_i^0(\boldsymbol{\beta})=a_i$, $A_i^T(\boldsymbol{\beta})=b_i$, $a^0(\boldsymbol{\beta})=a_0$, and $a^T(\boldsymbol{\beta})=b_0$, we have here problem (7.115) but without (7.115h) and (7.115i). The reader is invited to use (7.120) to obtain (7.116).

Special case 2. The control parameter does not appear in the initial and terminal conditions imposed on $s_i(t_0)$, $s_i(T)$, t_0, or T, $i=1,\ldots,n$. The problem then reduces to (7.119a)–(7.119d). We can obtain the necessary conditions (7.120) by introducing R new state variables $\beta_r(t)$, $r=1,\ldots,R$, each with the equation of motion

$$\dot{\beta}_r(t)=0 \tag{7.121}$$

and boundary conditions

$$\beta_r(t_0)\equiv\beta_r \text{ free} \quad \text{and} \quad \beta_r(T) \text{ free.} \tag{7.122}$$

Let the corresponding costate variables be $\Pi_r(t)$. The new Lagrangean is

$$\mathfrak{L}(t)=v(\mathbf{s},\mathbf{c},\boldsymbol{\beta},t)+\sum_{i=1}^{n}\pi_i f^i(\mathbf{s},\mathbf{c},\boldsymbol{\beta},t)+\sum_{r=1}^{R}\Pi_r\times 0+\sum_{j=1}^{m}\lambda_j g^j(\mathbf{s},\mathbf{c},\boldsymbol{\beta},t),$$

which is the same as in Theorem 7.11.1.

The only new equations are $(r=1,\ldots,R)$

$$\dot{\Pi}_r=-\frac{\partial\mathfrak{L}(t)}{\partial\beta_r}, \tag{7.123}$$

and the transversality conditions at initial and terminal time are

$$\Pi_r(t_0)=\frac{\partial K}{\partial\beta_r} \tag{7.124a}$$

because $\beta_r(t_0)\equiv\beta_r$ is free but appears in the initial cost $K(\beta)$, by convention

$$\Pi_r(T)=0 \tag{7.124b}$$

because $\beta_r(T)$ is free. Together they imply

$$\int_{t_0}^{T}-\frac{\partial\mathfrak{L}(t)}{\partial\beta_r}\,dt=0-\Pi_r(t_0) \quad \text{or} \quad \int_{t_0}^{T}\frac{\partial\mathfrak{L}}{\partial\beta_r}\,dt-\frac{\partial K}{\partial\beta_r}=0,$$

which is (7.120) in this special case. (Note that we could have equally well taken the convention that $\beta_r(T)\equiv\beta_r$ appears in K while $\beta_r(t_0)$ does not, but although the same result follows, this appears very artificial.)

We now offer an example of the second special case.

Example 7.11.1: peakload policy in continuous time. We follow the notation of Section 1.5: X denotes plant capacity, $x(t)$ is the supply at time t, $R_t(x_t)$ and $C_t(x_t)$ are the instantaneous revenue and cost functions, respectively, $K(X)$ is the cost of capital at time 0, and δ is the discount rate. The control parameter X is dealt with by introducing a state variable $X(t)$. We must maximize

$$\int_0^T [R_t(x(t)) - C_t(x(t))]e^{-\delta t}\,dt - K(X)$$

subject to

$$\dot{X}(t) = 0, \quad X(0) \equiv X \text{ free}, \quad X(T) \text{ free}, \quad 0 \le x(t) \le X(t),$$
$$\mathcal{L}(t) = [R_t(x(t)) - C_t(x(t))]e^{-\delta t} + \Pi(t) \times 0 + \lambda(t)[X(t) - x(t)].$$

The necessary conditions are

$$\frac{\partial \mathcal{L}(t)}{\partial x(t)} = [\mathrm{MR}_t(x(t)) - \mathrm{MC}_t(x(t))]e^{-\delta t} - \lambda(t) \le 0,$$
$$x(t) \ge 0, \quad \frac{\partial \mathcal{L}(t)}{\partial x(t)} x(t) = 0,$$

$$\dot{\Pi}(t) = -\lambda(t), \quad \Pi(0) = K'(X), \quad \Pi(T) = 0,$$
$$X(t) - x(t) \ge 0, \quad \lambda(t) \ge 0, \quad \lambda(t)[X(t) - x(t)] = 0.$$

From the differential equation in Π we obtain

$$\Pi(T) - \Pi(0) = -\int_0^T \lambda(t)\,dt,$$

or

$$0 = -K'(X) + \int_0^T \lambda(t)\,dt,$$

which we recognize as a special case of (7.120). The economic interpretation of the other conditions is as in Section 1.5. Note that the function $K(X)$ might also be interpreted as the initial cost net of the discounted scrap value of X after T periods of use.

Example 7.11.2: optimal choice of the quality and size of a mine. Suppose that you can buy a mine with known reserves, β_1 say, and a known quality of ore, β_2 say. If the rate of extraction is $x(t)$, this ore will provide $c(t) = f(\beta_2)x(t)$ units of output, which is nonstorable and yields a net benefit $v(c(t))$; the discount rate is $\delta > 0$. Let the purchase price of the mine be $p(\beta_2)$ per unit of reserves. After eliminating $c(t)$ the problem is to choose β_1, β_2, and $x(t)$ to maximize

$$\int_0^T v(f(\beta_2)x(t))e^{-\delta t}\,dt - p(\beta_2)\beta_1$$

subject to

$$\dot{s}(t) = -x(t), \quad s(0) = \beta_1, \quad s(T) = 0. \tag{7.125}$$

The function v is defined on positive c values only, and v and f are increasing functions with enough concavity to yield a concave Hamiltonian (e.g., $v(c)=\ln c$ and f strictly concave). The Hamiltonian is

$$H(t)=v(f(\beta_2)x)e^{-\delta t}-\pi x,$$

and the necessary conditions are

$$v'(f(\beta_2)x)f(\beta_2)e^{-\delta t}=\pi, \tag{7.126}$$

$$\dot{\pi}=0, \tag{7.127}$$

$$-\pi(0)\frac{\partial\beta_1}{\partial\beta_1}=-\frac{\partial}{\partial\beta_1}p(\beta_2)\beta_1, \quad \text{or} \quad \pi(0)=p(\beta_2), \tag{7.128}$$

$$0=-\frac{\partial}{\partial\beta_2}p(\beta_2)\beta_1+\int_0^T\frac{\partial H(t)}{\partial\beta_2}\,dt, \quad \text{or}$$

$$p'(\beta_2)\beta_1=\int_0^T v'(f(\beta_2)x)f'(\beta_2)xe^{-\delta t}\,dt. \tag{7.129}$$

From (7.126) $v'(f(\beta_2)x)e^{-\delta t}=\pi/f(\beta_2)$, a constant, and (7.129) can be integrated to yield, with (7.125),

$$\begin{aligned}p'(\beta_2)\beta_1&=v'(f(\beta_2)x)f'(\beta_2)e^{-\delta t}\int_0^T x\,dt\\&=\frac{\pi}{f(\beta_2)}f'(\beta_2)[\beta_1-0]\\&=\frac{p(\beta_2)}{f(\beta_2)}f'(\beta_2)\beta_1 \quad \text{by (7.128)},\end{aligned} \tag{7.130}$$

and finally we obtain

$$\frac{p'(\beta_2)}{p(\beta_2)}=\frac{f'(\beta_2)}{f(\beta_2)}, \tag{7.131}$$

which defines the optimal quality of the ore, say β_2^*, which we will assume is uniquely defined. Once this is known (7.126) yields

$$v'(f(\beta_2^*)x)f(\beta_2^*)e^{-\delta t}=p(\beta_2^*),$$

from which $x^*(t)$ can be calculated. Finally, $\beta_1^*=\int_0^T x^*(t)\,dt$ indicates the optimal size of the mine. Note that (7.131) requires that the proportional rate of gain in productivity, $f'(\beta_2)/f(\beta_2)$, equal the proportional rate of increase in price, $p'(\beta_2)/p(\beta_2)$.

This concludes our presentation of control parameters. For further details see Long and Vousden (1977) and Hestenes (1966).

Exercises

1. (Free endpoint) Find the time path of the control variable $c(t)$ and the terminal value of the state variable, $s(1)$, that maximize the integral of net benefit from

time 0 to time 1. Maximize $\int_0^1 [\alpha s(t) - \beta(c(t))^2]\,dt$ subject to $\dot{s}(t) = c(t)$, $s(0) = 0$, $s(1)$ free, where α and β are specified positive constants.

2. (Free endpoint with scrap value)
   (a) Derive the necessary conditions for the following problem:

$$V_1 \equiv \max_{c(t),\, s(T_1)} \int_0^{T_1} \ln(c(t))\,dt + \phi(s(T_1), T_1)$$

subject to $\dot{s}(t) = rs(t) - c(t)$, $s(0) = s_0$, $s(T_1)$ free, where

$$\phi(s(T_1), T_1) = (T_2 - T_1)\ln[s(T_1)e^{-rT_1}/(T_2 - T_1)] + r((T_2)^2 - (T_1)^2)/2,$$

and s_0, T_1, and T_2 are specified positive constants, $T_1 < T_2$.
   (b) Show that $s^*(T_1) = s_0(T_2 - T_1)e^{rT_1}/T_2$.
   (c) Show that over the interval $[0, T_1]$ the solution to the problem is identical to the solution to the following problem:

$$V_2 \equiv \max_{c(t)} \int_0^{T_2} \ln(c(t))\,dt$$

subject to $\dot{s}(t) = rs(t) - c(t)$, $s(0) = s_0$, and $s(T_2) = 0$. Show that $V_1 = V_2$. Can you guess the exact form of

$$V_{12} \equiv \max_{c(t)} \int_{T_1}^{T_2} \ln(c(t))\,dt$$

subject to $\dot{s}(t) = rs(t) - c(t)$, $s(T_1) = s^*(T_1)$, and $s(T_2) = 0$? Verify that your guess is correct.

3. In a study of the "political business cycle," Nordhaus assumes that the political party in power seeks to maximize its popularity index V before the next election takes place, at date T. Maximize $V = \int_0^T v(u(t), p(t))e^{\mu t}\,dt$ subject to $\dot{s}(t) = \beta(p(t) - s(t))$, $s(0) = s_0$, and $p(t) = m - nu(t) + bs(t)$, where s_0, μ, β, m, n, and b are specified positive constants and $b < 1$. The control is the rate of unemployment $u(t)$; $p(t)$ is the rate of inflation; μ is the fixed rate of decay of voters' memories, and the state variable $s(t)$ represents the expected rate of inflation. We take $v(u, p) = \delta - u^2 - Kp$, where δ and K are positive constants. Note that $s(T)$ is free.
   (a) Eliminate $p(t)$, apply the maximum principle, and solve explicitly for $u(t)$. Show that the optimal rate of unemployment is

$$u^*(t) = -B/A + [Kn/2 + B/A]e^{A(t-T)},$$

where $A \equiv \beta(1-b) - \mu$ and $B \equiv (\mu - \beta)Kn/2$. Prove that $u^*(t)$ is a monotone-decreasing function of time.
   (b) Construct a phase diagram in the (u, s) space for each of the three cases: $A > 0$ and $B < 0$; $A < 0$ and $B < 0$; $A < 0$ and $B > 0$. (*Hint:* Show that the transversality condition implies $u(T) = Kn/2$.)

4. (Free endpoint, free time) Modify the problem of exercise 3 by relaxing the assumption that T is exogenously fixed. Suppose that the government can choose T within some upper and lower bounds: $T_L \le T \le T_M$. What are the additional transversality conditions associated with the optimal choice of T (to be denoted

by T^*)? Show that if $T_L < T^* < T_M$, the transversality conditions uniquely determine $s(T^*)$.

5. The following problem is sometimes called doomsday or the gold miner problem. We prefer the fable of an economist marooned on a desert island with no hope of rescue, but in possession of a supply of cans of macaroni and cheese. After assuming that he has a can opener, we can address the following question: what is the rate of consumption that will maximize the economist's total utility before food runs out and the inevitable occurs? It is supposed that he must maintain at least some minimum intake $\bar{c}$, at which his utility level is zero. Formally, he must choose $c(t)$ and T to maximize $\int_0^T u(c(t))e^{-\delta t}\,dt$ subject to $\dot{s}(t) = -c(t)$, $c(t) \geq \bar{c}$; s_0 and δ are specified positive constants; u is strictly increasing and concave, and $u(\bar{c}) = 0$.
   (a) Apply the maximum principle, not omitting the transversality conditions relating to the choice of T and $s(T)$.
   (b) Show that the consumption at time T must exceed $\bar{c}$. Use the transversality condition and the first-order condition to characterize the optimal value $c^*(T)$; depict it on a graph of $u(c)$. Show that $s^*(T) = 0$. On a (c, s) phase diagram can you now represent the optimal path? Does $c(t)$ ever equal $\bar{c}$? Why does this make sense? Is the optimal value of T finite? How would a zero discount rate ($\delta = 0$) affect your findings?

6. Consider now a more pleasant variation on the theme of exercise 5. Suppose that you have come in possession of some amount of capital K_0 at time 0. You may keep it as long as you wish, say time T, when you must give back an amount at least as large as K_T, a specified positive constant. You have no other means of support. All other assumptions are the same as in exercise 5 except that there is a return on capital; thus, $\dot{K}(t) = rK(t) - c(t)$, where r is a specified positive constant and $K(T) \geq K_T$. In addition we assume $rK_0 > \bar{c} > rK_T$.

   Specify the problem and apply the maximum principle. Show by contradiction that $c^*(T) > \bar{c}$; use this result to derive an equation that defines $c^*(T)$; illustrate this on the graph of $u(c)$. Draw phase diagrams in the (c, K) plane, distinguishing various cases according to the relative values of δ and r. Give a description of the optimal policies. Does $c(t)$ always exceed $\bar{c}$? Show that even when $\delta < r$ the path of K is a monotone function of time (because at any instant, terminal time is freely chosen).

7. You have taken control of an established private club, which has a number of faithful old members. You are considering advertising as a means of increasing revenue. This brings in a flow of new but ephemeral visitors at the time of advertising. Unfortunately, it also drives away some old members, thereby diminishing their numbers permanently. There is also a cost to advertising. You want to maximize the total present value of net revenue over some specified planning horizon, at which date the club must close down; the number of members at that date is of no consequence.

   Let $s(t)$ be the stock of old members at time t, $x(t)$ be the flow of advertising that generates $\alpha x(t)$ ephemeral visitors at time t. The total number of customers at time t is $z(t) = s(t) + \alpha x(t)$. The rate of change in the stock of members is $\dot{s}(t) = -\gamma x(t)$. The revenue function $R(z(t))$ is increasing and concave; the unit cost of advertising is c, and we assume that there exists a positive value $\bar{z}$ such that $\alpha R'(\bar{z}) = c$. You must choose $x(t) \geq 0$ to maximize the total present value of profits (the discount rate is δ) subject to $s(0) = s_0$, $s(T) \geq 0$, and the above restrictions. s_0, T, α, γ and c are specified positive constants.

Formulate the problem (eliminate z). Apply the maximum principle and interpret the necessary conditions. Are they also sufficient for an optimum? Draw phase diagrams in the (ϕ, s) space where ϕ is the current-value costate - distinguish between $\alpha\delta - \gamma > 0$ and $\alpha\delta - \gamma < 0$. Identify the regions where $x > 0$ and $x = 0$, respectively. Describe in words the optimal policy for various assignments of T, s_0, and $\bar{z}$. What is the interpretation of $\bar{z}$? When is it optimal never to advertise?

8. Reconsider the commercial fishing fleet problem in exercise 7 of Chapter 6. The basic model is unchanged but parts (a), (b), and (c) now become the following:
   (a) Suppose that you own this fleet now. You intend to retire at time T and sell it. Thus $s(T)$ is free but nonnegative, and you must add to the maximand $e^{-\delta T} P_T s(T)$, where $P_T > 0$ is the current unit price for the fleet at time T. Apply the maximum principle and interpret all conditions, in particular the transversality condition.
   (b) Draw a phase diagram in the (ϕ, s) space, identifying the region where $x = 0$. Describe the optimal policy for various selected values of P_T and S_0.
   (c) Repeat part (b) with $p = 1$, $\delta = 0.1$, $m = 0.4$, $F(s) = 2(s + 0.5)^{1/2} - \sqrt{2}$, and $C(x) = (x+1)^2 - 1$. Find the intercepts of $\dot{\phi} = 0$ with the axes ($\bar{\phi}$ and $\bar{s}$, say).

CHAPTER 8

Discontinuities in the optimal controls

It was stated at the end of Chapter 4 and again in Section 6.1 that control variables are required only to be piecewise-continuous. This means that they can exhibit jump discontinuities at a finite number of dates along the horizon. These discontinuities in the control may in turn result in discontinuities for the time derivatives of the state and costate variables, but the state and costate variables are themselves piecewise-differentiable (i.e., there may exist a finite number of points where the left- and right-hand-side derivatives differ from one another). We claimed that this feature greatly enlarged the variety of problems that optimal control theory could handle and proved our point with a simple example in Section 6.1. In electrical engineering such discontinuities result in the operation of various circuit switches; in economics this takes the form of policy switches.

In the examples studied so far we have restricted the Hamiltonian to be strictly concave in the controls and the optimal trajectories have been continuous. Difficulties arise when we deal with problems that are linear in the controls (or can be made so), at least over some ranges; in these cases there are often bounds on the control variables, either imposed exogenously or generated endogenously. Except for a few theorems on time-optimal problems (see Pontryagin, 1962, pp. 120–4), there are no general results for dealing with these problems, so we have chosen to illustrate them with various examples. The reader should be aware that linearity in the controls and/or bounds on the control variables may generate discontinuities and be alert for them.

8.1 A classical bang–bang example

This example is found in Pontryagin et al. (1962) and has been reproduced by many writers; in our treatment we rely on phase diagram analysis and transversality conditions instead of the traditional algebraic solution, because it is simpler and makes the optimality of the bang–bang solution clear.

The problem is to reach an equilibrium position in minimum time when controlling not the speed of movement but the acceleration only; moreover, there are lower and upper bounds to the values the acceleration can take. The optimal solution turns out to be: set the acceleration at one of the bounds and then the other, never in between – hence the name "bang-bang." The problem is to maximize

$$\int_0^T [-1]\, dt \tag{8.1a}$$

subject to

$$\dot{s}_1 = c, \tag{8.1b}$$

$$\dot{s}_2 = s_1, \tag{8.1c}$$

$$-1 \leq c \leq 1, \tag{8.1d}$$

$$s_1(0),\ s_2(0) \text{ fixed}, \quad s_1(T) = s_2(T) = 0. \tag{8.1e}$$

If we interpret $s_2(t)$ as the distance to the origin at time t, then $s_1(t)$ is the speed and $c(t)$ the acceleration at that time. Note that c may be negative here, so that it is possible to decelerate by putting the engine in reverse, as in a ship, and this may eventually make the ship go backward. The requirement that both s_1 and s_2 be zero at time T means that we must reach the origin and have a zero speed at that time. If s_1 were not required to be zero, the system would still move since we do not control s_1 directly. Note that the integral is $-T$ and that the value of T is obviously free; note also that this is an autonomous problem in which time does not appear as an independent argument; these two observations determine the transversality condition below. The Hamiltonian is

$$H = -1 + \pi_1 c + \pi_2 s_1. \tag{8.2}$$

The necessary conditions are

$$\text{Maximize } H \text{ of (8.2) subject to } -1 \leq c \leq 1, \tag{8.3a}$$

$$\dot{\pi}_1 = -\partial H/\partial s_1 = -\pi_2, \tag{8.3b}$$

$$\dot{\pi}_2 = -\partial H/\partial s_2 = 0, \tag{8.3c}$$

$$H = -1 + \pi_1(t)c(t) + \pi_2(t)s_1(t) = 0, \quad \forall t \in [0, T], \tag{8.3d}$$

plus (8.1b) and (8.1c). We obtain condition (8.3d) by noting that the Hamiltonian is constant over the horizon (because $dH/dt = \partial H/\partial t = 0$ by equation (6.17) for an autonomous problem without discounting), while free terminal time implies that the Hamiltonian is nil at the end of the horizon (equation (7.116c)); hence, it is nil at all times. Condition (8.3a) can be expanded by taking the partial derivative of H with respect to c, $\partial H/\partial c = \pi_1$, and taking the bounds on c into account. If $\partial H/\partial c$ is positive (resp.

negative), we go to the upper (resp. lower) bound; only $\partial H/\partial c=0$ is compatible with an interior solution. Formally,

$$\partial H/\partial c=\pi_1>0 \Rightarrow c=1, \tag{8.4a}$$

$$\partial H/\partial c=\pi_1<0 \Rightarrow c=-1, \tag{8.4b}$$

$$\partial H/\partial c=\pi_1=0 \Leftarrow 0<c<1. \tag{8.4c}$$

Suppose that (8.4c) prevails over some interval of time. Then $\pi_1=0$; thus, $\dot{\pi}_1=0$ as well, and hence $\pi_2=0$ also by (8.3b). Substituting these values into (8.3d) we obtain $H=-1=0$ over that interval, a clear contradiction. Therefore, (8.4c) never occurs and either $c=1$ or $c=-1$ at any one time. This indicates that discontinuities may arise. We now show that there can be at most one such discontinuity. At that time, say t^*, the Hamiltonian is still zero, whether $c=1$ or $c=-1$. Therefore, we have both

$$H(t^*)=-1+\pi_1+\pi_2 s_1=0$$

and

$$H(t^*)=-1-\pi_1+\pi_2 s_1=0.$$

It follows that $\pi_1(t^*)=0$. However, we know from (8.3c) that π_2 is constant; hence, (8.3b) implies that π_1 is monotone. It can therefore be equal to zero only once and t^* is unique.

In order to investigate further we construct a phase diagram in the (s_1,s_2) space, where we distinguish between trajectories with $c=1$ and those with $c=-1$. This is done in Figure 8.1. Begin with $c=1$; then from (8.1) $\dot{s}_1>0$ and $\dot{s}_2$ has the sign of s_1, with $s_1=0$ forming the $\dot{s}_2=0$ locus. Those trajectories are drawn as full lines. The case of $c=-1$ is similar, but now $\dot{s}_1<0$ throughout; these trajectories are drawn as dashed lines.

The boundary conditions dictate that the optimal trajectory end at the origin. There are two and only two trajectories that reach the origin, $P0$ and $N0$. Therefore, with an arbitrary starting point (not on either one of these two trajectories) there must be a way to reach one of these trajectories. Consider, for instance, the initial point A. The only way to reach a stable branch leading to the origin is to follow the full line until N, whence the dashed line to the origin ($AN0$ trajectory). Therefore, in this case, as with all starting points below the $P0N$ line, which is called the *switch line,* it is optimal to set $c=1$ until the switch line is reached, then $c=-1$ until the origin is reached. The opposite applies to points above the switch line, such as B; the optimal policy is then $c=-1$ initially and $c=1$ at the end ($BP0$). Clearly, the discontinuities in the control occur at the switch points. Each optimal trajectory has at most one switch point and some have none (when the starting point is on the switch line), as already demonstrated.

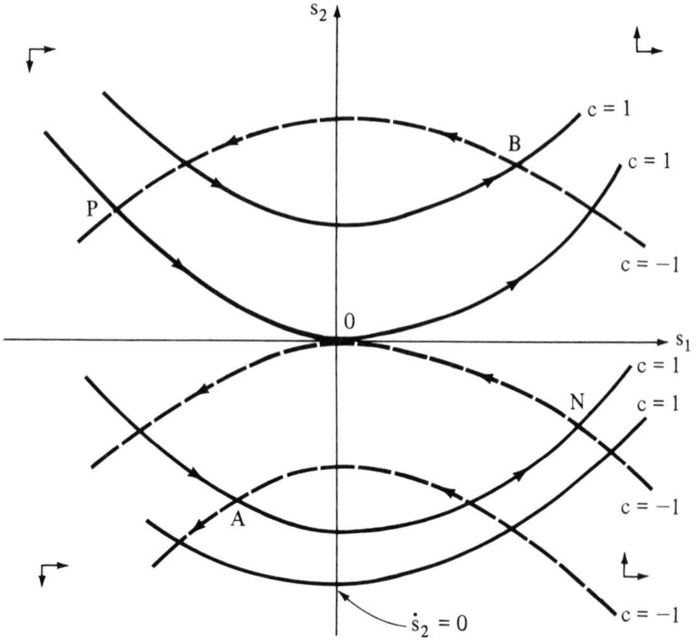

Figure 8.1

We now modify this example in order to illustrate the statement made earlier that discontinuities may occur when the problem is not linear in the control but can be made so. We maximize

$$\int_0^T [-1]\, dt$$
$$\text{subject to } \dot{s}_1 = f_1(s_2),\ \dot{s}_2 = f_2(c),\ -1 \le c \le 1, \tag{8.5}$$

where $f_i' > 0$, $f_i(0) = 0$, $i = 1, 2$, and the boundary conditions of (8.1e) apply. This problem is clearly not linear in the control variable. The Hamiltonian is

$$H = -1 + \pi_1 f_1(s_2) + \pi_2 f_2(c).$$

Some of the necessary conditions are

$$\dot{\pi}_1 = 0, \quad \dot{\pi}_2 = -\pi_1 f_1'(s_2), \quad \text{and} \quad H = 0 \text{ at all times.}$$

If an interior solution is optimal for some time, then $\partial H/\partial c = \pi_2 f_2'(c) = 0$, which implies $\pi_2 = 0$, which again implies $\dot{\pi}_2 = 0$ and thus $\pi_1 = 0$. Substituting, we have $H = -1 = 0$, a contradiction, and again the only optimal choices are $c=1$ or $c=-1$: a bang–bang solution emerges, although the problem was not linear in the control. Because the f functions are strictly increasing and go through the origin, it would be possible to transform (8.5) into a linear problem by redefining a new control variable $C=f_2(c)$. If this feature is not detected at the outset, a clue to the existence of a discontinuous solution is the fact that the first-order condition corresponding to the control variable (such as $\pi_2 f_2'(c)$) cannot be set to zero by choosing the control. Then changes in other variables, state or costate, will dictate whether the control should be set at the upper or lower bound. We now turn to a more complex example.

8.2 The beekeeper's problem

In this section we consider a bee population as a renewable resource that is exploited for its honey production. The apicultural process is perforce simplified, and the model can be interpreted in other ways. A bee population $s(t)$ produces a flow of honey $f(s(t))$; $q(t)$ of this honey is harvested and $x(t)$ of it is left for the bees and their young to feed on. The bee population grows naturally at an exponential rate n, but growth can be increased or decreased depending on whether $x(t)$ exceeds or is below some fixed ration $\bar{x}$. The beekeeper's aim is to obtain the largest possible revenue from the sale of honey during the fixed season $[0,T]$. The price of honey is unity, and there is a positive discount rate δ that reflects market forces. We assume that the bee population must be returned to its initial value at the end of the season; any other fixed level would result in similar policies. Finally, both $x(t)$ and $q(t)$ are required to be nonnegative; since x cannot be negative, q cannot exceed $f(s)$ (because $q+x=f(s)$), which means that one cannot gather more honey than is produced at any one time. The nonnegativity of q means that x cannot exceed $f(s)$; thus, one cannot feed the bees more honey than they produce at any time. This implies that once honey has been harvested it cannot be fed back to the bees, nor can commercial honey be bought from outside to feed them. We now state the problem formally as finding $x(t)$ and $q(t)$ that maximize

$$\int_0^T q(t)e^{-\delta t}\,dt \tag{8.6a}$$

subject to

$$\dot{s}(t)=ns(t)+x(t)-\bar{x}, \tag{8.6b}$$

$$x(t)+q(t)=f(s(t)), \tag{8.6c}$$

$$x(t)\geq 0, \quad q(t)\geq 0, \tag{8.6d}$$

$$s(0)=s(T)=s_0. \tag{8.6e}$$

We assume $f'>0$, $f''<0$; some other restrictions will be placed on the slope of f in order to distinguish several outcomes, but they are not essential at this stage. Some restrictions will also be placed on s_0 to make the problem feasible, but these will arise naturally in the course of the analysis. Eliminating q from the problem and skipping all time arguments, we maximize

$$\int_0^T [f(s)-x]e^{-\delta t}\,dt \tag{8.7a}$$

subject to

$$\dot{s}=ns+x-\bar{x}, \tag{8.7b}$$

$$0\leq x\leq f(s), \tag{8.7c}$$

$$s(0)=s(T)=s_0. \tag{8.7d}$$

The upper bound on x reflects the nonnegativity of q and (8.6c). The Hamiltonian and the Lagrangean are, respectively,

$$H=[f(s)-x]e^{-\delta t}+\pi[ns+x-\bar{x}], \tag{8.8}$$

$$\mathcal{L}=[f(s)-x]e^{-\delta t}+\pi[ns+x-\bar{x}]+\lambda[f(s)-x]. \tag{8.9}$$

Applying the maximum principle we obtain (8.7b) and

$$\dot{\pi}=-(e^{-\delta t}+\lambda)f'(s)-n\pi, \tag{8.10}$$

$$\text{Maximize } H \text{ of (8.8) subject to (8.7c).} \tag{8.11}$$

Consider the possibility of an interior solution in (8.11). If $0<x<f(s)$, then $\lambda=0$ since the upper bound is slack, and the first-order condition is $\partial H/\partial x=-e^{-\delta t}+\pi=0$. Differentiating this, we get $\dot{\pi}=-\delta e^{-\delta t}$ and (8.10) becomes

$$-\delta e^{-\delta t}=-e^{-\delta t}f'(s)-ne^{-\delta t},$$

or

$$f'(s)=\delta-n. \tag{8.12}$$

Assuming for the time being that $\delta\leq n$, it follows that (8.12) is never satisfied; hence, an interior solution never occurs, indicating a possible discontinuity. The two remaining possibilities are

$$\begin{aligned}\partial H/\partial x&=\pi-e^{-\delta t}>0 \Rightarrow x=f(s),\\ \partial H/\partial x&=\pi-e^{-\delta t}<0 \Rightarrow x=0.\end{aligned} \tag{8.13}$$

In the first eventuality $\dot{x}=f'[\dot{s}]$ and $\dot{x}$ has the sign of $\dot{s}$, while in the second one $\dot{x}=0$. We now proceed to construct a phase diagram in the (x,s)

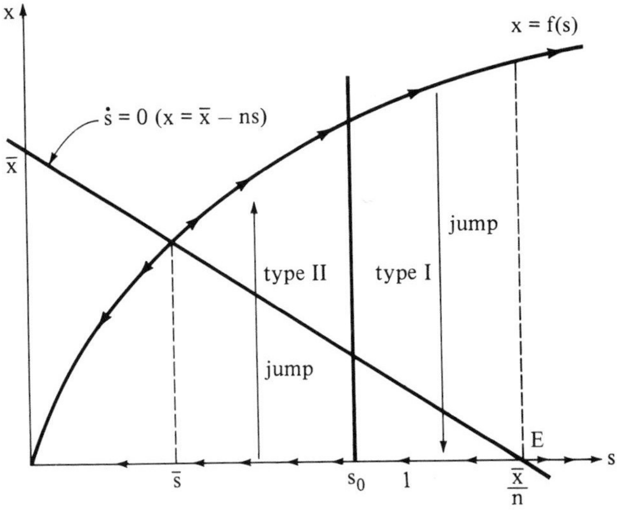

Figure 8.2

space for further analysis; this is done in Figure 8.2. The locus of $\dot{s}=0$ is obtained from (8.7b). It is a straight line intersecting the axes at $\bar{x}$ and $\bar{x}/n$. We also plot the graph of $x=f(s)$. There are only two possibilities: either the trajectory is on the graph of $x=f(s)$ or it is on the s axis. Above the $\dot{s}=0$ line, s increases (and so does x when following the $x=f(s)$ graph); under that line, s decreases (and so does x unless it keeps to zero values). The area above the $f(s)$ graph is ruled out by (8.7c), and anywhere between the graph and the s axis is ruled out by our previous argument. (Thus, the presumed $\dot{s}=0$ line turns out to be a misnomer.) The equilibrium point is at E. We now proceed to restrict the values s_0 may take to ensure the existence of a solution with the specific boundary conditions (8.7d). If $s_0>\bar{x}/n$, s will always increase and we cannot satisfy (8.7d). Define the value $\bar{s}$ by $n\bar{s}+f(\bar{s})=\bar{x}$ that marks the intersection of the $\dot{s}=0$ line with $x=f(s)$. Again if $s_0<\bar{s}$ we cannot satisfy (8.7d) as s always decreases. Hereafter, we assume $\bar{s}<s_0<\bar{x}/n$. There are two distinct possibilities. The optimal policy for the indicated s_0 value could be to first set $x=0$ and after some time to jump up onto $x=f(s)$ and return to the s_0 line (type II). Alternatively, we could begin with $x=f(s)$ and, after a jump down to $x=0$, proceed to the s_0 line (type I). Such jumps are sketched in Figure 8.2. In order to resolve this dilemma we must turn to another phase diagram, in the (ψ, s) space, where $\psi=e^{\delta t}\pi$ is the current-

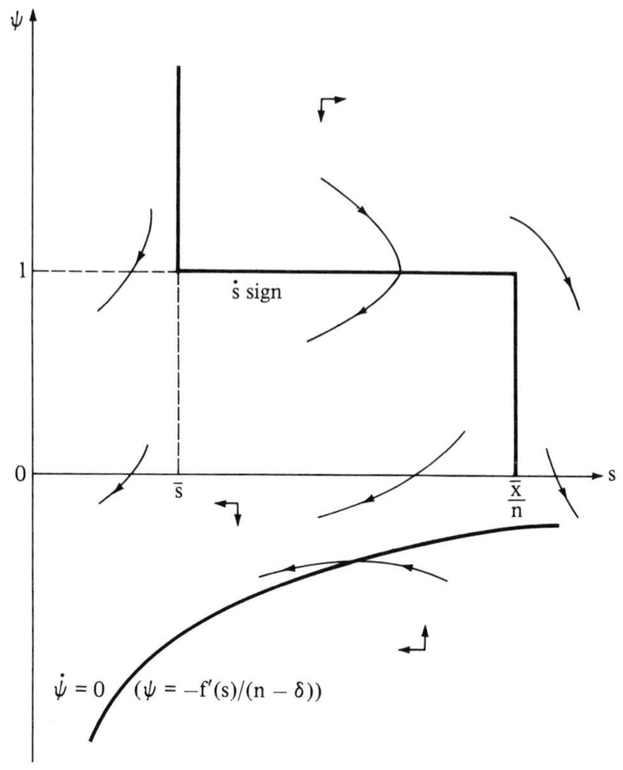

Figure 8.3

value costate (Figure 8.3). Following the technique of Section 4.4 we obtain a differential equation for ψ by differentiating ψ and using (8.10),

$$\dot{\psi} = (\delta - n)\psi - (1+\mu)f'(s), \tag{8.14}$$

where $\mu = e^{\delta t}\lambda$ is the current value of the multiplier. Equation (8.13) becomes

$$\psi > 1 \Rightarrow x = f(s) \quad \text{and} \quad \psi < 1 \Rightarrow x = 0. \tag{8.15}$$

This takes the place of the first-order condition when discontinuities occur, and it seems difficult to use it to replace the x term by a ψ term in the $\dot{s}$ equation (8.7b). However, careful inspection reveals this to be unnecessary. In the relevant region (between $s = \bar{s}$ and $s = \bar{x}/n$) it is always true that $x = f(s)$ implies $\dot{s} > 0$ and $x = 0$ implies $\dot{s} < 0$. Therefore, using (8.15) we have

$$\psi > 1 \Rightarrow \dot{s} > 0 \quad \text{and} \quad \psi < 1 \Rightarrow \dot{s} < 0.$$

Note that when $\psi > 1$, $x = f(s)$ and $\dot{s} = ns + f(s) - \bar{x} > 0$, and this changes abruptly to $\dot{s} = ns - \bar{x} < 0$ when $\psi < 1$. There is a discontinuity in $\dot{s}$ due to the one in x, and the line $\psi = 1$ is not really an $\dot{s} = 0$ locus. We call it "$\dot{s}$ sign" and draw the trajectories with a kink when they cross it since their slope is $d\psi/ds = \dot{\psi}/\dot{s}$ and exhibits a discontinuity when $\dot{s}$ does.

We now turn to $\dot{\psi} = 0$; this yields $\psi = -(1+\mu)f'(s)/(n-\delta)$; the denominator is nonnegative by the assumption $\delta \leq n$ (to be relaxed later). If $\delta = n$, then $\dot{\psi} < 0$ by (8.14). If $\delta < n$, the denominator is positive, so that ψ as given by this equation is negatively valued; hence, $\psi < 1$ and $\mu = 0$. The $\dot{\psi} = 0$ locus has become $\psi = -f'(s)/(n-\delta)$. Given our assumptions on f ($f' > 0$, $f'' < 0$), this has the shape depicted in Figure 8.3. From (8.14) $\dot{\psi}$ is positive only when ψ is negative enough, that is, below the $\dot{\psi} = 0$ locus, and $\dot{\psi}$ is negative above the locus. This and the information gathered on $\dot{s}$ enable us to draw the diagram, at least between $\bar{s}$ and $\bar{x}/n$. This is sufficient to resolve the dilemma we had. Clearly, any trajectory starting below $\psi = 1$, whether at ψ positive or ψ negative, entails s decreasing throughout; this conflicts with (8.7d). Therefore, we must start above $\psi = 1$, and this means $x = f(s)$ at first, followed by a jump to $x = 0$. In Figure 8.2 policies of type I are optimal. Note that no matter how large T is, it is possible to find a trajectory of type I that lasts long enough, because after the downward jump the path can be arbitrarily close to the equilibrium point E and its motion very slow.

We now relax our assumption $\delta \leq n$ and examine the case where $\delta > n$. Equation (8.12) is no longer inconsistent but defines some value s^* by $f'(s^*) = \delta - n$, where the slope of f is equal to $\delta - n$. Thus, an interior solution may occur, but during the time that it prevails s is fixed at s^* and $\dot{s} = 0$; hence, by (8.7b) x is also fixed at $x^* = \bar{x} - ns^*$. The point (x^*, s^*) is, of course, on the presumed $\dot{s} = 0$ line in an (x, s) phase diagram. The possibility of an interior solution thus does not rule out discontinuities, since jumps are required in order to pass from $x = 0$ to $x = x^*$ to $x = f(s)$. To discover the nature of the optimal solution requires once again the use of the (ψ, s) diagram, which we have constructed in Figure 8.4. (In that figure we have assumed $\bar{s} < s^* < \bar{x}/n$; if s^* falls outside the relevant area, we obtain somewhat different outcomes, which will be briefly examined shortly.) From equation (8.14) the $\dot{\psi} = 0$ locus is given by the expression $\psi = (1+\mu)f'(s)/(\delta - n) > 0$ under our current assumption. While $\psi \leq 1$, $\mu = 0$ and this expression simplifies to $\psi = f'(s)/(\delta - n)$. When $\psi \geq 1$, we have $x > 0$; hence, $\partial \mathcal{L}/\partial x = 0$ and $\psi = 1 + \mu$. This and (8.14) imply that the $\dot{\psi} = 0$ locus above $\psi = 1$ is the vertical line $s = s^*$. In the relevant area, $\psi = 1$ is the $\dot{s}$-sign line that separates $\dot{s} > 0$ from $\dot{s} < 0$, and we have a new "equilibrium" at I with two stable arms and two unstable arms. This equilibrium is somewhat peculiar. It is true that $\dot{\psi}$ approaches 0 smoothly but

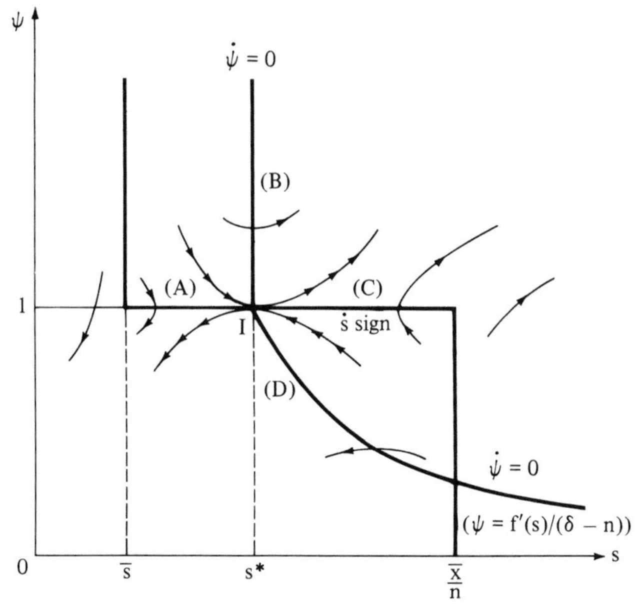

Figure 8.4

$\dot{s}$ does not: as we jump from $x = f(s^*)$ or $x = 0$ to $x = x^*$, there is a discontinuity in $\dot{s}$. The arms (stable and unstable) have been drawn with a horizontal slope at I since $\dot{\psi}$ – hence the slope $d\psi/ds = \dot{\psi}/\dot{s}$ – tends to zero around I. This is an important point, because equilibrium I can then be reached in finite time; we shall need this observation later. While at I itself, $s = s^*$, $x = x^*$, and of course $\dot{s} = 0$. We use these arms to define regions (A) through (D) in the relevant area of the plane, as in Figure 8.4. Any trajectory that begins in region (B) or (D) entails a monotone motion for s; hence, it is ruled out. Trajectories in region (A) have the population s increasing at first and then decreasing and are of type I, as in Figure 8.2. However, trajectories in region (C) are of type II, and if s_0 is larger than s^*, type II policies are expected to emerge. The economic intuition behind this result is that when the discount rate was low, it paid to let the bee population increase at first so that full harvesting would take place on the larger bee population. However, if the discount rate is large enough, the preference for earlier harvesting reverses the argument.

We now illustrate the consequences of this analysis in the (x, s) space in Figure 8.5 for several eventualities. With an initial value such as s_{01}, trajectories will be of type I, moving from G_1 to H_1, down to J_1 with a

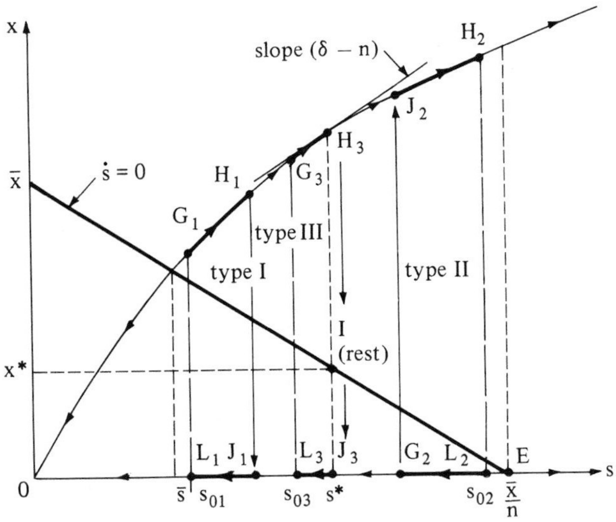

Figure 8.5

jump and on to L_1. With an initial value such as s_{02}, trajectories will be of type II: $L_2G_2J_2H_2$. Other possibilities also emerge because of the peculiarities of equilibrium point I discussed earlier. Since $\dot{s}$ does not tend to zero when approaching I, it is possible to reach this equilibrium in finite time, and this gives rise to new types of policies when s_0 is close enough to s^* and T is long enough. Suppose the initial value is s_{03} and suppose that we ride the stable arm separating regions (A) and (B) in Figure 8.4; there is still much time left when I is reached, so that leaving I immediately along the unstable arm between regions (A) and (D) would force us to reach s_{03} too soon. Then there must be a rest period at I itself in between travel along the arms. This is a new type (type III) of policy which is depicted as $G_3H_3IJ_3L_3$ in Figure 8.5. This involves two jumps from H_3 to I and from I to J_3 with a rest period at I in between. A type IV policy with a long horizon and s_{04} just above s^* could also be described; we leave this to the reader. Let us now give the economic intuition behind these new types of policies. At point I, we have $f'(s^*) = \delta - n$ and $\dot{s} = ns^* + x^* - \bar{x} = 0$. The first expression is the golden rule for this problem: the marginal physical product of the resource plus its natural rate of growth is matched with the rate of discount, and this could be maintained forever with $x = x^*$. Thus, when initial (and terminal) stock is close enough to the golden rule level and there is enough time, it becomes optimal to reach that point and remain there as long as possible. Finally, note that we have not exhausted all eventualities in this problem. For instance, δ could be close enough to n so that s^* may be well above the relevant region (or not exist at all) and all policies would be of type I. Conversely, if δ is very large, s^* could be below $\bar{s}$ and only type II policies would prevail. Finally, freeing the terminal bee population from the constraint of being equal to the initial one would further broaden the range of optimal policies encountered here. The bulk of the analysis remains unchanged; the drawing of conclusions is left to the reader as an exercise. To facilitate this task the directions of trajectories outside the relevant $[\bar{s}, \bar{x}/n]$ interval have been drawn in all the diagrams.

In this section we used a simple renewable resource problem to generate a rich array of discontinuous policies for the analysis of which we needed to study the interaction between the (state, control) and the (state, costate) phase diagrams. Next we shall analyze a problem with two controls, two state variables, and two constraints.

8.3 One-sector optimal growth with reserves

This model is similar to the ones analyzed in Sections 4.4 and 6.4, but it has some additional features. We first state the problem and then proceed to describe it. Find c and x that maximize

$$\int_0^T u(c)e^{-\delta t}\,dt + e^{-\delta T}p_T s_T \tag{8.16}$$

subject to

$$\dot{s} = F(s) - ms + x - c, \tag{8.17}$$

$$\dot{X} = -x, \tag{8.18}$$

$$0 \le x \le 1; \tag{8.19}$$

$$p_T, T, s_0, X_0 \text{ exogenously specified;} \quad s_T \text{ free;} \quad X_T \ge 0. \tag{8.20}$$

We assume $u' > 0$, $u'' < 0$, and $u'(0) = \infty$, and the same for F. The notation is as in Section 6.4, but now we have a stock X of the good on which we can draw; there are bounds of 0 and 1 on the flow of good from this source. Note that we are disregarding potential constraints on the use of capital stock for consumption ($c \le F(s) + x$) or on the use of reserves to augment capital stock ($c \ge x$). Although these are interesting features, they would bring the number of inequality constraints to four and complicate the analysis so as to redirect our main focus away from the discontinuities. As another simplification we have not included a scrap value for X_T. This would only lengthen the analysis without altering its substance. A rationale for this assumption is that the reserves are firm-specific and of no value to others outside it. The Hamiltonian is

$$H = u(c)e^{-\delta t} + \pi[F(s) - ms + x - c] - \lambda x. \tag{8.21}$$

The necessary conditions are (8.17)–(8.20) plus

$$u'(c)e^{-\delta t} = \pi; \tag{8.22}$$

$$\begin{aligned} \partial H/\partial x &= \pi - \lambda > 0 \Rightarrow x = 1, \\ \partial H/\partial x &= \pi - \lambda < 0 \Rightarrow x = 0, \\ \partial H/\partial x &= \pi - \lambda = 0 \Leftarrow 0 < x < 1; \end{aligned} \tag{8.23}$$

$$\dot{\pi} = \pi(m - F'(s)), \quad \pi_T = p_T e^{-\delta T}, \tag{8.24}$$

$$\dot{\lambda} = 0, \quad X_T \geq 0, \quad \lambda \geq 0, \quad \lambda X_T = 0. \tag{8.25}$$

First suppose that $\lambda = 0$; then $X_T \geq 0$. By (8.22) we know that $\pi > 0$; hence, $\pi > \lambda$ at all times and (8.23) implies $x = 1$ for the whole interval. This will happen if $X_0 \geq T$. Hereafter, we analyze the case $X_0 < T$. Note also that (8.24) and (8.22) imply $u'(c_T) = p_T$, which fixes the optimal value of c_T.

Suppose now that $0 < x < 1$ for some time; then by (8.23) $\pi = \lambda$ and π is constant; (8.24) implies $\pi(m - F'(s)) = 0$, and we distinguish two cases:

(a) In the case where $F'(s) > m$ for all s, the preceding equation implies $\pi = 0$, which contradicts (8.22). In this case interior solutions for x are ruled out and $x = 1$ and $x = 0$ are the only possibilities. Since also $\pi > 0$ implies $\dot{\pi} < 0$ in this case, we must have $x = 1$ at first and $x = 0$ later. The optimal policy is clear: set $x = 1$ until date X_0, where the reserves are exhausted; thereafter, set $x = 0$. Differentiating (8.22) and using (8.24) yield

$$\dot{c} = \frac{-u'}{u''}[F'(s) - m - \delta]. \tag{8.26}$$

Therefore, the consumption path may or may not be monotone. This case will be illustrated after the analysis of case (b).

(b) We now turn to the case where there exists some value s^* such that $m = F'(s^*)$. Then $0 < x < 1$ is compatible with $s = s^*$; of course, $\dot{s} = 0$ during that time interval. Therefore, the optimal controls in this interval are given by $c^*(t)$ and $x^*(t)$ defined by

$$u'(c^*(t))e^{-\delta t} = \lambda \quad \text{(from (8.22) and } \pi = \lambda) \tag{8.27}$$

and

$$0 = F(s^*) - ms^* + x^*(t) - c^*(t) \quad \text{(from (8.17))}. \tag{8.28}$$

Outside this interval equation (8.26) applies with x set at either 0 or 1.

We can now construct the phase diagram in Figure 8.6. Equation (8.26) defines the $\dot{c} = 0$ locus as $s = \bar{s}$, where $F'(\bar{s}) = m + \delta$; hence, $\bar{s} < s^*$. Two

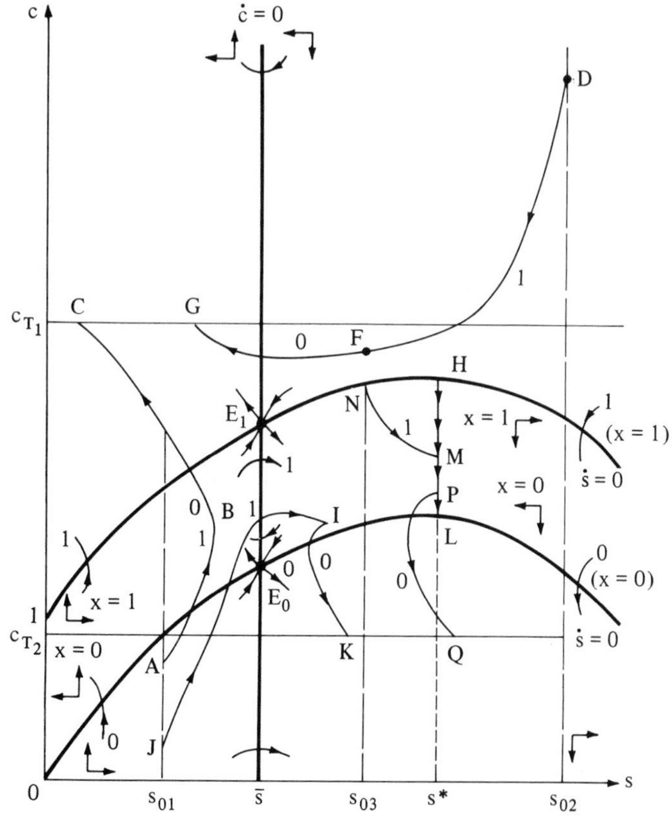

Figure 8.6

$\dot{s}=0$ loci are drawn, one for $x=0$ and the other for $x=1$, and so labeled, using (8.17). This equation and (8.26) enable us to find the directions of trajectories in the usual manner. The only complication occurs in the region between the two $\dot{s}=0$ loci: we must now label trajectories as $x=0$ or $x=1$ (or 0 and 1) since they go in different directions. Note that at any point of intersection of two trajectories of a different type, the $\dot{s}$ component in the $x=1$ path is higher than in the $x=0$ path; therefore, its slope in the (c, s) plane is algebraically larger. There are two saddle-point equilibria E_1 and E_0, neither of which is attainable in finite time. We now turn to the interior solution. This can happen only when $s=s^*$; hence, from (8.28), $\dot{x}^*=\dot{c}^*$, which is negative by (8.27). Furthermore, this trajectory is feasible only between the $\dot{s}=0$ loci since any other point on the $s=s^*$ line would violate (8.19). For instance, at a point above H, we would have

$c = F(s^*) - ms^* + 1 + \alpha$, $(\alpha > 0)$ say, and (8.28) would yield $x^* = 1 + \alpha$, a contradiction. Similar reasoning rules out sections under point L.

There are many possible policies, depending on the values of T, X_0, s_0, and so on, and it would take too much space to categorize them all. We have instead selected a sample and invite readers to augment it so as to increase their understanding of the problem. Some of these policies keep well away from $s = s^*$ and will serve as illustrations of case (a). Consider the initial value s_{01} and a scrap value p_T such that terminal consumption is at c_{T_1}. One possible optimal path is ABC, although the switch from $x = 1$ to $x = 0$ may occur earlier (smaller X_0) or later (larger X_0). A different initial capital stock such as s_{02} may yield the path DFG with the switch at F. Consider now an initial capital s_{01} with terminal consumption c_{T_2}; then the optimal path could be JIK, which exhibits several changes in the signs of $\dot{c}$ and $\dot{s}$. All of these examples could apply to case (a) since s^* plays no role. Consider now an initial capital of s_{03} and terminal consumption c_{T_2}, the optimal path could be $NMPQ$ with MP section corresponding to an interior solution. It is advantageous to follow this path rather than pursue the NM trajectory further to meet with an $x = 0$ path to the right of s^*, because it is wasteful to go past s^* (the net production function $F(s) - ms$ is decreasing there); if it is possible to keep π constant until a feasible $x = 0$ path can be followed to the left of s^*, this will be done. Therefore, the policy, which applies only to case (b), is: set $x = 1$ at first, then stay at s^*, adjusting x to keep $\dot{s} = 0$, and finally move off to $x = 0$. This policy involves two jump discontinuities because in the interior of the segment HL, x is strictly between 0 and 1 and over the whole horizon x takes on the values 1, $[\alpha, \beta]$, and 0 with $1 > \alpha > \beta > 0$. This policy could be called a bang–slide–bang solution; it shows that interior solutions can coexist with discontinuities. It is also worth noting that the special features of this model have enabled us to describe the solution of a two-state, two-control problem in a two-dimensional phase diagram. Thus, this is a hint that sometimes adding to a model rigidities that yield bang–bang types of solutions may be a first step toward analyzing otherwise intractable problems.

8.4 Highest consumption path

This is a special case of Example 6.4.1, with a linear utility function. Because of the linearity of the utility function, a golden rule equilibrium attainable in finite time appears. This is in contrast with the results of Section 6.4, in which the equilibrium could be reached only in infinite time. Therefore, our analysis here will highlight the fact that the equilibrium of Section 6.4 was not reached in finite time, not because this would have been unfeasible, but because it was suboptimal to do so. We now state the problem; the notation is identical to that of Section 6.4. Find c that maximizes

$$\int_0^T ce^{-\delta t}\,dt \tag{8.29}$$

subject to

$$\dot{s}=F(s)-ms-c, \tag{8.30}$$

$$0\leq c\leq F(s); \tag{8.31}$$

$$T, s_0, s_T \text{ exogenously specified.} \tag{8.32}$$

We assume $F(0)=0$, $F'>0$, $F''<0$, and $F'(0)>\delta+m>F'(\infty)$. The Hamiltonian is

$$H=ce^{-\delta t}+\pi[F(s)-ms-c], \tag{8.33}$$

and the Lagrangean is

$$\mathcal{L}=H+\mu[F(s)-c]. \tag{8.34}$$

The necessary conditions are (8.30)–(8.32) plus

$$\begin{aligned} \partial H/\partial c &= e^{-\delta t}-\pi<0 \Rightarrow c=0,\\ \partial H/\partial c &= e^{-\delta t}-\pi>0 \Rightarrow c=F(s),\\ \partial H/\partial c &= e^{-\delta t}-\pi=0 \Leftarrow 0<c<F(s) \end{aligned} \tag{8.35}$$

(the last two cases of (8.35) can also be stated as $e^{-\delta t}=\pi+\mu$), as well as

$$\dot{\pi}=m\pi-(\pi+\mu)F'(s), \tag{8.36}$$

$$\mu\geq 0,\quad F(s)-c\geq 0,\quad \mu[F(s)-c]=0. \tag{8.37}$$

We shall also use the current-value costate variable $\psi=e^{\delta t}\pi$ when convenient.

First suppose that $0<c<F(s)$ over some time interval. Then by (8.35) and (8.37) $\pi=e^{-\delta t}$ (or $\psi=1$) and $\mu=0$. Substituting these in (8.36) yields $\dot{\pi}=-\delta e^{-\delta t}=e^{-\delta t}[m-F'(s)]$, or

$$F'(s^*)=m+\delta, \tag{8.38}$$

which defines the value s^*. Therefore, over that time interval, $s=s^*$ and $\dot{s}=0$, which with (8.30) gives

$$c^*=F(s^*)-ms^*. \tag{8.39}$$

Equations (8.38) and (8.39) define the golden rule equilibrium, which is the only point compatible with an interior solution for (8.35). Also recall that $\psi^*=1$.

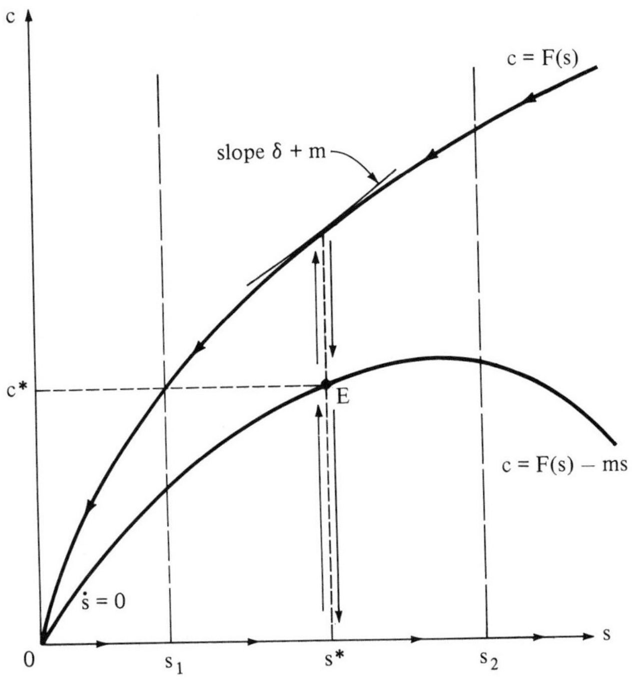

Figure 8.7

We now turn to boundary solutions. Suppose that $c=0$; then by (8.35)–(8.37) we have $\mu=0$, $\pi>e^{-\delta t}>0$ (or $\psi>1$), $\dot{\pi}=\pi[m-F'(s)]$, or

$$\dot{\psi}=\psi[\delta+m-F'(s)] \tag{8.40}$$

and

$$\dot{s}=F(s)-ms. \tag{8.41}$$

We take this last expression to be positive since the possibility that, at some high level of capitalization, depreciation overtakes productivity in absolute terms does not seem very sensible – it has little effect on our analysis in any case. Finally, consider the case in which $c=F(s)$. Then by (8.35)–(8.37) we obtain $\pi<e^{-\delta t}$ (or $\psi<1$); indeed, $\pi+\mu=e^{-\delta t}$, $\dot{\pi}=m\pi-e^{-\delta t}F'(s)$, or

$$\dot{\psi}=(\delta+m)\psi-F'(s) \tag{8.42}$$

and

$$\dot{s}=-ms<0. \tag{8.43}$$

We are now ready to construct the phase diagrams that will enable us to select optimal policies. The (c,s) diagram is represented in Figure 8.7. We

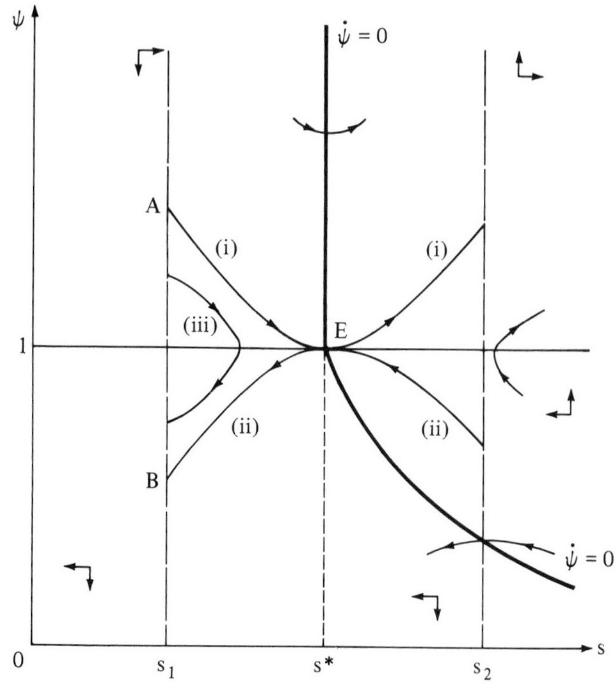

Figure 8.8

have the upper bound $c = F(s)$ and the lower bound $c = 0$. The $\dot{s} = 0$ locus, $c = F(s) - ms$, is also represented, although only the equilibrium point E on it is ever used. As indicated, $\dot{s} > 0$ along the s axis and $\dot{s} < 0$ along $c = F(s)$. We expect trajectories going past $s = s^*$ to jump (up or down) to the golden rule equilibrium, but we need the (ψ, s) diagram to confirm our guess. This is constructed in Figure 8.8. There are two cases to consider: above $\psi = 1$, where by (8.41) s increases, and below $\psi = 1$, where by (8.43) s decreases. Thus, the $\psi = 1$ line delineates positive and negative values of $\dot{s}$, although it is not an $\dot{s} = 0$ locus because of the discontinuity in c, hence in $\dot{s}$. Above $\psi = 1$, equation (8.40) gives the $\dot{\psi} = 0$ locus at $s = s^*$. Below $\psi = 1$, the $\dot{\psi} = 0$ locus is given by (8.42), which yields $\psi = F'(s)/(\delta + m)$. Recalling that F' is a positively valued and decreasing function, we have the $\dot{\psi} = 0$ locus as represented. The sign of $\dot{\psi}$ can easily be everywhere ascertained, and this is indicated in Figure 8.8 by the usual arrows. Note that trajectories reaching or leaving E have a horizontal slope at E because $d\psi/ds = \dot{\psi}/\dot{s} = 0$, since $\dot{\psi} = 0$ and $\dot{s} \neq 0$ wherever $\psi \neq 1$. (This is similar to the beekeeper's problem.) This explains why E can be reached in finite time. We now examine a few optimal trajectories with various endpoint conditions:

(i) Let $s_0 = s_1$ and $s_T = s_2$ with $s_1 < s^* < s_2$; then with T large enough the optimal policy is to set $c = 0$ until s^* is reached, jump to E, stay there as long as possible, and jump down to $c = 0$ just in time to reach s_2 at time T. Clearly, if T is too small, the problem is unfeasible, whereas if T is very large, most of the time is spent at the equilibrium. This policy is reminiscent of "turnpike" results. In Figure 8.8 the optimal trajectory is (i). Therefore, the equilibrium is approached by the stable arm and abandoned by following the unstable arm in the appropriate region. Whenever the optimal path involves E, only the stable or unstable arms are used; no other trajectory is involved.

(ii) When the boundary conditions are reversed, that is, $s_0 = s_2$ and $s_T = s_1$, trajectory (ii) in Figure 8.8 is used. In the (c, s) plane this involves setting $c = F(s)$ until s^* is reached, staying at s^*, and jumping again to $c = F(s)$ for the last stretch.

(iii) Let $s_0 = s_T = s_1$. Then the optimal policy could be (iii) in Figure 8.8 if time is short. This involves setting $c = 0$ and then switching to $c = F(s)$. If, however, T is very large, the optimal policy could be AEB in Figure 8.8 with some time spent at E. This would involve two jump discontinuities.

8.5 Concluding comments

In this chapter we have illustrated a class of problems that can be handled only by the methods of optimal control theory. Conditions under which discontinuities arose were bounded controls and the inability to vary the control variable so as to keep the first-order condition balanced at zero. In many cases the Hamiltonian was a linear function of some of the control variables, but we saw that this was not necessary. A variety of optimal policies have emerged with controls set to one of their bounds for some time and then, after a jump discontinuity, moving smoothly along an interior path, or resting temporarily at an equilibrium, or even moving to another bound. The presence of discontinuities has in particular led to the appearance of equilibria that can be reached in finite time and are indeed part of the optimal policy. Thus, discontinuities should not be looked upon as mere oddities that can always be smoothed away (they cannot always be, as in Sections 8.1 and 8.3) but should be seen as contributing richness to the behavior exhibited in control problems. Finally, as in Section 8.3 they sometimes allow the analysis of otherwise intractable problems. We have also used the (state, costate) and the (state, control) phase diagrams jointly in order to elicit the precise optimal path.

Exercises

1. Consider a very simple problem in which the aim is to maximize the present value of the sales from a finite stock of a resource. The flow of sales is bounded, and a scrap value exists for any stock remaining at the end of the horizon. Formally, choose a piecewise-continuous control $c(t)$ to maximize

$$V=\int_0^T c(t)e^{-\delta t}\,dt+e^{-\delta T}s(T)$$

subject to $\dot{s}(t)=-c(t)$, $0\leq c(t)\leq 1$, $s(T)\geq 0$, and $s(0)=s_0$, where δ, T, and s_0 are specified positive constants. Apply the maximum principle to this problem; pay particular attention to the conditions characterizing the choice of the control that maximizes the Hamiltonian subject to the constraints on c and also to the transversality conditions. Prove by contradiction that an interior solution (i.e., $0<c(t)<1$) cannot persist for any finite time interval. Show that when $T>s_0$ the solution is bang–bang, and when $T<s_0$, the control is continuous. In each case derive the complete solution, including the value of the switch point when appropriate, and ensure that the transversality condition is satisfied. Give a verbal account of the optimal solution.
2. Repeat exercise 1 when the constraint $0\leq c(t)\leq 1$ is replaced by $-1\leq c(t)\leq 1$. In other words, buying as well as selling is now permissible. Again show that an interior solution is never optimal and that the control is bang–bang when $T>s_0$ but continuous when $T<s_0$. Give a verbal account of your results.
3. Here we examine the optimal manner of consuming a resource when the current stock of the resource is an argument of the utility function. This is perhaps because mere ownership pleases the individual or perhaps because the quality of the resource declines with the level of stocks. Find $c(t)$ to maximize

$$\int_0^T c(t)s(t)e^{-\delta t}\,dt$$

subject to $\dot{s}(t)=-c(t)$, $0\leq c(t)\leq 1$, $s(T)\geq 0$; s_0 is a specified positive constant, as is δ. Show that it is never optimal to have $0<c(t)<1$ for any length of time. Draw the (state, costate) phase diagram; divide the space in two regions: one where $c=0$ and one where $c=1$. Describe the optimal solution; show that c is continuous when $s_0>T$ but discontinuous when $s_0<T$. Show that it is never optimal to enter the interior of the region where $c=0$.
4. Reconsider the problem of exercise 3 modified by the introduction of a scrap value: the maximand is now

$$\int_0^T c(t)s(t)e^{-\delta t}\,dt+e^{-\delta T}ps(T),$$

where p is a positive constant. Show that the results are similar to those of exercise 3 except that it now may be optimal to reach the interior of the $c=0$ region.

Give a verbal account of the optimal policy and explain why it might be optimal to consume no resource for some time while a positive stock is available.

5. Modify the problem of exercise 4 by introducing depreciation. The state equation is now $\dot{s}(t) = -c(t) - ms(t)$, where m is a specified positive constant. How are the results of exercise 4 modified, if at all?

6. In this exercise there is a most preferred value of the state variable with a penalty applying when the system is not at it. The rate of change in the state variable is the control and it is bounded. Find $c(t)$ to maximize

$$\int_0^2 -(s(t)-1)^2\, dt$$

subject to $\dot{s}(t) = c(t)$, $-1 \leq c(t) \leq 1$, and $s(t) = s_0$, where s_0 is a specified constant and $s(2)$ is free. Show that $-1 < c(t) < 1$ never occurs unless $s(t) = 1$. Draw a phase diagram in the (state, costate) space. Derive the optimal solution when $s_0 = 0$ and when $s_0 = 2$. What is the sign of the costate variable in these two cases? Explain your result. If a discount factor $e^{-\delta t}$ is applied, how are the results altered?

7. In this exercise we examine the consumption of a growing resource that might suffer from overcrowding. We seek to choose $c(t)$ to maximize

$$\int_0^T c(t)\, dt$$

subject to $\dot{s}(t) = s(t)(100 - s(t)) - c(t)$, $0 \leq c(t) \leq 3{,}000$, $s(0) = s_0$, and $s(T) = s_T$, where s_0 and s_T are specified positive constants. Show that an interior solution $(0 < c < 3{,}000)$ is compatible with only one pair of values for c and s. Draw the phase diagrams in the (c, s) space and the (π, s) space, where π is the costate variable. Identify a steady state with positive consumption that can be reached in finite time and where it is optimal to stay as long as possible provided that T is large enough. Choose a few values for T, s_0, and s_T and illustrate the various types of optimal paths.

8. Modify the problem of exercise 7 by imposing a smaller upper bound on the consumption for the resource, namely, $0 \leq c(t) \leq 900$. All other data remain unchanged. Show that an interior solution $0 < c(t) < 900$ is never optimal now. Draw the phase diagrams. How many steady states with a positive consumption are there now? Are they attainable in finite time? Choose some s_0 and s_T values so that the optimal control has two jump discontinuities. Describe the optimal path when $s_0 = 70$, $s_T = 95$, and T is relatively large.

9. Reconsider the model of Section 8.3. Construct a phase diagram in the (φ, s) space in which $\varphi(t) = \pi(t)e^{\delta t}$ and match the trajectories with those of Figure 8.6. Show that shifts from $x = 1$ to $x = 0$ can occur only when $s(t) < s^*$ and shifts from $x = 0$ to $x = 1$ can occur only when $s(t) > s^*$. Discuss why this is so.

10. Reconsider the model of Section 8.4. Derive the exact solution when $\delta = 0.1$, $F(s) = 2\sqrt{s}$, $s_0 = 64$, $s_T = 144$, $T = 12$, and $m = 0$. Identify the intervals over which $c > 0$ and those over which $c = 0$. Suppose now that $\delta = 0.06$, $m = 0.04$, $F(s) = 2\sqrt{s}$, $T = 75$, $s_0 = 100e$, and $s_T = 100e^{-1.5}$, where e is the exponential number. What is the optimal solution now?

CHAPTER 9

Infinite-horizon problems

In economics it is often convenient to postulate that the time horizon is infinite. One of the reasons for adopting this assumption is that one avoids the problem of specifying the end-of-horizon stocks or a scrap value function. Also, this formulation quite often leads to simplified formulas and appealing results; for example, the idea of a long-run equilibrium can be given a precise treatment. To critics who point out that the world is predicted to end in finite time, one may offer the following defense: provided that the optimal path for an infinite-horizon problem does not differ significantly from the solution of a control problem with a very large but finite horizon, the convenience of working with an infinite-horizon model is worth the loss of "realism."

There are, however, certain technical difficulties associated with optimal control problems having an infinite time horizon. In particular, the finite-horizon transversality conditions do not carry over to the infinite-horizon case. Furthermore, it is possible that the integral does not converge for all feasible paths. If this case arises, how would one rank alternative paths? These, and other issues, will be examined in this chapter.

9.1 Optimality criteria

Consider the problem of attempting to find the control vector $\mathbf{c}(t)$ that maximizes the integral

$$V=\int_0^\infty v(\mathbf{s}(t),\mathbf{c}(t),t)\,dt \tag{9.1}$$

subject to

$$\dot{s}_i=f^i(\mathbf{s}(t),\mathbf{c}(t),t),\quad i=1,2,\ldots,n, \tag{9.1a}$$

$$g^j(\mathbf{s}(t),\mathbf{c}(t),t)\geq 0,\quad j=1,2,\ldots,m', \tag{9.1b}$$

$$g^h(\mathbf{s}(t),\mathbf{c}(t),t)=0,\quad h=m'+1,\ldots,m, \tag{9.1b$'$}$$

$$s_i(0)=s_{i0},\quad i=1,2,\ldots,n. \tag{9.1c}$$

We assume that there are no restrictions on the limiting behavior of the state variables, but alternative specifications are possible; see equation

(9.4) in the next section. The constraints (9.1b) and (9.1b′) are assumed to satisfy the rank condition stated in Chapter 6.

A solution $(\mathbf{s}(t), \mathbf{c}(t))$ of (9.1a)–(9.1c) is called a feasible path. If for all feasible paths $(\mathbf{s}(t), \mathbf{c}(t))$ the integral (9.1) converges, then the optimal path is clearly one that yields the highest value for the integral. (Note that convergence is ensured if $v(\mathbf{s}, \mathbf{c}, t)$ takes the form $v(\mathbf{s}, \mathbf{c}, t) = e^{-\delta t}u(\mathbf{s}, \mathbf{c}, t)$, where $u(\mathbf{s}, \mathbf{c}, t)$ is bounded and $\delta > 0$.)

Suppose the integral (9.1) does not converge for all feasible paths. How would one compare any two feasible paths $(\mathbf{s}^*(t), \mathbf{c}^*(t))$ and $(\mathbf{s}(t), \mathbf{c}(t))$? In order to answer this question, let us consider, for any finite t, the difference of the cumulative performances up to that time:

$$z(t) \equiv \int_0^t v(\mathbf{s}^*, \mathbf{c}^*, \tau)\, d\tau - \int_0^t v(\mathbf{s}, \mathbf{c}, \tau)\, d\tau. \tag{9.2}$$

(We have skipped the time arguments for simplicity.) Clearly, if there exists some T such that for all $t \geq T$, $z(t) \geq 0$, then $(\mathbf{s}^*, \mathbf{c}^*)$ is better (or no worse) than $(\mathbf{s}, \mathbf{c})$. This leads to the following definition.

Definition 9.1.1: overtaking criterion. A path $(\mathbf{s}^*, \mathbf{c}^*)$ is said to be "no worse" than path $(\mathbf{s}, \mathbf{c})$ under the overtaking criterion if the difference in cumulative performances $z(t)$ is nonnegative for all t sufficiently large. A path is optimal under the overtaking criterion if under that criterion it is no worse than any other feasible path $(\mathbf{s}, \mathbf{c})$.

The overtaking criterion, while intuitively appealing, is in practice very difficult to apply, as we shall see in Section 9.3. Furthermore, it fails to rank any pair of paths whose difference $z(t)$ changes sign periodically. Consider the following example, in which $z(t)$ takes the form $z(t) = (\sin t)/t$. In this example, the difference in cumulative performances oscillates around zero but vanishes (approaches zero) as t becomes very large. This suggests another definition of "no worse" in the sense that

$$\lim_{t \to \infty} z(t) \geq 0.$$

However, such a definition would still be too restrictive, because in general $z(t)$ may not approach a limit, as the following example illustrates:

$$\begin{aligned} z(t) &= \sin t \ (\geq 0) \text{ for } 2m\Pi \leq t < (2m+1)\Pi \\ &= (\sin t)/t \ (\leq 0) \text{ for } (2m+1)\Pi \leq t < (2m+2)\Pi \end{aligned} \tag{9.3}$$

for all $m = 0, 1, 2, \ldots$. In this more complicated example, $z(t) = 1$ periodically, but for all t in $[(2m+1)\Pi, (2m+2)\Pi]$, $m = 0, 1, 2, \ldots$, $z(t)$ is nonpositive; thus, a limit does not exist. However, in a sense $(\mathbf{s}^*, \mathbf{c}^*)$ is no worse (or indeed strictly better) than $(\mathbf{s}, \mathbf{c})$, because in intervals of time when $z(t)$ is negative, its value is bounded below, and the lower bound approaches zero as t tends to infinity. To formalize this idea, let $\inf z(t)$ denote the greatest lower bound of the graph of $z(t')$, for all $t' \geq t$. By definition $\inf z(t)$ is a nondecreasing function of t, so that while $z(t)$ may oscillate, $\inf z(t)$ does not, and hence the limit

$$\lim_{t \to \infty} [\inf z(t)]$$

exists (it may, of course, be infinite). We can now define optimality according to the catching-up criterion.

Definition 9.1.2: catching-up criterion. A path (s^*, c^*) is said to be "no worse" than path (s, c) under the catching-up criterion if

$$\lim_{t \to \infty} [\inf z(t)] \geq 0.$$

A path is optimal according to the catching-up criterion if under that criterion it is no worse than any other feasible path.

One advantage of the catching-up criterion is that sufficient conditions are easily verified, as we shall see in Section 9.3.

A moment's reflection will convince the reader that if a path is optimal under the overtaking criterion, then it is also optimal under the catching-up criterion, and that both criteria are equivalent to the maximization of the integral (9.1) if it converges for all feasible paths. The choice of optimality criteria depends on personal tastes. In practice, it is recommended that one proceed under the assumption that the integral converges. If convergence does not occur, one may look for paths that satisfy the overtaking criterion. If none exists, one can fall back on the catching-up criterion. One can even set up weaker criteria (see, e.g., Seierstad and Sydsaeter, 1977).

9.2 Necessary conditions

It is clear that if $(\mathbf{s}^*, \mathbf{c}^*)$ is optimal under any one of the criteria mentioned in the preceding section, then the "truncated path" $(\mathbf{s}^*(t), \mathbf{c}^*(t))$, $t \leq T$ must be an optimal path for the truncated problem of finding $\mathbf{c}(t)$ that maximizes

$$V = \int_0^T v(\mathbf{s}, \mathbf{c}, t)\, dt$$

subject to (9.1a)–(9.1c) and the terminal condition

$$\mathbf{s}(T)=\mathbf{s}^*(T).$$

It follows that all the necessary conditions (Theorem 6.5.1) for finite-horizon problems (with the exception of transversality conditions) are also necessary for the infinite-horizon problems (for a formal proof, see Halkin, 1974). It remains to find out the appropriate transversality conditions. Suppose that the following terminal conditions are imposed on the state variables:

$$\lim_{t\to\infty} s_i(t)=\bar{s}_i, \quad i=1,2,\dots,k, \tag{9.4a}$$

$$\lim_{t\to\infty} s_i(t)\geq\bar{s}_i, \quad i=k+1,\dots,k+p, \tag{9.4b}$$

and no condition on $s_i(t)$, $i=k+p+1,\dots,n$, as $t\to\infty$.

One might expect that the following "transversality conditions" are necessary:

$$\lim_{t\to\infty} \pi_i(t)\geq 0, \quad i=k+1,\dots,k+p, \tag{9.5a}$$

$$\lim_{t\to\infty} \pi_i(t)=0, \quad i=k+p+1,\dots,n. \tag{9.5b}$$

Unfortunately, in general, conditions (9.5) are not necessary conditions. Only with considerable restrictions on the functions v, f^i, g^j does one obtain transversality conditions (9.5) or something like them; see Seierstad (1977b), Benveniste and Scheinkman (1982), and Michel (1982). Since the restrictions are too strict to be applicable to most problems of economic interest, we will not reproduce them here. Instead, we recommend the use of sufficiency theorems to identify optimal paths. This is the subject matter of the next section.

9.3 Sufficient conditions

The following theorem states the sufficient conditions for optimality under the catching-up criterion (and also for cases in which convergence occurs).

Theorem 9.3.1: sufficiency. Let $(\mathbf{s}^*, \mathbf{c}^*, \boldsymbol{\pi}^*, \boldsymbol{\lambda}^*)$ satisfy the necessary conditions of Theorem 6.5.1 and the terminal conditions (9.4a)–(9.4b). Then $(\mathbf{s}^*, \mathbf{c}^*)$ is optimal if the Lagrangean is concave in $(\mathbf{s}, \mathbf{c})$ and if

$$\lim_{t\to\infty} \boldsymbol{\pi}^*(t)\cdot[\mathbf{s}(t)-\mathbf{s}^*(t)]\geq 0, \tag{9.6}$$

where $\mathbf{s}(t)$ is any other feasible path satisfying the terminal conditions (9.4) and the expression in (9.6) is the inner product of $\boldsymbol{\pi}^*$ and $\mathbf{s}-\mathbf{s}^*$. (Note that (9.6) can be replaced by a weaker condition, where "lim" is replaced by "lim inf.")

To prove this result, observe that for any given T the sufficiency arguments in Chapter 6 yield

$$
\begin{aligned}
z(T) &\equiv \int_0^T v(\mathbf{s}^*, \mathbf{c}^*, t)\,dt - \int_0^T v(\mathbf{s}, \mathbf{c}, t)\,dt \\
&= \int_0^T [(H^* - \boldsymbol{\pi}^* \cdot \dot{\mathbf{s}}^*) - (H - \boldsymbol{\pi}^* \cdot \dot{\mathbf{s}})]\,dt \\
&= \int_0^T [(H^* + \dot{\boldsymbol{\pi}}^* \cdot \mathbf{s}^*) - (H + \dot{\boldsymbol{\pi}}^* \cdot \mathbf{s})]\,dt + \boldsymbol{\pi}^*(T) \cdot [\mathbf{s}(T) - \mathbf{s}^*(T)] \\
&\geq \boldsymbol{\pi}^*(T) \cdot [\mathbf{s}(T) - \mathbf{s}^*(T)].
\end{aligned}
$$

Hence, if (9.6) is satisfied, then $(\mathbf{s}^*, \mathbf{c}^*)$ is optimal according to the catching-up criterion.

Corollary 9.3.1. For optimality under the overtaking criterion, (9.6) must be modified to a more stringent condition: there exists some T such that for all $t > T$,

$$\boldsymbol{\pi}^*(t) \cdot [\mathbf{s}(t) - \mathbf{s}^*(t)] \geq 0. \tag{9.7}$$

This result follows clearly from the proof of Theorem 9.3.1.

There are many different cases in which condition (9.6) is satisfied. The following corollary displays some of these. In applying them it is important to keep in mind the ever-present possibility of making a change of variable from $s_i(t)$ to $\tilde{s}_i(t) = -s_i(t)$, which results in a costate $\tilde{\pi}_i(t) = -\pi_i(t)$.

Corollary 9.3.2. In Theorem 9.3.1 condition (9.6) can be replaced by (9.8a)–(9.8c):

(i) For $i = 1, 2, \ldots, k$, either $|\pi_i^*(t)| < N$ for some N, or

$$\lim_{t \to \infty} \pi_i^*(t)[s_i(t) - s_i^*(t)] \geq 0. \tag{9.8a}$$

(ii) For $i = k+1, \ldots, n$,

$$\lim_{t \to \infty} \pi_i^*(t) s_i^*(t) = 0, \tag{9.8b}$$

$$\lim_{t \to \infty} \pi_i^*(t) \geq 0, \quad \text{and} \quad 0 \leq s_i(t) < M \text{ for some } M. \tag{9.8c}$$

9.4 Autonomous problems

A special class of infinite-horizon problems often encountered is that in which none of the functions f^i and g^h, g^j contains t as an argument and in which $v(\mathbf{s}, \mathbf{c}, t)$ takes the form

$$v(\mathbf{s}, \mathbf{c}, t) = e^{-\delta t} u(\mathbf{s}, \mathbf{c}),$$

where $\delta \geq 0$ is called the discount rate. These problems are called "autonomous problems" because, as described in Chapter 4, the necessary conditions, stated with the use of current-value costate variables, do not contain an independent time argument and the resulting system of differential equations is autonomous in the sense defined in Chapter 2. When the time horizon is infinite, an additional feature of this class of problems is that the optimal value of each control variable at any time depends only on the values of the state variables. Before formalizing this idea, let us consider an example.

Example 9.4.1. Find $c(t)$ that maximizes

$$V(b, t_0) = \int_{t_0}^{\infty} e^{-\delta t} (\ln c(t))\, dt \tag{9.9}$$

subject to

$$\dot{s} = rs - c, \tag{9.9a}$$

$$s(t_0) = b > 0, \tag{9.9b}$$

$$\lim_{t \to \infty} s(t) = 0, \tag{9.9c}$$

where we assume $\delta > r \geq 0$.

Since the problem is autonomous, we can form the current-value Hamiltonian

$$\tilde{H} = \ln c + \psi(rs - c)$$

and obtain the necessary conditions (9.9a)–(9.9c) and

$$\partial \tilde{H}/\partial c = 1/c - \psi = 0, \tag{9.9d}$$

$$\dot{\psi} = \delta\psi - \partial \tilde{H}/\partial s = (\delta - r)\psi. \tag{9.9e}$$

Differentiate (9.9d) with respect to t:

$$(1/c^2)\dot{c} = -\dot{\psi},$$

or

$$\dot{c}/c = -\dot{\psi}c = -\dot{\psi}/\psi = r - \delta. \tag{9.10}$$

Solving (9.10),

$$c(t) = Ke^{(r-\delta)t}, \quad K = \text{const} = c(t_0)e^{(\delta - r)t_0}. \tag{9.11}$$

Substituting (9.11) into (9.9a):

$$\dot{s} - rs = -Ke^{(r-\delta)t},$$

or

$$e^{-rt}(\dot{s}-rs) = -Ke^{-\delta t}. \tag{9.12}$$

Integrating both sides from t_0 to any arbitrary T yields

$$\int_{t_0}^{T} e^{-rt}(\dot{s}-rs)\,dt = e^{-rT}s(T) - e^{-rt_0}s(t_0) = -\frac{K}{\delta}(e^{-\delta t_0})[1-e^{-\delta(T-t_0)}].$$

Rearranging terms after multiplying both sides by e^{rt_0}, we get

$$s(T)e^{-r(T-t_0)} = [s(t_0) - c(t_0)/\delta] + [c(t_0)/\delta]e^{-\delta(T-t_0)},$$

or

$$s(T) = [s(t_0) - c(t_0)/\delta]e^{r(T-t_0)} + [c(t_0)/\delta]e^{(r-\delta)(T-t_0)}. \tag{9.13}$$

Taking the limit $T \to \infty$ and noting that the last term vanishes because we assumed $r-\delta<0$, we must have

$$c(t_0) = s(t_0)\delta = b\delta \tag{9.14}$$

in order to satisfy (9.9c). Substituting this into (9.11) and (9.13), respectively, we get

$$c^*(t) = b\delta e^{(r-\delta)(t-t_0)}, \tag{9.15a}$$

$$s^*(t) = be^{(r-\delta)(t-t_0)}. \tag{9.15b}$$

Thus,

$$c^*(t) = \delta s^*(t) \quad \text{for all } t \geq t_0. \tag{9.16}$$

We are able to express the optimal control as a function of the current value of the state variable. This property holds for all autonomous infinite-horizon problems as we shall see shortly. First, however, let us calculate the value function $V(b, t_0)$. Substitute (9.15a) into (9.9):

$$\begin{aligned} V(b, t_0) &= \int_{t_0}^{\infty} e^{-\delta t}[(\ln b\delta) + (r-\delta)(t-t_0)]\,dt \\ &= [\delta^{-1}e^{-\delta t}((r-\delta)t_0 - \delta^{-1}(r-\delta) - \ln b\delta - t(r-\delta))]_{t_0}^{\infty} \\ &= \delta^{-1}e^{-\delta t_0}\left(\frac{r-\delta}{\delta} + \ln b\delta\right). \end{aligned}$$

This is true in particular when $t_0 = 0$; hence,

$$V(b, 0) = \delta^{-1}\left[\frac{r-\delta}{\delta} + \ln b\delta\right].$$

It follows that

$$V(b, t_0) = e^{-\delta t_0}V(b, 0). \tag{9.17}$$

In other words, the value of this autonomous infinite-horizon program starting at t_0 with $s(t_0)=b$ is equal to $e^{-\delta t_0}$ times the value of the same infinite-horizon program starting at $t=0$ with $s(0)=b$. Let us generalize this property to any autonomous infinite-horizon problem.

A typical autonomous problem

Let

$$V(\mathbf{b}, t_0) = \max_{\mathbf{c}} \int_{t_0}^{\infty} e^{-\delta t} u(\mathbf{s}, \mathbf{c})\, dt \tag{9.18}$$

subject to

$$\dot{s}_i = f^i(\mathbf{s}, \mathbf{c}), \quad i = 1, 2, \ldots, n, \tag{9.19a}$$

$$g^j(\mathbf{s}, \mathbf{c}) \geq 0, \quad j = 1, 2, \ldots, m', \tag{9.19b}$$

$$g^h(\mathbf{s}, \mathbf{c}) = 0, \quad h = m'+1, \ldots, m, \tag{9.19c}$$

$$s_i(t_0) = b_i, \quad i = 1, 2, \ldots, n, \tag{9.20a}$$

$$\lim_{t\to\infty} s_i(t) = \bar{s}_i, \quad i = 1, 2, \ldots, k, \tag{9.20b}$$

$$\lim_{t\to\infty} s_i(t) \geq \bar{s}_i, \quad i = k+1, \ldots, k+p, \tag{9.20c}$$

no restriction on the terminal behavior of

$$s_i(t), \quad i = k+p+1, \ldots, n. \tag{9.20d}$$

(Note that the cases $k=0$ and $p=0$ are admitted.) It is immediately clear that

$$\begin{aligned} V(\mathbf{b}, t_0) &= e^{-\delta t_0} \max_{c} \int_{t_0}^{\infty} e^{-\delta(t-t_0)} u(\mathbf{s}, \mathbf{c})\, dt \\ &= e^{-\delta t_0} \max_{c} \int_{0}^{\infty} e^{-\delta \tau} u(\mathbf{s}, \mathbf{c})\, d\tau, \text{ where } \tau \equiv t - t_0, \\ &= e^{-\delta t_0} V(\mathbf{b}, 0), \end{aligned} \tag{9.21}$$

where

$$V(\mathbf{b}, 0) = \max_{c} \int_{0}^{\infty} e^{-\delta t} u(\mathbf{s}, \mathbf{c})\, dt$$

subject to (9.19a)–(9.19c), (9.20b)–(9.20d), and

$$s_i(0) = b_i, \quad i = 1, 2, \ldots, n.$$

Thus, we have proved (9.17) for any autonomous infinite-horizon problem. Let us proceed a step further and define the (current-value) return function

$$W(\mathbf{b}, t_0) \equiv e^{\delta t_0} V(\mathbf{b}, t_0). \tag{9.22a}$$

Then from (9.21)

$$W(\mathbf{b}, t_0) = V(\mathbf{b}, 0); \tag{9.22b}$$

in other words, W is independent of t_0 and can be written as $W(\mathbf{b})$. Thus, we have

$$W(\mathbf{b}) = e^{\delta t_0} V(\mathbf{b}, t_0), \quad \text{where } \mathbf{s}(t_0) = \mathbf{b}. \tag{9.22c}$$

Since this is true for any starting time and any starting capital stock, we can write (9.22c) more generally as

$$W(\mathbf{s}_t) = e^{\delta t} V(\mathbf{s}_t, t). \tag{9.23a}$$

Now recall that in Chapter 4 under the assumption of differentiability of V, we proved a relationship between the costates and the function V,

$$\boldsymbol{\pi}^*(t) = V_{\mathbf{s}}(\mathbf{s}_t, t). \tag{9.23b}$$

If $\boldsymbol{\psi}(t)$ denotes the current-value costates, it follows from (9.23a) and (9.23b) that

$$\boldsymbol{\psi}^*(t) = W_{\mathbf{s}}(\mathbf{s}_t). \tag{9.23c}$$

Thus, for autonomous infinite-horizon problems, the optimal current-value costates at any time depend only on the values of the state variables at that time. This implies that the optimal values of the control variables depend on the current states alone. Thus, we have provided an informal proof of the following result.

Theorem 9.4.1. In an autonomous infinite-horizon problem such as (9.18)–(9.20), the maximum present-value return function has the following property:

$$V(\mathbf{b}, t_0) = e^{-\delta t_0} V(\mathbf{b}, 0). \tag{9.24a}$$

It follows that the maximum current-value return function depends only on the initial values of the state variables and not on the starting date; hence, it can be expressed as $W(\mathbf{b})$. Consequently, the current-value vector $\boldsymbol{\psi}^*(t)$ and the optimal control vector $\mathbf{c}^*(t)$ can be expressed as functions of the current-state vector $\mathbf{s}^*(t)$:

$$\boldsymbol{\psi}^*(t) = \boldsymbol{\theta}(\mathbf{s}^*(t)) = W_{\mathbf{s}}(\mathbf{s}^*(t)), \tag{9.24b}$$

$$\begin{aligned} \mathbf{c}^*(t) &= \boldsymbol{\omega}(\boldsymbol{\psi}^*(t), \mathbf{s}^*(t)) \\ &= \boldsymbol{\omega}(\boldsymbol{\theta}(\mathbf{s}^*(t)), \mathbf{s}^*(t)) \equiv \mathbf{G}(\mathbf{s}^*(t)). \end{aligned} \tag{9.24c}$$

It is important to stress that this result cannot be obtained in the case of autonomous finite-horizon problems. (For example, in problem (4.38), equations (4.44), (4.46), and (4.48) yield

$$c^*(t)=\delta[s^*(t)-e^{rt}(s_T e^{-rt}-s_0 e^{-\delta T})(1-e^{-\delta T})^{-1}],$$

showing that $c^*(t)$ depends on t.) The reason is that in the finite-horizon case the step preceding (9.21) fails.

9.5 Steady states in autonomous infinite-horizon problems

For autonomous problems, it is more convenient to work with the current-value Hamiltonian and current-value costates. The resulting differential equations in $(\mathbf{s}, \boldsymbol{\psi})$ are autonomous (i.e., do not contain t as an independent argument). Quite often these differential equations yield an equilibrium point $(\mathbf{s}^\infty, \boldsymbol{\psi}^\infty)$ with the saddle-point property. If the discount rate is positive and the Lagrangean is concave in $(\mathbf{s}, \mathbf{c})$, then the sufficiency theorem stated in Section 9.3 implies that the path leading to $(\mathbf{s}^\infty, \boldsymbol{\psi}^\infty)$ is an optimal path, provided that all feasible paths for $\mathbf{s}$ are bounded (or $\mathbf{s}$ is required to be nonnegative and $\boldsymbol{\psi}^\infty \geq 0$). In Chapter 4 we studied various versions of the optimal growth model that have the saddle-point property. It turns out that for the general autonomous problem (9.18) with the discount rate $\delta > 0$, if a steady state $(\mathbf{s}^\infty, \boldsymbol{\psi}^\infty)$ exists, then it cannot be locally stable in the (state, costate) space; that is, at least one of the roots is positive (or has positive real part), so that we have either only conditional stability (in the sense of saddle point) or only complete instability; see Kurz (1968) for a proof. In the special case where there is only *one* variable, it can also be shown that the optimal path of the state variable is monotone. An informal proof of this result is offered below.

Recall that along an optimal path, the control vector can be expressed as a function of the state variables. When there is only one state variable, using (9.24) we can describe the evolution of the optimal path of the state variable by a single first-order differential equation:

$$\dot{s}_1^*(t)=f^1[s_1^*(t), G(s_1^*(t))]. \tag{9.25}$$

If the optimal path of s_1 were nonmonotone, there would exist t_1 and t_2 such that $s_1(t_1)=s_1(t_2)$ and the sign of $\dot{s}_1(t_2)$ would be opposite to that of $\dot{s}_1(t_1)$. But this is impossible, because we know that f^1 in (9.25) takes on the same value at $s_1(t_1)$ and $s_1(t_2)$ for $s_1(t_1)=s_1(t_2)$. This argument establishes that $s_1^*(t)$ must be monotone. Notice that the argument relies on the assumption that there is only *one* state variable; if there were several state variables, then s_j $(j \neq 1)$ would in general appear in (9.25), so that $s_1(t_1)=s_1(t_2)$ is consistent with $\dot{s}_1(t_1) \neq \dot{s}_1(t_2)$. The following theorem summarizes these results.

Theorem 9.5.1. Steady states of autonomous infinite-horizon problems with a positive discount rate either are unstable or exhibit the saddle-point property. In the special case where there is only one state variable, the optimal path for the state variable must be monotone.

To reinforce the readers' grasp of this theorem, we provide a simple example.

Example 9.5.1: saddle point in an infinite-horizon autonomous problem. Let $s(t)$ denote the biomass of a fish species. Without commercial exploitation, the rate of growth of $s(t)$ is assumed to take the following form:

$$\dot{s} = s(1-s).$$

Introducing exploitation, let $x(t)$ denote the rate of landing (harvest). We assume the following relationship between the output x and the inputs s and n (where n stands for "effort"):

$$x = 2s^{1/2}n^{1/2}. \tag{9.26}$$

The fish are sold at the fixed price $p=1$, and effort costs w dollars per unit. The optimization problem is to find $n(t)$ that maximizes

$$V(s_0) = \int_0^\infty e^{-rt}(2s^{1/2}n^{1/2} - wn)\,dt \tag{9.27}$$

subject to

$$n \geq 0, \tag{9.27a}$$

$$\dot{s} = s(1-s) - 2s^{1/2}n^{1/2}, \tag{9.27b}$$

$$s(0) = s_0 > 0 \tag{9.27c}$$

(r and w are positive constants, and we assume $r<1$). There is no restriction on the limiting behavior of $s(t)$. Note that (9.27b) implies that $s(t)$ can never become negative.

The current-value Hamiltonian is

$$\tilde{H} = 2s^{1/2}n^{1/2} - wn + \psi(s - s^2 - 2s^{1/2}n^{1/2}), \tag{9.28}$$

and the necessary conditions are

$$\partial\tilde{H}/\partial n = (1-\psi)s^{1/2}n^{-1/2} - w \leq 0,$$
$$n \geq 0, \quad [(1-\psi)s^{1/2}n^{-1/2} - w]n = 0, \tag{9.29a}$$

$$\dot{\psi} = r\psi - \partial\tilde{H}/\partial s = \psi(r-1+2s) - (1-\psi)s^{-1/2}n^{1/2}, \tag{9.29b}$$

$$\dot{s} = \partial\tilde{H}/\partial\psi = s(1-s) - 2s^{1/2}n^{1/2}. \tag{9.29c}$$

From (9.29a) if $\psi \geq 1$ or $s=0$, then $n=0$. If $\psi<1$ and $s>0$, then n is given by

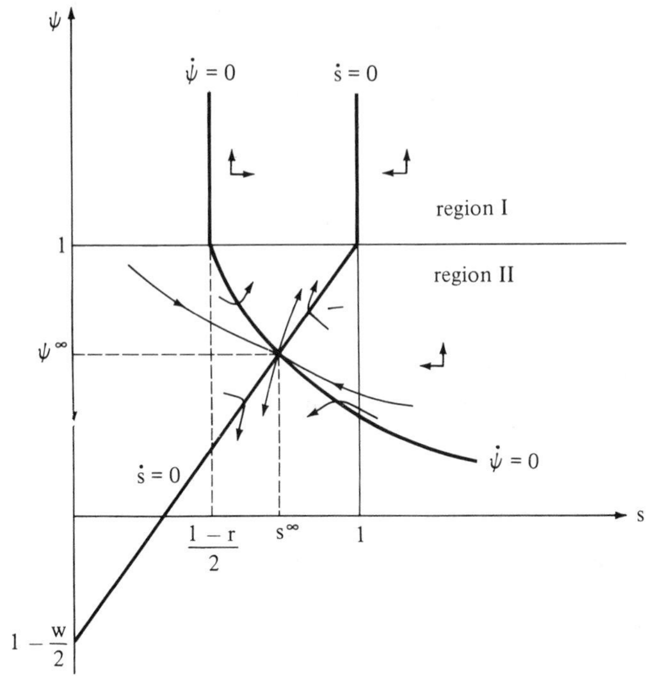

Figure 9.1

$$n = w^{-2}s(1-\psi)^2. \tag{9.30}$$

Thus, the optimal value of the control is a function of s and ψ:

$$n = G(s,\psi) = \begin{cases} 0 & \text{if } \psi \geq 1 \text{ or } s = 0, \\ s(1-\psi)^2/w^2 & \text{if } \psi < 1 \text{ and } s > 0. \end{cases} \tag{9.31}$$

Substituting (9.31) into (9.29b) and (9.29c), we have a pair of autonomous differential equations in (s, ψ); that is, there is no independent time term. Figure 9.1 is the phase diagram for this system. There are two regions. In region I, which consists of the area defined by $\psi \geq 1$ and $s \geq 0$, we have $n = 0$. In region II ($\psi < 1$ and $s > 0$), n is positive. Let us examine the properties of the $\dot{\psi} = 0$ locus first. In region I, this is the vertical line $s = (1-r)/2$ (we have assumed $r < 1$). In region II, substitution of (9.31) into (9.29b) yields, for $s > 0$,

$$\dot{\psi} = \psi(r-1+2s) - (1-\psi)^2/w \equiv M(s,\psi). \tag{9.32a}$$

Thus, the locus $\dot{\psi} = 0$ is a downward-sloping curve, starting at $(s, \psi) =$

$((1-r)/2, 1)$ and approaching the horizontal axis asymptotically. (To see this, note that when ψ goes to zero in (9.32a) with $\dot{\psi}=0$, s must become very large.) $M(s,\psi)$ takes on positive values to the right of that curve, because $\partial M/\partial s = 2\psi > 0$ for all $\psi > 0$. Thus, $\dot{\psi} > 0$ (resp. < 0) to the right (resp. left) of the $\dot{\psi}=0$ locus.

Turning to the $\dot{s} > 0$ locus, we note that in region I, $n=0$, so that $\dot{s} > 0$ if and only if $0 < s < 1$, and $\dot{s}=0$ if and only if $s=0$ or $s=1$. In region II, substitution of (9.30) into (9.29c) yields

$$\dot{s} = s(1-s) - 2s(1-\psi)/w \equiv N(s,\psi), \tag{9.32b}$$

so that $\dot{s}=0$ if $s=0$ or

$$\psi = 1-(1-s)w/2. \tag{9.33}$$

(9.33) is the equation of a straight-line segment joining $(s,\psi)=(0, 1-w/2)$ with $(s,\psi)=(1,1)$. To the right of this line, $\dot{s}<0$, because $\partial N/\partial s = -1 < 0$.

The intersection of the $\dot{s}=0$ locus (9.32b) with the curve $M(s,\psi)=0$ yields a unique equilibrium point, denoted by (s^∞, ψ^∞), where $0 < \psi^\infty < 1$ and $(1-r)/2 < s^\infty < 1$. There is another equilibrium point, namely, the origin, but starting from any positive s there is no path leading to it.

The point (s^∞, ψ^∞) is a saddle point, as can be seen from the phase diagram. To confirm this, we linearize (9.32a) and (9.32b):

$$\begin{bmatrix} \dot{s} \\ \dot{\psi} \end{bmatrix} = \begin{bmatrix} N_s & N_\psi \\ M_s & M_\psi \end{bmatrix} \begin{bmatrix} s-s^\infty \\ \psi-\psi^\infty \end{bmatrix},$$

where

$$N_s = [1-s^\infty - 2(1-\psi^\infty)/w] - s^\infty = -s^\infty < 0,$$
$$N_\psi = 2s^\infty/w > 0,$$
$$M_s = 2\psi^\infty > 0,$$
$$M_\psi = (r-1+2s^\infty) + 2(1-\psi^\infty)/w > 0,$$

and where the simplification of N_s comes from setting $\dot{s}=0$ in (9.32b) and the sign of M_ψ is obtained by recalling that $s^\infty > (1-r)/2$. The trace is $N_s + M_\psi = r > 0$, and the determinant is negative. Therefore, the roots are real and have opposite signs; hence, the equilibrium is a saddle point.

Starting from any $s_0 > 0$, the optimal policy is to choose an appropriate $\psi(0)$ so that $(s_0, \psi(0))$ is located on the stable branch of the saddle point. The system will approach the equilibrium point along the stable branch. Notice that the state variable $s^*(t)$ is monotone, thus verifying Theorem 9.5.1. However, one should bear in mind that in models with two or more state variables, it is possible that their time paths are nonmonotone; in fact, limit cycles may sometimes be optimal (see Ryder and Heal, 1973; Benhabib and Nishimura, 1979).

9.6 Further properties of autonomous infinite-horizon problems

In Theorem 9.4.1 we have shown that autonomous infinite-horizon problems have the property

$$V(\mathbf{b}, t_0) = e^{-\delta t_0} V(\mathbf{b}, 0) = e^{-\delta t_0} W(\mathbf{b}).$$

Hence, for any t_1 and $\mathbf{s}(t_1) \equiv \mathbf{s}_1$

$$V(\mathbf{s}_1, t_1) = e^{-\delta t_1} V(\mathbf{s}_1, 0) \equiv e^{-\delta t_1} W(\mathbf{s}_1). \tag{9.34}$$

We now use this identity to prove an important relationship between the value function V and the Hamiltonian, assuming that the former is differentiable.

Consider problem (9.18). Let $(\mathbf{s}^*, \mathbf{c}^*)$ denote the optimal path. Then for any t_1,

$$V(\mathbf{b}, t_0) = \int_{t_0}^{t_1} e^{-\delta t} u(\mathbf{s}^*, \mathbf{c}^*)\, dt + V(\mathbf{s}^*(t_1), t_1). \tag{9.35}$$

Use (9.34) in (9.35):

$$V(\mathbf{b}, t_0) = \int_{t_0}^{t_1} e^{-\delta t} u(\mathbf{s}^*, \mathbf{c}^*)\, dt + V(\mathbf{s}^*(t_1), 0) e^{-\delta t_1}. \tag{9.36}$$

Since (9.36) is true for all t_1, differentiating both sides with respect to t_1 yields

$$\begin{aligned} 0 &= e^{-\delta t_1} u(\mathbf{s}^*(t_1), \mathbf{c}^*(t_1)) - \delta V(\mathbf{s}^*(t_1), 0) e^{-\delta t_1} \\ &\quad + e^{-\delta t_1} [\partial V(\mathbf{s}^*(t_1), 0)/\partial \mathbf{s}] \cdot (d\mathbf{s}^*/dt_1). \end{aligned} \tag{9.37}$$

But

$$\frac{d\mathbf{s}^*(t_1)}{dt_1} = f(\mathbf{s}^*(t_1), \mathbf{c}^*(t_1))$$

and

$$e^{-\delta t_1} \partial V(\mathbf{s}^*(t_1), 0)/\partial \mathbf{s} = \partial [V(\mathbf{s}^*(t_1), t_1)]/\partial \mathbf{s} = \boldsymbol{\pi}^*(t_1).$$

Hence, (9.37) becomes

$$\delta V(\mathbf{s}^*(t_1), t_1) = e^{-\delta t_1} u(\mathbf{s}^*(t_1), \mathbf{c}^*(t_1)) + \boldsymbol{\pi}^*(t_1) \cdot \mathbf{f}(\mathbf{s}^*(t_1), \mathbf{c}^*(t_1)) \equiv H(t_1), \tag{9.38}$$

or equivalently, by (9.34),

$$\tilde{H}(t_1) \equiv u(\mathbf{s}^*(t_1), \mathbf{c}^*(t_1)) + \boldsymbol{\psi}^*(t_1) \cdot \mathbf{f}(\mathbf{s}^*(t_1), \mathbf{c}^*(t_1)) = \delta V(\mathbf{s}^*(t_1), 0), \tag{9.39}$$

where $\boldsymbol{\psi}^*(t_1)$ is the vector of current-value costates.

If $\delta = 0$, equation (9.39) implies that along an optimal path the current-value Hamiltonian of an autonomous infinite-horizon problem is identically zero. If $\delta > 0$, then (9.34) and (9.38) imply that

$$\lim_{t \to \infty} H(t) = 0. \tag{9.40}$$

(We can prove condition (9.40) without relying on the assumed differentiability of V; see Michel, 1982. However, note also that in our definition of the Hamiltonian we ignore anomalous cases in which the multiplier associated with the integrand cannot be set at unity.) We summarize these results in the following theorem.

Theorem 9.6.1. The Hamiltonian and the value function of the autonomous infinite-horizon problem (9.18) satisfy the following properties:

$$\tilde{H}(\mathbf{s}^*, \mathbf{c}^*, \boldsymbol{\psi}^*) = \delta V(\mathbf{s}^*, 0) \equiv \delta W(\mathbf{s}^*), \tag{9.41}$$

$$\text{If } \delta > 0 \text{ then } \lim_{t \to \infty} H(t) = 0. \tag{9.42}$$

When $\delta > 0$, (9.41) has an interesting economic interpretation. Recall that $V(\mathbf{s}^*(t_1), 0)$ is the value of the maximization problem

$$\int_{t_1}^{\infty} e^{-\delta(t - t_1)} u(\mathbf{s}, \mathbf{c})\, dt$$

subject to $\mathbf{s}(t_1) = \mathbf{s}^*(t_1)$ (9.19a–9.19c) and (9.20b–9.20d). It is thus the "stock of total wealth," measured in utility units. If the discount rate is interpreted as an interest rate, then $\tilde{H}$ is the income earned as interest on total wealth (see Weitzman, 1976; Kemp and Long, 1982).

Remark (a). It is instructive to relate (9.41) to the Hamilton–Jacobi–Bellman equation (5.63) of Chapter 5. Since for autonomous infinite-horizon problems $V(\mathbf{s}, t)$ can be written as $V(\mathbf{s}, 0)e^{-\delta t} \equiv W(\mathbf{s})e^{-\delta t}$, that equation becomes

$$0 = \max_{\mathbf{c}(t)} [e^{-\delta t} u(\mathbf{s}, \mathbf{c}) + e^{-\delta t} W_{\mathbf{s}}(\mathbf{s}) \cdot \mathbf{f}(\mathbf{s}, \mathbf{c}) - \delta e^{-\delta t} W(\mathbf{s})], \tag{9.43}$$

or equivalently,

$$\delta W(\mathbf{s}) = \max_{\mathbf{c}(t)} [u(\mathbf{s}, \mathbf{c}) + W_{\mathbf{s}}(\mathbf{s}) \cdot \mathbf{f}(\mathbf{s}, \mathbf{c})]. \tag{9.44}$$

But recalling that $W_{\mathbf{s}}(\mathbf{s})$ is equal to the current-value costate (equation (9.23c)), it can be seen that (9.44) is identical to (9.41).

Remark (b). Equation (9.41) is consistent with the transversality condition for free-terminal-time problems with a scrap value function (see Section 7.6). Consider problem (9.18). Suppose we have solved the problem and obtain the optimal time path $\mathbf{s}^*(t)$ and the value function $V(\mathbf{s}_t, t)$ and hence $W(\mathbf{s}_t)$; see (9.23a). For any t_1 and $\hat{\mathbf{s}}$, let $V_0(\mathbf{s}_0, \hat{\mathbf{s}}, t_1)$ denote the value of the following fixed-time, fixed-endpoint problem:

$$\max \int_{t_0}^{t_1} u(\mathbf{s}, \mathbf{c}) e^{-\delta t}\, dt \tag{9.45}$$

subject to (9.19a)–(9.19c) and $\mathbf{s}(t_0) = \mathbf{s}_0$, $\mathbf{s}(t_1) = \hat{\mathbf{s}}$. Then by definition, $V(\mathbf{s}_0, t_0)$ must be the value of the following free-time, free-endpoint problem with the scrap value function $e^{\delta t} W(\mathbf{s}_t) = V(\mathbf{s}_t, t)$:

$$\max_{\hat{\mathbf{s}}, t_1} [V_0(\mathbf{s}_0, \hat{\mathbf{s}}, t_1) + e^{-\delta t_1} W(\hat{\mathbf{s}})]. \tag{9.46}$$

Clearly, because the scrap value function in (9.46) is not any arbitrary function, but is the value function of the remaining portion of the original autonomous infinite-horizon problem, any t_1 will solve (9.46), provided that that $\hat{\mathbf{s}}$ is chosen to be $\mathbf{s}^*(t_1)$, the optimal value of the state variable at time t_1. The necessary conditions for (9.46) are

$$\partial V_0 / \partial t_1 - \delta e^{-\delta t_1} W(\mathbf{s}^*(t_1)) = 0, \tag{9.47a}$$

$$\partial V_0 / \partial \hat{\mathbf{s}} + e^{-\delta t_1} W_{\mathbf{s}}(\mathbf{s}^*(t_1)) = 0. \tag{9.47b}$$

Condition (9.47a) is a special case of the transversality condition (7.92) of the free-terminal-time problems discussed in Section 7.6. Recalling that $\partial V_0 / \partial t_1 = H(t_1)$, we see that (9.47a) is identical to (9.41). Condition (9.47b) simply says that the costate variables are continuous (recall that $\partial V_0 / \partial \hat{\mathbf{s}} = -\boldsymbol{\pi}(t_1^-)$ and $e^{-\delta t_1} W_{\mathbf{s}}(\mathbf{s}(t_1)) = \boldsymbol{\pi}(t_1^+)$).

The result of Theorem 9.6.1 can also be useful in deriving the solution to some problems. We now present one such problem, which also incorporates a control parameter. (Necessary conditions that characterize the optimal choice of control parameters are stated in Section 7.11.)

Example 9.6.1: optimal resource depletion under the maximin criterion. An economy produces an output q using a stock of capital K and a flow of extracted resource x. The production function is

$$q(t) = [x(t)]^{\alpha} [K(t)]^{1-\alpha}, \tag{9.48}$$

where we assume $0 < \alpha < 0.5$; we shall see later why this assumption is crucial. Gross output is allocated between consumption $c(t)$ and investment $I(t)$. There is no capital depreciation, so that

$$\dot{K}(t) = I(t). \tag{9.49}$$

We are dealing with an exhaustible resource; denoting the current stock of resource by $R(t)$, we have

$$\dot{R}(t) = -x(t). \tag{9.50}$$

We require $I(t) \geq 0$, $x(t) \geq 0$, and $c(t) \geq b$ for all t, where b is a control parameter. The planner's objective is to select the highest possible value for the lower bound on consumption, b; this is why it is called the maximin criterion. (Another name for it is the *Rawlsian criterion* because it equates society's welfare with that of the poorest generation.) The objective can be stated as choosing the largest constant b or maximizing

$$W = \int_0^\infty \delta b e^{-\delta t}\, dt = b, \quad \delta > 0, \tag{9.51}$$

because $\int_0^\infty \delta e^{-\delta t}\, dt = 1$ for any $\delta > 0$. The maximization is subject to (9.49), (9.50), and

$$I(t) + c(t) = [x(t)]^\alpha [K(t)]^{1-\alpha}, \tag{9.52a}$$

$$c(t) \geq b, \tag{9.52b}$$

$$I(t) \geq 0, \quad x(t) \geq 0, \tag{9.52c}$$

$$K(0) = K_0 > 0, \quad R(0) = R_0 > 0, \quad \text{and} \quad \lim_{t \to \infty} R(t) = 0. \tag{9.52d}$$

We can use (9.52a) to eliminate $c(t)$ and restate (9.52b) as

$$[x(t)]^\alpha [K(t)]^{1-\alpha} - I(t) \geq b. \tag{9.52e}$$

The Hamiltonian of this problem is

$$H = \delta b e^{-\delta t} - \pi_1 x + \pi_2 I, \tag{9.53a}$$

and the Lagrangean is

$$\mathcal{L} = H + \lambda (x^\alpha K^{1-\alpha} - I - b). \tag{9.53b}$$

The necessary conditions are (9.49), (9.50), (9.52c)–(9.52e), and

$$\frac{\partial \mathcal{L}}{\partial I} = \pi_2 - \lambda \leq 0, \quad I \geq 0, \; I \frac{\partial \mathcal{L}}{\partial I} = 0, \tag{9.54a}$$

$$\frac{\partial \mathcal{L}}{\partial x} = -\pi_1 + \alpha \lambda x^{\alpha - 1} K^{1-\alpha} \leq 0, \quad x \geq 0, \; x \frac{\partial \mathcal{L}}{\partial x} = 0, \tag{9.54b}$$

$$\frac{\partial \mathcal{L}}{\partial \lambda} = x^\alpha K^{1-\alpha} - I - b \geq 0, \quad \lambda \geq 0, \; \lambda \frac{\partial \mathcal{L}}{\partial \lambda} = 0, \tag{9.54c}$$

$$\dot{\pi}_1 = 0, \tag{9.54d}$$

$$\dot{\pi}_2 = -(1-\alpha) \lambda x^\alpha K^{-\alpha}, \tag{9.54e}$$

$$\int_0^\infty \frac{\partial \mathcal{L}}{\partial b}\, dt = 0. \tag{9.54f}$$

Conditions (9.54a)–(9.54e) are the familiar ones; condition (9.54f) relates to the optimal choice of the control parameter b according to Theorem 7.11.1. It can be rewritten as

$$\int_0^\infty (\delta e^{-\delta t} - \lambda(t))\, dt = 0,$$

or

$$\int_0^\infty \lambda(t)\, dt = 1. \tag{9.55}$$

At this stage the existence of a solution to the problem appears by no means assured. Our strategy will be to suppose that there exists a positive ($x > 0$, $I > 0$, $b > 0$) solution, use the necessary conditions to derive it, and establish its optimality with the sufficiency results of Theorem 9.3.1. Although it is possible to use only the necessary conditions stated above, it is much more efficient to make use of Theorem 9.6.1, which states

$$e^{\delta t} H = \delta W,$$

or, using (9.51) and (9.53a),

$$\delta b - e^{\delta t}\pi_1 x + e^{\delta t}\pi_2 I = \delta b;$$

hence,

$$\frac{\pi_1}{\pi_2} = \frac{I}{x}. \tag{9.56}$$

When $x > 0$ and $I > 0$, (9.54a) and (9.54b) yield

$$\frac{\pi_1}{\pi_2} = \alpha x^{\alpha-1} K^{1-\alpha}; \tag{9.57}$$

hence,

$$I = \alpha x^{\alpha} K^{1-\alpha}. \tag{9.58}$$

Investment is seen to equal a constant fraction α of gross output. We suppose further that $c(t) = b$ for all t. (Recall that any consumption in excess of b contributes nothing to the objective criterion of (9.49).) Then (9.58) and (9.54c) yield

$$b = (1-\alpha) x^{\alpha} K^{1-\alpha}, \tag{9.59}$$

and this enables us to express $I(t)$ in terms of the constant b:

$$I(t) = \frac{\alpha}{1-\alpha} b. \tag{9.60}$$

Therefore, $\dot{K} = \alpha b/(1-\alpha)$ and

$$K(t)=K_0+\frac{\alpha}{1-\alpha}bt. \tag{9.61}$$

Manipulating (9.59) and (9.61) yields

$$x(t)=\left(\frac{\alpha}{1-\alpha}bt+K_0\right)^{(\alpha-1)/\alpha}\left(\frac{b}{1-\alpha}\right)^{1/\alpha}. \tag{9.62}$$

To determine the value of b we note that (9.50) and (9.52d) imply $R_0=\int_0^\infty x(t)\,dt$, which, with (9.62), gives

$$R_0=\left[\left(\frac{\alpha}{1-\alpha}bt+K_0\right)^{(2\alpha-1)/\alpha}\right]_0^\infty\left(\frac{1-\alpha}{b}\right)^{(\alpha-1)/\alpha}\frac{1}{2\alpha-1}$$

$$=(K_0)^{(2\alpha-1)/\alpha}\left(\frac{1-\alpha}{b}\right)^{(\alpha-1)/\alpha}\frac{1}{1-2\alpha},$$

because $\alpha<0.5$ implies that $(2\alpha-1)/\alpha$ is negative. Finally, we obtain

$$b=(1-\alpha)[(1-2\alpha)R_0]^{\alpha/(1-\alpha)}(K_0)^{(1-2\alpha)/(1-\alpha)}, \tag{9.63}$$

which we can substitute into (9.60)–(9.62) to obtain the precise solutions for I, K, and x.

We now turn to the task of determining the solutions for the multiplier and the costates. From (9.56) and (9.60) we deduce that

$$\pi_2(t)=\pi_1\frac{1-\alpha}{\alpha b}x(t). \tag{9.64}$$

Since b and π_1 are constant and $\int_0^\infty x(t)\,dt=R_0$, we can integrate both sides of (9.64) and use (9.55) and (9.54a) (with $I>0$) to obtain

$$1=\int_0^\infty \lambda(t)\,dt=\int_0^\infty \pi_2(t)\,dt=\pi_1\frac{1-\alpha}{\alpha b}\int_0^\infty x(t)\,dt=\pi_1\frac{1-\alpha}{\alpha b}R_0;$$

hence,

$$\pi_1=\frac{\alpha b}{(1-\alpha)R_0}. \tag{9.65}$$

This can be substituted into (9.64) and used in conjunction with (9.62) to get, after simplification,

$$\lambda(t)=\pi_2(t)=\frac{1}{R_0}\left(\frac{\alpha}{1-\alpha}bt+K_0\right)^{(\alpha-1)/\alpha}\left(\frac{b}{1-\alpha}\right)^{1/\alpha}. \tag{9.66}$$

The expressions in (9.65) and (9.66) with b given by (9.63) complete the solution to the problem.

It remains to verify that this solution satisfies the condition of Theorem 9.3.1. First consider $\lim_{t\to\infty}\pi_1^*(t)[R(t)-R^*(t)]\equiv L_1$, where $\pi_1^*(t)$ is the constant of (9.65), $R^*(t)$ is the optimal path determined by (9.62) and (9.50), and $R(t)$ is any other feasible path. Since π_1 is constant and $R(t)$ is required to satisfy (9.52d), we have

$$L_1 = \pi_1 \lim_{t\to\infty} [R(t) - R^*(t)] = 0.$$

Next consider $L_2 \equiv \lim_{t\to\infty} \pi_2^*(t)[K(t) - K^*(t)]$, where $\pi_2^*(t)$ and $K^*(t)$ are given by (9.66) and (9.61) and $K(t)$ is any feasible path:

$$\begin{aligned} L_2 &= \lim_{t\to\infty} \pi_2^*(t)K(t) - \lim_{t\to\infty} \pi_2^*(t)K^*(t) \\ &= \lim_{t\to\infty} \pi_2^*(t)K(t) - \lim_{t\to\infty} \frac{1}{R_0}\left(\frac{b}{1-\alpha}\right)^{1/\alpha}\left[\frac{1}{1-\alpha}bt + K_0\right]^{(2\alpha-1)/\alpha} \\ &\qquad \text{by (9.61) and (9.66)} \\ &= \lim_{t\to\infty} \pi_2^*(t)K(t) \quad \text{since } \frac{2\alpha-1}{\alpha} < 0 \\ &\geq 0, \end{aligned}$$

as $\pi_2^*(t) \geq 0$, and any feasible $K(t)$ must be positive because $\dot{K}(t) = I(t) \geq 0$ and $K(0) = K_0 > 0$. Therefore, $L_1 + L_2 \geq 0$, and the solution gives a maximum. It is interesting that the optimal path does not tend to a steady state, since $K^*(t)$ increases without bound.

Exercises

1. The problem of optimal income transfer in a growing economy was first considered by Hamada (1967). The government's objective is to maximize the integral of discounted utility of the representative worker, $V = \int_0^\infty [u(c(t))e^{-\delta t}]\,dt$, where $c(t) \geq 0$ is consumption per worker. The capitalists' income is $Y(t) = N(t)[f(k(t)) - c(t)]$, where $N(t)$ is the number of workers and $k(t)$ is the capital/labor ratio $K(t)/N(t)$. We require $Y(t) \geq 0$. The rate of growth of the number of workers is exogenous, $\dot{N}(t) = nN(t)$, and the capitalists' propensity to save is constant; hence, $\dot{K}(t) = sY(t)$. We take δ, n, and s to be specified positive constants.
   (a) Show that this problem can be reduced to a single-state variable problem in which V is maximized subject to $\dot{k}(t) = s[f(k(t)) - c(t)] - nk(t)$, $f(k(t)) - c(t) \geq 0$, $c(t) \geq 0$.
   (b) Assume that $k(0) = k_0 > 0$, $\lim_{t\to+\infty} k(t) \geq 0$, and u and f have the following properties:

$$f'(0) = +\infty,\ f'(s) > 0,\ f'(+\infty) = 0,\ f''(s) < 0,$$

$$u'(0) = +\infty,\ u'(c) > 0,\ u'(+\infty) = 0,\ u''(c) < 0.$$

Show that there exists a steady state at k^*, where $f'(k^*)=(\delta+n)/s$, and construct a phase diagram in the (φ, k) space. Show that the steady state is a saddle point and the long-run optimum.

(c) Rework the problem under the assumption that the utility function is linear $(u(c)=c)$.

2. The rate of change of a country's net worth $A(t)$ is given by

$$\dot{A}(t)=R(A(t))-c(t)+y,$$

where $R(A(t))$ denotes interest income, $c(t)$ is aggregate consumption, and y is an exogenous flow of income. It is assumed that $R(0)=0$, $R'(A)>0$, and $R''(A)<0$. A negative net wealth implies that the country is a debtor. The country's aim is to maximize welfare represented by $W=\int_0^\infty U(c(t))e^{-\delta t}\,dt$ subject to the above constraints; $U''(c)<0$, and $U'(0)=+\infty$. The following assumptions are made. Let A_M be the negative number defined by $R(A_M)+y=0$; assume that there exists a value $A^*>A_M$ such that $R'(A^*)=\delta$ and that $A(0)=A_0>A_M$; we require $\lim_{t\to\infty} A(t)\geq A_M$. Derive the necessary conditions, construct a phase diagram, and show that the optimal path converges to (A^*, ψ^*), where $\psi^*=U'(y+R(A^*))$. Verify that this is a saddle point.

3. A country uses capital to extract a resource, which is used as an input in the production of a final output. The two production functions are $R(t)=K_2(t)$ and $Q(t)=[R(t)K_1(t)]^{1/2}$. $R(t)$ denotes the rate of extraction of the resource. $K_1(t)$ and $K_2(t)$ are the amounts of capital used in the two industries; they are control variables and can be freely chosen at any time, subject to the constraint $K_1(t)+K_2(t)=K(t)$, where $K(t)$ is the total stock of capital, a state variable. We have $\dot{K}(t)=I(t)$, $\dot{S}(t)=-R(t)$, and $C(t)=Q(t)-I(t)$. $S(t)$ is the existing stock of resource; $Q(t)$ denotes final output, which is shared between consumption $C(t)$ and investment $I(t)$; we assume that $I(t)$ is unrestricted in sign. The utility function is $U(C(t))=(1-\eta)^{-1}(C(t))^{1-\eta}$ $(\eta>0, \eta\neq 1)$. The country's objective is to maximize $\int_0^\infty U(C(t))e^{-\delta t}\,dt$ subject to the above constraints. In addition, it is required that $\lim_{t\to\infty} K(t)\geq 0$ and $\lim_{t\to\infty} S(t)\geq 0$; $S(0)$ and $K(0)$ are specified. Set up the problem as a control problem and eliminate by substitution the controls $Q(t)$, $I(t)$, $K_1(t)$, and $K_2(t)$. Show that if $S(0)\geq 0.5K(0)\eta/(\delta-0.5)$ and $\delta>0.5$, the following solution path satisfies all the necessary and sufficient conditions:

(i) The costate variable associated with S is zero for all t; hence, $R(t)=0.5K(t)$;

(ii) $\dot{\pi}(t)=-0.5\pi(t)$, where $\pi(t)$ is the costate for capital stock;

(iii) $C(t)=C(0)\exp((0.5-\delta)t/\eta)$;

(iv) $I(t)=0.5K(t)-C(t)$;

(v) $C(0)=K(0)[0.5-(0.5-\delta)/\eta]$.

4. Reconsider the yabbies problem of exercises 8 and 9 in Chapter 6. Let $T=+\infty$ and require $\lim_{t\to\infty} s(t)\geq 0$. Determine the optimal path in all three cases. For the special case $\delta=5$, $R(c)=c(1-0.5c)$, $f(s)=10s(1-s)$, find all steady-state equilibria and show that one of the equilibria is an unstable focus, which is not part of the optimal path because the Hamiltonian is strictly convex in the

state variable in that region. (What is the sign of the costate variable at that point?)

5. Reconsider the model of exercise 6 in Chapter 6. Describe the optimal policy when $E(0)=E_0>0$ is specified, $E(T)\geq 0$ is free, and $T=+\infty$. Carry out the exercise again under the added assumption that equipment stock depreciates at the rate $m>0$ (i.e., $\dot{E}(t)=b(t)-mE(t)$).

CHAPTER 10

Three special topics

10.1 Problems with two-state variables

Nearly all the models hitherto encountered in this book have contained a single state variable. (Exceptions are the models of Sections 8.1, 8.3, and 9.6.) We have relied very heavily on phase diagrams in shedding light on the optimal solution. When there are two state variables, however, the (state, costate) space is four-dimensional and cannot be represented straightforwardly. It must be understood that, given the usual regularity conditions, we have in the maximum principle a set of necessary and sufficient conditions for an optimum, whatever the size of the problem, and if all functional forms and other restrictions were fully specified, we could - possibly using numerical methods - provide an explicit solution to the problem. However, since most models of interest in economic theory involve some unspecified functional forms, an explicit solution is normally unobtainable. This is why phase diagrams are such a useful device for pulling together all the pieces of information contained in the maximum principle.

Since they fail us here, we must devise other means of synthesizing the information. Unfortunately, this is often quite difficult, and in many cases a complete characterization of the solution escapes us. This is not to say that we cannot offer a partial characterization of the solution. It is the aim of this section to illustrate what can indeed be done. First note that in the models of Sections 8.1 and 8.3, the analysis was reduced to a two-dimensional phase diagram. The reader is referred to those sections.

Reduction of a two-sector growth model to a one-state-variable model

This is a straightforward generalization of the one-sector growth model introduced in Chapter 4 and further analyzed in subsequent chapters. There are now two industries (or plants, or regions, etc.) that can independently produce the consumption good, each with its own technology. The problem is to find $c_1(t) \geq 0$ and $c_2(t) \geq 0$ that maximize

$$\int_0^\infty u(c_1+c_2)e^{-\delta t}\,dt \tag{10.1a}$$

subject to

$$\dot{s}_1 = F_1(s_1) - m_1 s_1 - c_1, \tag{10.1b}$$

$$\dot{s}_2 = F_2(s_2) - m_2 s_2 - c_2; \tag{10.1c}$$

$$s_1(0),\, s_2(0) \text{ exogenously specified.} \tag{10.1d}$$

Using the current-value costate variables ψ_1 and ψ_2 the Hamiltonian is

$$H = u(c_1+c_2) + \psi_1[F_1(s_1) - m_1 s_1 - c_1] + \psi_2[F_2(s_2) - m_2 s_2 - c_2]. \tag{10.2}$$

As long as the usual strict concavity requirements are met, with derivatives ranging from $+\infty$ to 0, and nonnegativity conditions are ignored, the following conditions are optimal: (10.1b), (10.1c), and

$$u' - \psi_1 = 0, \tag{10.3a}$$

$$u' - \psi_2 = 0, \tag{10.3b}$$

$$\dot{\psi}_1 = \psi_1[\delta + m_1 - F_1'], \tag{10.3c}$$

$$\dot{\psi}_2 = \psi_2[\delta + m_2 - F_2'], \tag{10.3d}$$

where the prime denotes a derivative.

These conditions would normally require a four-dimensional phase diagram, but this problem exhibits some particular features. Clearly, $\psi_1 = \psi_2$; hence, the net marginal products are the same in both industries: $F_1' - m_1 = F_2' - m_2$. This suggests that the two industries could be operated as one, maximizing their joint net product subject to the total availability of capital. This aggregation would not be possible if the two kinds of capital could not be measured in the same units. Define

$$F(s) = \max_{s_1, s_2}[F_1(s_1) - m_1 s_1 + F_2(s_2) - m_2 s_2 \mid s_1 + s_2 = s]. \tag{10.4}$$

The envelope theorem (Theorem 1.2.8) and the optimality conditions immediately yield

$$F'(s) = F_1'(s_1) - m_1 = F_2'(s_2) - m_2. \tag{10.5}$$

Consider the aggregated problem of maximizing

$$\int_0^\infty u(c)e^{-\delta t}\,dt$$

subject to

$$\dot{s} = F(s) - c,$$

$$s(0) = s_1(0) + s_2(0).$$

It has the Hamiltonian $H=u(c)+\psi[F(s)-c]$ and the optimality conditions

$$u'=\psi, \quad \dot{s}=F(s)-c, \quad \text{and} \quad \dot{\psi}=\psi(\delta-F'(s)). \tag{10.6}$$

Letting $\psi=\psi_1=\psi_2$ and $F(s)$ as defined by (10.4), we see that conditions (10.6) along with (10.5) can be used to duplicate the optimality conditions of problem (10.1). The analysis can be carried out in the (s, ψ) space, and (10.5) is used to get the optimal path of s_1 and s_2. The phase diagram in the (s, ψ) space is similar to Figure 6.5 without region A and presents no difficulties. Suppose $s(0)<\tilde{s}$; we then follow the stable arm to the equilibrium. However, problems arise when the path is to be represented in the (s_1, s_2) space. Equation (10.5) defines what we shall term an efficient locus by its second equality; it is an upward-sloping curve in the (s_1, s_2) plane along which the net marginal products of both industries are the same. The equilibrium values of s_1 and s_2 are found at the intersection of the efficient locus with the line of equation $s_1+s_2=\tilde{s}$. While an interior solution (both c_1 and c_2 positive) prevails, the optimal path follows the efficient locus to the equilibrium point. A difficulty arises if the initial values $s_1(0)$ and $s_2(0)$ are not on the efficient locus. Then clearly a corner solution must be considered (a diagram is helpful). Suppose, for instance, that the initial point is below the efficient locus; that is, $s_2(0)$ is relatively too small. Then in order to correct this imbalance we shall set $c_2=0$ at first. The optimality conditions become (10.1b), (10.3a), (10.3c), (10.3d), and

$$\dot{s}_2=F_2(s_2)-m_2 s_2, \tag{10.7a}$$

$$u'<\psi_2. \tag{10.7b}$$

We can use (10.7a) and (10.3d) to determine the path of s_2 and ψ_2, and (10.1b), (10.3a), and (10.3c) can yield a phase diagram in the (s_1, ψ_1) space. The resulting optimal trajectory in the (s_1, s_2) space now has an arm beginning at the initial point and reaching up to the efficient locus. A set of initial conditions with $s_1(0)$ relatively small would result in a similar initial phase with $c_1=0$; the possibility that $c_1=c_2=0$ can be ruled out by assuming infinite marginal utility of nil consumption.

The reduction of a two-state-variable problem to a single-state-variable one is often far more difficult than in the preceding example. One very interesting treatment is that of Hadley and Kemp (1971, ch. 6), which analyzes an economy with a consumption good industry and a capital good industry, with nontransferable capital between the two industries. Many two-state-variable models, in anticipation of difficulties with their solution, have been structured in such a way that several distinct phases can be distinguished, each with a solution simpler than the whole problem. Examples of such models are Takayama (1985, pp. 627–37) and Pitchford (1977). We find other examples in this book (Sections 8.1 and 8.3) where the linearities of the model were exploited for that purpose.

10.2 Trade in capital goods: jumps in the state variables

In the models studied thus far capital goods have been essentially home-grown: no trade in capital took place within the horizon, although the inclusion of a scrap value implied the sale of capital at the end of the horizon. We maintained the requirement that state variables be continuous, indeed differentiable, albeit not continuously so. This is an appropriate assumption in cases where the planning unit does not have access to exterior capital markets, but not otherwise. For instance, there is no reason that an individual firm may not purchase or sell some capital asset at any time. In order to take this possibility into account we must allow jump discontinuities to occur for the state variables. This is a departure from the control problem format expounded here and requires a new result. This was provided by Vind (1967).

10.2.1 *The general case*

We first state the result for a problem with one state variable, no constraints, and upward jumps only. Many state variables and constraints introduce no new complications; downward jumps will be treated later.

We alter the problem stated at the beginning of Chapter 4 to take into account the possibility of jumps in the state variable. Find $c(t)$, θ_j, $s(\theta_j^-)$, and $s(\theta_j^+)$, $j=1,\dots,J$, that maximize

$$V=\int_0^T u(s(t),c(t),t)\,dt+\sum_{j=1}^{J} p(\theta_j)[s(\theta_j^-)-s(\theta_j^+)] \tag{10.8}$$

subject to

$$\dot{s}(t)=f(s(t),c(t),t), \quad \text{except at } \theta_j,\ j=1,\dots,J, \tag{10.9}$$

$$s(\theta_j^+)\geq s(\theta_j^-), \quad j=1,\dots,J, \tag{10.10}$$

and

$$s(0)=s_0, \quad s(T)=s_T. \tag{10.11}$$

The dates θ_j at which jumps will occur are to be chosen optimally. The values $s(\theta_j^-)$ and $s(\theta_j^+)$ are, respectively, the values of the state variable immediately before and after a jump. A price $p(\theta_j)$ is paid for an extra unit of capital at time θ_j. Thus, the state variable follows the differential equation (10.9) at all times except when jumps occur. Equation (10.10) reflects the fact that only upward jumps are allowed; the cost of purchases of capital has been subtracted from the objective function (10.8).

The Hamiltonian for this is

$$H = u(s, c, t) + \pi f(s, c, t). \tag{10.12}$$

Theorem 10.2.1: upward jumps in the interior of the horizon. An optimal solution to the above problem, with upward jumps at θ_j, $j = 1, \ldots, J$, must necessarily satisfy (10.9)–(10.11) plus

$$\max_{c(t)} H \tag{10.13}$$

$$\dot{\pi} = -\partial H / \partial s \tag{10.14}$$

and in addition

$$p(t) \geq \pi(t), \quad \text{all } t \in [0, T], \qquad p(\theta_j) = \pi(\theta_j), \quad j = 1, \ldots, J, \tag{10.15}$$

$$H(\theta_j^-) - H(\theta_j^+) + \dot{p}(\theta_j)[s(\theta_j^-) - s(\theta_j^+)] = 0 \quad \text{if } \theta_j \in (0, T),\ j = 1, \ldots, J, \tag{10.16}$$

where

$$H(\theta_j^-) = u(s(\theta_j^-), c(\theta_j^-), \theta_j) + \pi(\theta_j) f(s(\theta_j^-), c(\theta_j^-), \theta_j)$$

and

$$H(\theta_j^+) = u(s(\theta_j^+), c(\theta_j^+), \theta_j) + \pi(\theta_j) f(s(\theta_j^+), c(\theta_j^+), \theta_j).$$

The ingenious method devised by Vind and adopted by others (e.g., Arrow and Kurz, 1970) was to introduce an artificial time that coincides with natural-time outside jumps but keeps running while natural time stands still at jump points. We shall use instead a simpler method of derivation that relies on the economic content of jumps. To begin with, let us interpret equation (10.15). It indicates that the internal price of capital, π, never exceeds the outside price p; at jump points the two prices are equal. Presumably no purchase takes place when the outside price is higher and if a purchase occurs (at a jump), the two prices must be equal. To understand equation (10.16) it must be thought of as characterizing the choice of θ_j. The last term is the "gain" made by postponing the purchase of $[s(\theta_j^+) - s(\theta_j^-)]$ units by one instant (supposing $\dot{p}(\theta_j) < 0$). To understand the first two terms, recall that we showed in Section 7.5 that $H(T)$ was the marginal contribution of an increase in T to the maximand V. Here we use the difference $H(\theta_j^+) - H(\theta_j^-)$ to evaluate the loss of postponing the injection of $[s(\theta_j^+) - s(\theta_j^-)]$ units of capital by an instant.

This discussion brings out similarities between conditions (10.15) and (10.16) with transversality conditions. Indeed, problems with free end-point and a scrap value function actually allow a jump from s_T to zero at terminal time, and T may also be free. The same can be said about problems with free initial point and an initial purchase cost. These observations are the basis for the strategy we adopt in proving (10.15) and (10.16). We consider a problem with a single jump at time θ in order to simplify the notation (several jumps at θ_j, $j=1,\ldots,J$, simply require summation over j).

Let c, $s(\theta^-)$, $s(\theta^+)$, and θ maximize

$$V=\int_0^\theta u(c,s,t)\,dt+p(\theta)[s(\theta^-)-s(\theta^+)]+\int_\theta^T u(c,s,t)\,dt \tag{10.17}$$

subject to $\dot{s}=f(c,s,t)$ except at θ, s_0 and s_T are fixed, and $s(\theta^+)\geq s(\theta^-)$. Using integration by parts with an arbitrary π function, as in equation (4.77), we easily obtain

$$\begin{aligned}V=&\int_0^\theta (H+\dot{\pi}s)\,dt+\pi_0 s_0-\pi(\theta^-)s(\theta^-)+\pi(\theta^+)s(\theta^+)-\pi_T s_T\\&+\int_\theta^T (H+\dot{\pi}s)\,dt+p(\theta)[s(\theta^-)-s(\theta^+)].\end{aligned} \tag{10.18}$$

For an optimal choice of θ we partially differentiate V with respect to θ:

$$\begin{aligned}\frac{\partial V}{\partial\theta}=&H(\theta^-)+\dot{\pi}(\theta^-)s(\theta^-)+\int_0^\theta\left[(H_s+\dot{\pi})\frac{ds}{d\theta}+H_c\frac{dc}{d\theta}\right]dt-\frac{d\pi(\theta^-)}{d\theta}s(\theta^-)\\&-H(\theta^+)+\dot{\pi}(\theta^+)s(\theta^+)+\int_\theta^T\left[(H_s+\dot{\pi})\frac{ds}{d\theta}+H_c\frac{dc}{d\theta}\right]dt\\&+\frac{d\pi(\theta^+)}{d\theta}s(\theta^+)+\dot{p}(\theta)[s(\theta^-)+s(\theta^+)].\end{aligned}$$

We now substitute the optimal π, c, and s functions and use the other necessary conditions, $H_s+\dot{\pi}=0$, $H_c=0$. We also note that $\dot{\pi}(\theta^-)=d\pi(\theta^-)/d\theta$ and $\dot{\pi}(\theta^+)=d\pi(\theta^+)/d\theta$ and set the derivative to zero if $\theta\in(0,T)$:

$$\frac{\partial V}{\partial\theta}=H(\theta^-)-H(\theta^+)+\dot{p}(\theta)[s(\theta^-)-s(\theta^+)]=0. \tag{10.19}$$

This is condition (10.16).

In order to choose $s(\theta^+)$ and $s(\theta^-)$ to maximize V we need to form a Lagrangean, $\mathcal{L}=V+\lambda[s(\theta^+)-s(\theta^-)]$. We obtain

$$\frac{\partial\mathcal{L}}{\partial s(\theta^-)}=-\pi(\theta^-)+p(\theta)-\lambda=0,$$

$$\frac{\partial\mathcal{L}}{\partial s(\theta^+)}=\pi(\theta^+)-p(\theta)+\lambda=0,$$

and

$$s(\theta^+)-s(\theta^-)\geq 0,\quad \lambda\geq 0,\quad \lambda[s(\theta^+)-s(\theta^-)]=0.$$

Therefore, if a jump occurs, $\lambda=0$ and $\pi(\theta^+)=\pi(\theta^-)=p(\theta)$, whereas if no jump occurs, $p(\theta)-\pi(\theta)=\lambda\geq 0$. Indeed, we must never have $p(t)<\pi(t)$, since this is inconsistent with a maximum of V. To sum up, we have $p(t)\geq \pi(t)$ everywhere, and at a jump, where $s(\theta^+)>s(\theta^-)$, we have $p(\theta)=\pi(\theta)$; this is condition (10.15).

We now provide a simple example of a problem in which an upward jump is optimal within the interior of the horizon.

Example 10.2.1: an interior jump upward. A firm has a stock of an exhaustible resource $s(0)=2$. The rate of extraction is $c(t)$: $\dot{s}(t)=-c(t)$. The extraction is costless, and the extracted resource is transformed into a final output $y=2c^{1/2}N^{1/2}$, where N is the quantity of a fixed factor, say land. Output price is 1 (per unit) and $N=1$. The firm cannot sell its resource stock, but it can acquire additional resource at the (buying) price $p(t)$ per unit of stock, where $p(t)=(t-1)^2+1$, $t\geq 0$. Therefore, the lowest buying price ever is $p(1)=1$. Suppose also that the planning horizon is $[0,2]$ and that $s(2)$ must equal 1. The firm must solve the following problem: find c that maximizes

$$\int_0^2 2c^{1/2}\,dt+\sum_j p(\theta_j)[s(\theta_j^-)-s(\theta_j^+)]$$

subject to

$$\dot{s}=-c,$$
$$s(0)=2,\quad s(2)=1,$$
$$s(\theta_j^+)\geq s(\theta_j^-).$$

$H=2c^{1/2}-\pi c$ and the maximum principle yields $c=\pi^{-2}$ and $\dot{\pi}=0$. Thus, c and π are constant. This is expected: with a zero discount rate and a concave objective, it is optimal to spread extraction evenly. In the absence of jumps $\int_0^2 c\,dt=2-1=1$ yields $c=\frac{1}{2}$ and $\pi=(2)^{1/2}$. We now show that an upward jump at $\theta=1$ is optimal, using conditions (10.15) and (10.16). Recall that for an upward jump we require $p(t)\geq\pi(t)$ at all times and equality at the jump time. We have shown that π is constant, hence a jump will occur, if at all, at the minimum value of p, namely, $p(1)=1$. This determines the (constant) values of π and c: $\pi(t)=1$ and $c(t)=1$ for all $t\in[0,2]$. We can now evaluate the size of the jump: $c(t)=1$ for $0\leq t<1$, with $s(0)=2$, implies $s(1^-)=1$ and $c(t)=1$ for $1<t\leq 2$, with $s(2)=1$, requires $s(1^+)=2$. Hence, the firm purchases one unit of the resource. To check on the optimality of this policy we calculate

$$\begin{aligned}&H(1^-)-H(1^+)+\dot{p}(1)[s(1^-)-s(1^+)]\\&\quad=(2\times 1-1\times 1)-(2\times 1-1\times 1)+0[1-2]=0,\end{aligned}$$

as required by (10.16). The maximum revenue is now

$$\int_0^2 2\,dt + 1(-1) = 3$$

instead of

$$\int_0^2 2\frac{1}{\sqrt{2}}\,dt = 2\sqrt{2},$$

previously.

Let us now extend these results to allow for downward jumps as well. First note that if the same price $p(t)$ is used for buying (upward jump) or selling (downward jump) capital, we will necessarily have $p(t) = \pi(t)$ everywhere. This unrestrained opening of the model destroys it, since the path of the costate is then fixed at the outset. Under these circumstances no solution would exist for most problems.

There is, however, a more sensible way of modeling two-way trade in capital: we must introduce two sets of prices, a buying price and a selling price of capital, say $p^b(t)$ and $p^s(t)$, respectively, with $p^b(t) > p^s(t)$ to avoid infinitely large profits. Some distortion of a free market such as transaction costs is presumably the cause of this discrepancy. We must replace the expression in (10.17) by

$$\begin{aligned} V &= \int_0^{\theta_1} u(c,s,t)\,dt - p^b(\theta_1)[s(\theta_1^+) - s(\theta_1^-)] \\ &\quad + \int_{\theta_1}^{\theta_2} u(c,s,t)\,dt + p^s(\theta_2)[s(\theta_2^-) - s(\theta_2^+)] + \int_{\theta_2}^{T} u(c,s,t)\,dt \\ &= \int_0^{\theta_1} (H + \dot{\pi}s)\,dt + \pi_0 s_0 - \pi(\theta_1^-)s(\theta_1^-) - p^b(\theta_1)[s(\theta_1^+) - s(\theta_1^-)] \\ &\quad + \int_{\theta_1}^{\theta_2} (H + \dot{\pi}s)\,dt + \pi(\theta_1^+)s(\theta_1^+) - \pi(\theta_2^-)s(\theta_2^-) \\ &\quad + \int_{\theta_2}^{T} (H + \dot{\pi}s)\,dt + \pi(\theta_2^+)s(\theta_2^+) - \pi_T s_T \\ &\quad + p^s(\theta_2)[s(\theta_2^-) - s(\theta_2^+)], \end{aligned} \tag{10.20}$$

with

$$\begin{aligned} &s(\theta_1^+) - s(\theta_1^-) \geq 0, \\ &s(\theta_2^-) - s(\theta_2^+) \geq 0. \end{aligned}$$

(The form of the expression in (10.20) presumes, without loss, that the upward jump – at θ_1 – must occur before the downward jump – at θ_2.)

We apply the same technique as before with $\mathcal{L} = V + \lambda_1[s(\theta_1^+) - s(\theta_1^-)] + \lambda_2[s(\theta_2^-) - s(\theta_2^+)]$. If an upward jump occurs at θ_1, we differentiate $\mathcal{L}$ with respect to $s(\theta_1^-)$ and $s(\theta_1^+)$:

$$-\pi(\theta_1^-)+p^b(\theta_1)-\lambda_1=0,$$
$$\pi(\theta_1^+)-p^b(\theta_1)+\lambda_1=0,$$
$$\lambda_1\geq 0,\quad s(\theta_1^+)-s(\theta_1^-)\geq 0,\quad \lambda_1[s(\theta_1^+)-s(\theta_1^-)]=0.$$

If a downward jump occurs at θ_2, we differentiate $\mathcal{L}$ with respect to $s(\theta_2^-)$ and $s(\theta_2^+)$:

$$-\pi(\theta_2^-)+p^s(\theta_2)+\lambda_2=0,$$
$$\pi(\theta_2^+)-p^s(\theta_2)-\lambda_2=0,$$
$$\lambda_2\geq 0,\quad s(\theta_2^-)-s(\theta_2^+)\geq 0,\quad \lambda_2[s(\theta_2^+)-s(\theta_2^-)]=0.$$

Therefore, in either case at a jump, $\pi(\theta_j^-)=\pi(\theta_j^+)$, $j=1,2$, and we have

$$p^s(t)\leq\pi(t)\leq p^b(t)\quad \forall t\in[0,T]. \tag{10.21}$$

Moreover, differentiating (10.20) with respect to θ_1 and θ_2 yields two expressions similar to (10.19):

$$\begin{aligned}&H(\theta_1^-)-H(\theta_1^+)-\dot{p}^b(\theta_1)[s(\theta_1^+)-s(\theta_1^-)]=0,\\&H(\theta_2^-)-H(\theta_2^+)+\dot{p}^s(\theta_2)[s(\theta_2^-)-s(\theta_1^+)]=0.\end{aligned} \tag{10.22}$$

We can now gather our results. At an upward jump,

$$\begin{aligned}&s(\theta_1^+)>s(\theta_1^-)\quad\text{and}\quad p^b(\theta_1)=\pi(\theta_1)>p^s(\theta_1),\\&H(\theta_1^-)-H(\theta_1^+)-\dot{p}^b(\theta_1)[s(\theta_1^+)-s(\theta_1^-)]=0.\end{aligned} \tag{10.23}$$

At a downward jump,

$$\begin{aligned}&s(\theta_2^+)<s(\theta_2^-)\quad\text{and}\quad p^s(\theta_2)=\pi(\theta_2)<p^b(\theta_2),\\&H(\theta_2^-)-H(\theta_2^+)+\dot{p}^s(\theta_2)[s(\theta_2^-)-s(\theta_2^+)]=0.\end{aligned} \tag{10.24}$$

These results indicate that the internal valuation of capital is bracketed by the buying and selling prices. Clearly, if the internal valuation is strictly within that interval, no trade takes place, and if trade does take place, the internal price equals the buying or selling price, depending on whether a purchase or a sale takes place, respectively. These are no more than generalized market-clearing conditions which state that the demand price never exceeds the supply price and that both prices are equal when a trade takes place, with the internal valuation of capital playing the role of a demand price vis-à-vis the buying price and of a supply price vis-à-vis the selling price. We now state these results formally.

Theorem 10.2.2: upward and downward jumps in the interior. Consider the problem of maximizing

$$\begin{aligned}V=&\int_0^{\theta_1}u(c,s,t)\,dt-p^b(\theta_1)[s(\theta_1^+)-s(\theta_1^-)]\\&+\int_{\theta_1}^{\theta_2}u(c,s,t)\,dt+p^s(\theta_2)[s(\theta_2^-)-s(\theta_2^+)]+\int_{\theta_2}^{T}u(c,s,t)\,dt\end{aligned}$$

subject to

$$s(\theta_1^+)-s(\theta_1^-)\geq 0,$$
$$s(\theta_2^-)-s(\theta_2^+)\geq 0,$$
$$\dot{s}=f(s,c,t) \text{ except at } \theta_j,\ j=1,2,$$
$$s(0)=s_0 \quad\text{and}\quad s(T)=s_T,$$

while allowing the possibility of an upward jump at time θ_1 or a downward jump at time θ_2, $\theta_j\in(0,T)$, $j=1,2$. Let $H(t)=u(s,c,t)+\pi f(s,c,t)$ be the Hamiltonian evaluated along an optimal path. Assume $p^b(t)$ and $p^s(t)$ to be differentiable. If an upward jump in s is optimal at time θ_1, then (10.21) and (10.23) hold, as do the usual necessary conditions. If a downward jump in s is optimal at time θ_2, then (10.21) and (10.24) hold, as do the usual necessary conditions.

The generalization to several state variables and many jumps is straightforward, at the cost of adding some notation.

Notation

The dates of jumps are θ_j, $j=1,\ldots,J$.
The set of variables that exhibit upward jumps at date θ_j is Θ_j^1; $\Theta_j^1=\{i\mid s_i(\theta_j^-)<s_i(\theta_j^+)\}$.
The set of variables that exhibit downward jumps at date θ_j is Θ_j^2; $\Theta_j^2=\{i\mid s_i(\theta_j^-)>s_i(\theta_j^+)\}$.

Note that while we can have both Θ_j^1 and Θ_j^2 nonempty, their intersection is always empty because some variables may have upward jumps and other variables downward jumps at the same date, but any one variable cannot jump both upward and downward at the same time.

Corollary 10.2.1. If many jumps are optimal and follow the pattern described in the notation above and if $p_i^s(t)$ and $p_i^b(t)$ are everywhere differentiable, then in addition to the maximum principle, the following conditions apply:

$$p_i^s(t)\leq\pi_i(t)\leq p_i^b(t),\quad i=1,\ldots,I,\ t\in[0,T], \tag{10.25}$$

$$s_i(\theta_j^+)>s_i(\theta_j^-),\ p_i^b(\theta_j)=\pi_i(\theta_j)>p_i^s(\theta_j),\quad i\in\Theta_j^1,\ j=1,\ldots,J, \tag{10.26a}$$

$$s_i(\theta_j^-)>s_i(\theta_j^+),\ p_i^b(\theta_j)>\pi_i(\theta_j)=p_i^s(\theta_j),\quad i\in\Theta_j^2,\ j=1,\ldots,J, \tag{10.26b}$$

$$H(\theta_j^-)-H(\theta_j^+)-\sum_{i\in\Theta_j^1}\dot{p}_i^b(\theta_j)(s_i(\theta_j^+)-s_i(\theta_j^-))$$
$$+\sum_{i\in\Theta_j^2}\dot{p}_i^s(\theta_j)(s_i(\theta_j^-)-s_i(\theta_j^+))=0,\quad j=1,\ldots,J. \tag{10.27}$$

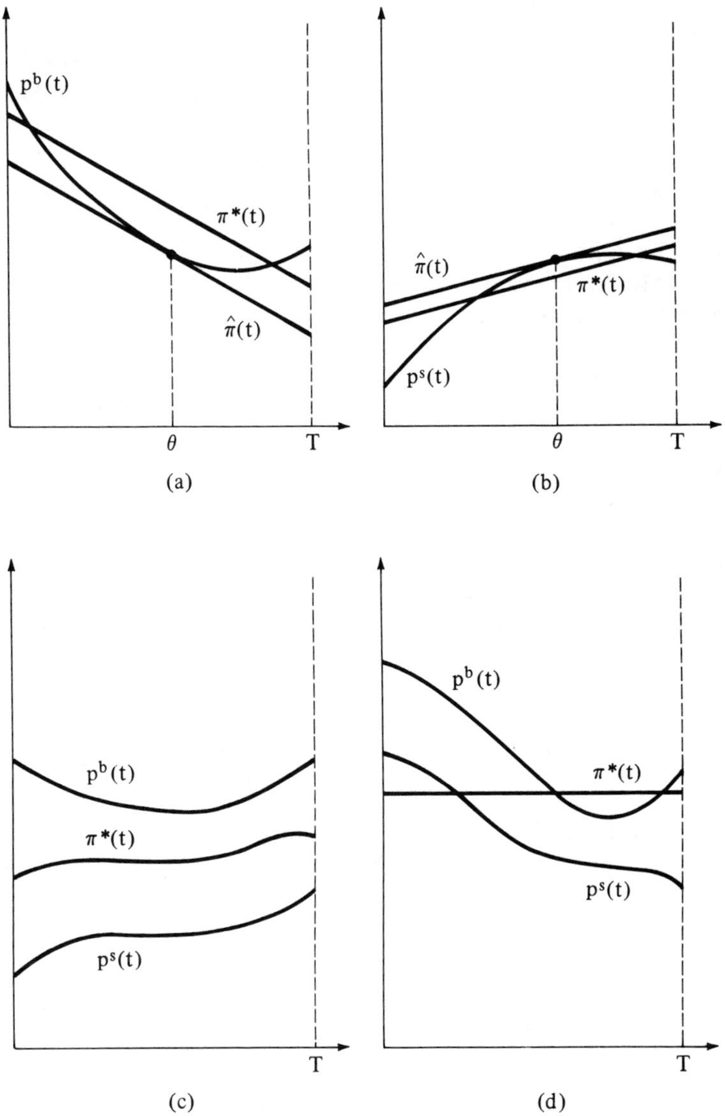

Figure 10.1

In Figure 10.1 we illustrate various cases. The costate variable before jumps were allowed is denoted by $\pi^*(t)$, while the variable when jumps are allowed is denoted by $\hat{\pi}(t)$; θ is the jump time and $p^b(t)$ and $p^s(t)$ are the buying price and the selling price of capital, respectively. Case (a) illustrates an upward jump; the previous path $\pi^*(t)$ is no longer optimal, for during some interval it exceeds the price at which stock can be procured from outside; $\hat{\pi}(t)$ is now the optimal path with $p^b(\theta)=\hat{\pi}(\theta)$. Case (b) illustrates a downward jump at time θ. Case (c) illustrates the possibility that although jumps are allowed, they are suboptimal: the original valuation of capital $\pi^*(t)$ is bracketed by the buying price and the selling price at all times; thus, no trade in capital takes place. Case (d) illustrates the possibility that the introduction of some capital price patterns may result in a problem without a solution. In the case illustrated, as in Example 10.2.1, the value of the costate must be constant, and this is incompatible with it remaining not above the buying price and not below the selling price.

This presentation makes it clear that while it is possible for a state variable to exhibit several jumps (some upward and some downward), this occurrence is highly unlikely for arbitrarily chosen buying and selling prices. We now illustrate downward jumps on a variant of Example 10.2.1.

Example 10.2.2: an interior jump downward. It is now assumed that no resource can be purchased but that instead it can be sold at price $p^s(t)=2-(t-1)^2$. This price path exhibits a maximum at $t=1$: $p(1)=2$. The previous optimality conditions still prevail; $c=\pi^{-2}$ is constant. If a sale takes place at $t=1$, we have $\pi(t)=p(1)=2$ for all $t\in[0,2]$ and $c=\frac{1}{4}$. Therefore, $\int_0^2 c\,dt=\frac{1}{4}(2)=\frac{1}{2}$ and 0.5 unit is sold at time 1. The costate variable remains above the price path $p^s(t)$ everywhere but at $t=1$. The reader is invited to verify that $H(1^-)-H(1^+)+\dot{p}^s(1)[s(1^-)-s(1^+)]=0$.

Remark: jumps at the boundary. In Section 10.2.1 we have hitherto restricted our attention to jumps occurring at some time $\theta\in(0,T)$. A slight modification is needed for jumps that occur at time 0 or time T. We must return to equation (10.19), which reflects the choice of the timing of the jump; that equation was valid if $\theta\in(0,T)$. If we wish to restrict θ properly to the closed interval $[0,T]$, we must add $\partial V/\partial\theta\le 0$ if $\theta=0$ and $\partial V/\partial\theta\ge 0$ if $\theta=T$. More precisely, letting

$$H(0^-)=u(c(0),s_0,0)+\pi(0)f(c(0),s_0,0)$$

and

$$H(T^+)=u(c(T),s_T,T)+\pi(T)f(c(T),s_T,T),$$

we require

$$\begin{aligned} &H(0^-)-H(0^+)+\dot{p}(0)[s_0-s(0^+)]\le 0 \quad \text{for a jump at } \theta=0,\\ &H(T^-)-H(T^+)+\dot{p}(T)[s(T^-)-s_T]\ge 0 \quad \text{for a jump at } \theta=T. \end{aligned} \tag{10.28}$$

The other conditions are unaltered. We now illustrate boundary jumps with an example.

Example 10.2.3: jumps at the boundary. In this example a naturally wasting resource can be harvested, but larger harvests hasten wastage. The problem is to choose c to maximize

$$V=\int_0^1 2\ln c\,dt,$$

subject to

$$\dot{s}=-2s^{-1/2}c,$$

$$s(0)=4,\quad s(1)=1.$$

The Hamiltonian $H=2\ln c-2\pi s^{-1/2}c$ is everywhere strictly concave in c, and maximizing it yields $c=\pi^{-1}s^{1/2}$. Substitution yields the maximized Hamiltonian $H(s,\pi)=\ln s-2(\ln\pi)-2$, which is strictly concave in s, given π. The state and costate obey the differential equations

$$\dot{\pi}=-s^{-1}\quad\text{and}\quad\dot{s}=-2\pi^{-1}.$$

To solve this system of nonlinear equations, differentiate the second one to get $\ddot{s}=2\pi^{-2}\dot{\pi}$ and by substitution $\ddot{s}=-0.5s^{-1}(\dot{s})^2$, or $\ddot{s}/\dot{s}=-0.5\dot{s}/s$, which can easily be integrated (twice). The general solution is

$$\begin{aligned}\pi(t)&=3\alpha^{-1}(\beta-\alpha t)^{1/3},\\ s(t)&=(\beta-\alpha t)^{2/3},\\ c(t)&=\alpha/3,\quad \alpha>0,\ \beta>0.\end{aligned}\tag{10.29}$$

In the first instance let no jumps be allowed; the boundary conditions $s(0)=4$ and $s(1)=1$ then yield $\alpha=7$, $\beta=8$, and the solution is $\pi^*(t)=\frac{3}{7}(8-7t)^{1/3}$, $s^*(t)=(8-7t)^{2/3}$, $c^*(t)=\frac{7}{3}$, with $\pi^*(t)$ ranging from $\frac{6}{7}$ to $\frac{3}{7}$ over the horizon and the value function $V=2\ln c\simeq 1.6945$.

In order to generate an upward jump at time $T=1$ say, all we need do is offer a buying price that is lower than the current internal valuation at time T. Since currently $\pi^*(1)=\frac{3}{7}$, let us use $p^b(t)=2.3-2t$ (we will also need to check that $p^b(t)$ remains above the new $\pi(t)$ path once the latter is determined). The general solution (10.29) is still valid for $t\in[0,1)$ and using the boundary condition $s(0)=4$ and $\pi(1)=p^b(1)=0.3$, we obtain $\beta=8$ and $\alpha\simeq 7.567$. The solution is

$$\hat{s}(t)=(8-7.567t)^{2/3},\quad \hat{\pi}(t)=0.3965(8-7.567t)^{1/3},\quad \hat{c}(t)=2.5223;$$

$\hat{s}(1^-)=0.57234$ and since $s(1)=1$, there is a purchase of 0.42766 unit of stock at $T=1$, at a price of 0.3 per unit. The value function is $V=2\ln c-0.3(0.42766)=1.722$, an improvement, and finally

$$\begin{aligned}H(1^-)-H(1^+)+\dot{p}(s(1^-)-s_1)&=\ln(0.57234)-\ln 1-2(-0.42766)\\ &=0.2973>0,\end{aligned}$$

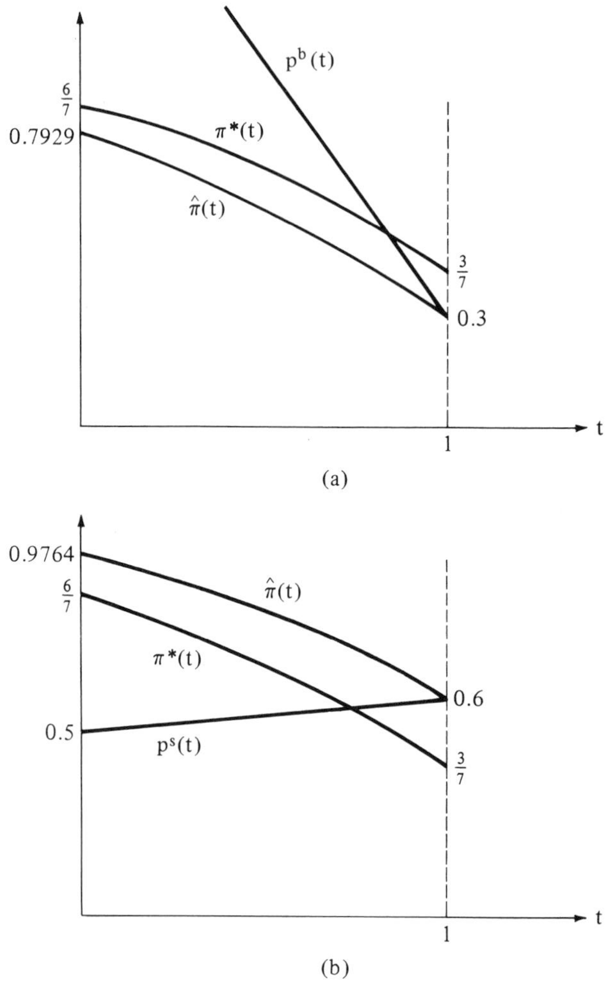

Figure 10.2

as required for jumps at the terminal time. The paths of $\pi^*(t)$, $\hat{\pi}(t)$, and $p^b(t)$ are plotted in Figure 10.2a; it can be verified that $\hat{\pi}(t) \le p^b(t)$ everywhere.

We now present a downward jump at $T = 1$. Let the selling price be $p^s(t) = 0.5 + 0.1t$; $p^s(1) = 0.6 > \pi^*(1) = \frac{3}{7}$, indicating that a downward jump at $T = 1$ is warranted. We calculate the solution on $t \in [0, 1)$ by using the general solution (10.29) and the boundary conditions $s(0) = 4$, $\pi(1) = 0.6$. We obtain $\beta = 8$ and $\alpha \simeq 6.145$. The solution is

$$\hat{s}(t)=(8-6.145t)^{2/3}, \quad \hat{\pi}(t)=0.4882(8-6.145t)^{1/3}, \quad \hat{c}(t)=2.0483;$$

$\hat{s}(1^-)=1.50972$ and since $s(1)=1$, there is a sale of 0.50972 unit of stock at $T=1$, at a price of 0.6 per unit. The value function is $V=2\ln c+0.6(0.50972)=1.7399$, an improvement over no jumps, and finally

$$\begin{aligned} H(1^-)-H(1^+)+\dot{p}(s(1^-)-s_1) &= \ln(1.50972)-\ln 1+0.1(0.50972) \\ &= 0.4629>0 \end{aligned}$$

as expected. The paths of $\pi^*(t)$, $\hat{\pi}(t)$, and $p^s(t)$ are plotted in Figure 10.2b; note that $\hat{\pi}(t)\geq p^s(t)$ everywhere.

10.2.2 *The case of the strictly concave Hamiltonian*

We again turn our attention to the possible occurrence of jumps within the interior of the planning horizon. We begin our discussion with a theorem that modifies a proposition of Arrow and Kurz (1970, p. 57).

Theorem 10.2.3. Let $H^0(\mathbf{s}, \boldsymbol{\pi}, t)=\max_{\mathbf{c}} H(\mathbf{c}, \mathbf{s}, \boldsymbol{\pi}, t)$. If $H^0(\mathbf{s}, \boldsymbol{\pi}, t)$ is strictly concave in $\mathbf{s}$ and if the exogenous price paths of capital, $\mathbf{p}(t)$, are continuously differentiable, jumps in the state variables may be optimal only if they take place at the initial time or at the terminal time.

Proof. Consider a hypothetical jump point θ where variables s_i, $i\in I$, exhibit a jump. For those doing an upward jump ($s_i(\theta^+)>s_i(\theta^-)$) we must have $p_i^b(t)\geq\pi_i(t)$, all t and $p_i^b(\theta)=\pi_i(\theta)$. This implies that p_i^b must decrease toward π_i before time θ and must increase away from π_i after time θ. In terms of slopes these restrictions can be expressed as

$$\dot{p}_i^b(\theta^-)\leq\dot{\pi}_i(\theta^-) \quad \text{and} \quad \dot{p}_i^b(\theta^+)\geq\dot{\pi}_i(\theta^+);$$

hence,

$$\dot{\pi}_i(\theta^+)-\dot{\pi}_i(\theta^-)\leq\dot{p}_i^b(\theta^+)-\dot{p}_i^b(\theta^-)=0 \quad \text{by differentiability of } p_i^b.$$

Therefore,

$$[\dot{\pi}_i(\theta^+)-\dot{\pi}_i(\theta^-)][s_i(\theta^+)-s_i(\theta^-)]\leq 0 \tag{10.30}$$

if s_i exhibits an upward jump at instant θ.

For a downward jump ($s_i(\theta^+)<s_i(\theta^-)$), we must have $p_i^s(t)\leq\pi_i(t)$, all t and $p_i^s(\theta)=\pi_i(\theta)$; therefore, p_i^s must increase toward π_i before time θ and decrease away from it after time θ. This leads to

$$\dot{\pi}_i(\theta^+)-\dot{\pi}_i(\theta^-)\geq\dot{p}_i^s(\theta^+)-\dot{p}_i^s(\theta^-)=0 \quad \text{by differentiability of } p_i^s.$$

Hence, (10.30) is also valid at a downward jump, and we can sign the inner product

$$[\dot{\pi}(\theta^+)-\dot{\pi}(\theta^-)]\cdot[\mathbf{s}(\theta^+)-\mathbf{s}(\theta^-)]\le 0, \tag{10.31}$$

where the elements of $\mathbf{s}(\theta^+)$ and $\mathbf{s}(\theta^-)$ are such that

$$s_i(\theta^+)\neq s_i(\theta^-),\ i\in I,\quad \text{and}\quad s_j(\theta^+)=s_j(\theta^-),\ j\notin I.$$

We know from the maximum principle that $\dot{\pi}_i=-H^0_{s_i}$; hence,

$$\dot{\pi}_i(\theta^+)-\dot{\pi}_i(\theta^-)=-H^0_{s_i}(\mathbf{s}(\theta^+),\pi,\theta)+H^0_{s_i}(\mathbf{s}(\theta^-),\pi,\theta).$$

Substituting this into (10.31), we obtain

$$-[H^0_{\mathbf{s}}(\mathbf{s}(\theta^+),\pi,\theta)-H^0_{\mathbf{s}}(\mathbf{s}(\theta^-),\pi,\theta)]\cdot[\mathbf{s}(\theta^+)-\mathbf{s}(\theta^-)]\le 0.$$

This inequality contradicts the assumed strict concavity of H in s, which requires

$$[H^0_{\mathbf{s}}(\mathbf{s}(\theta^+),\pi,\theta)-H^0_{\mathbf{s}}(\mathbf{s}(\theta^-),\pi,\theta)]\cdot[\mathbf{s}(\theta^+)-\mathbf{s}(\theta^-)]<0$$

for any values of $\mathbf{s}(\theta^+)$ and $\mathbf{s}(\theta^-)$, thus proving the theorem. □

We can present a geometric interpretation of the preceding proof in the case where a single state variable, say s_1, jumps at time θ. Consider first an upward jump, that is, $s_1(\theta^+)>s_1(\theta^-)$, for which $p_1^b(t)$ must just "touch" $\pi(t)$ from above at time θ. Yet we know that $\dot{\pi}_1(\theta^-)=-H^0_{s_1}(s_1(\theta^-),\ldots,\theta)$ and $\dot{\pi}_1(\theta^+)=-H^0_{s_1}(s_1(\theta^+),\ldots,\theta)$. The implication of the strict concavity of H^0 in s_1 is that $\dot{\pi}_1$ increases at an upward jump – because the first derivative of H^0 with respect to s_1 decreases. The geometric interpretation of this configuration is that there is a kink in π_1 at time θ and that, at the kink, the two branches of π_1 form an angle of less than 180° (from above). Therefore, it is impossible for a smooth p_1^b curve to just touch π_1 at θ from above since this would require the graph of π_1 to be below the tangent to p_1^b at θ.

A similar argument can be made for downward jumps where p_1^s must touch π_1 from below and where π_1 exhibits a kink forming an angle of less than 180° (from below).

Remark. The conditions of applicability of Theorem 10.2.3 could be weakened to the requirement that H^0 be strictly concave in each s_i, individually, if we are considering jumps in one state variable at a time, a most likely occurrence.

Under the differentiability assumption for outside prices, Theorem 10.2.3 severely restricts the applicability of our previous results on jumps. It is relatively seldom that one encounters problems where $H^0(\mathbf{s},\pi,t)$ is not strictly concave in $\mathbf{s}$, as we recall that concavity is the most common restriction ensuring sufficiency of the maximum principle. Note, however, that a problem with a nonstrictly concave Hamiltonian (it was linear) was encountered in Examples 10.2.1 and 10.2.2, and jumps occurred in the interior of the horizon with differentiable price paths.

We wish to offer two other approaches to the modeling of the important phenomenon of trade in capital goods. The first approach introduces nondifferentiable price paths, while the second pursues the implications of smoothness and does away with jumps.

Nondifferentiable price paths. Our proof of Theorem 10.2.3 makes it clear that we can devise nondifferentiable price paths that induce jumps in the interior of the horizon even with a strictly concave Hamiltonian. There is no particular reason to insist on differentiability since it is mainly an assumption of convenience without any real economic content here. For concreteness and to investigate the anatomy of jumps in such problems we will present an example. Before we do this, however, we must state precisely the optimality conditions corresponding to jumps with nondifferentiable price paths. This will entail a modification of Theorem 10.2.2.

Theorem 10.2.4. Consider the problem described in Theorem 10.2.2 but allow $p^b(t)$ and $p^s(t)$ to be piecewise-differentiable. Then the usual necessary conditions hold and

$$p^s(t) \le \pi(t) \le p^b(t), \quad \forall t \in [0, T]. \tag{10.32}$$

Furthermore, if an upward jump in s is optimal at time θ_1,

$$s(\theta_1^+) > s(\theta_1^-), \quad p^b(\theta_1) = \pi(\theta_1) \ge p^s(\theta_1); \tag{10.33}$$

$$H(\theta_1^-) - H(\theta_1^+) - \dot{p}^b(\theta_1^-)[s(\theta_1^+) - s(\theta_1^-)] \ge 0 \tag{10.34a}$$

and

$$H(\theta_1^-) - H(\theta_1^+) - \dot{p}^b(\theta_1^+)[s(\theta_1^+) - s(\theta_1^-)] \le 0. \tag{10.34b}$$

If a downward jump is optimal at time θ_2,

$$s(\theta_2^+) < s(\theta_2^-), \quad p^s(\theta_2) = \pi(\theta_2) \le p^b(\theta_2); \tag{10.35}$$

$$H(\theta_2^-) - H(\theta_2^+) + \dot{p}^s(\theta_2^-)[s(\theta_2^-) - s(\theta_2^+)] \ge 0 \tag{10.36a}$$

and

$$H(\theta_2^-) - H(\theta_2^+) + \dot{p}^s(\theta_2^+)[s(\theta_2^-) - s(\theta_2^+)] \le 0. \tag{10.36b}$$

Note that $\dot{p}(\theta^-)$ and $\dot{p}(\theta^+)$ represent left- and right-hand-side derivatives of the appropriate p function in the event that it is not differentiable at θ. If p is differentiable at θ, these derivatives are identical and the (a) and (b) parts of (10.34) and (10.36) reduce to an equality as in Theorem 10.2.2.

Proof. We can transform the objective function into expression (10.20). Conditions (10.32), (10.33), and (10.35) are obtained as before from the optimal choice of $s(\theta_i^+)$ and $s(\theta_i^-)$. In order to obtain (10.34) and (10.36), which are new, we must take the derivative of (10.20) with respect to θ_i and evaluate it at θ_i^- and at θ_i^+, noting that only for the nondifferentiable prices will this distinction be relevant. Since we seek a maximum, we must let the derivative evaluated at θ_i^- be nonnegative while the right-hand-side derivative is nonpositive. We obtain

$$H(\theta_i^-)-H(\theta_i^+)+\dot{p}^j(\theta_i^-)[s(\theta_i^-)-s(\theta_i^+)]\geq 0, \quad i=1,2, \ j=s,b,$$

and

$$H(\theta_i^-)-H(\theta_i^+)-\dot{p}^j(\theta_i^+)[s(\theta_i^+)-s(\theta_i^-)]\leq 0, \quad i=1,2, \ j=s,b,$$

where $H(\theta_i^-)=H(s(\theta_i^-),c(\theta_i^-),\pi(\theta_i),\theta_i)$, and so on. These conditions can be specialized to (10.34) and (10.36). □

Theorem 10.2.4 can be extended to several variables and many jump dates, as was done in Corollary 10.2.1.

Corollary 10.2.2. Suppose that it is optimal for variables to exhibit jumps according to the notation given on page 316 but $p_i^s(t)$ and $p_i^b(t)$ are only piecewise-differentiable. Then, in addition to the maximum principle, the following conditions are also necessary: (10.25), (10.26), and

$$H(\theta_j^-)-H(\theta_j^+)-\sum_{i\in\Theta_j^1}\dot{p}_i^b(\theta_j^-)[s_i(\theta_j^+)-s_i(\theta_j^-)] + \sum_{i\in\Theta_j^2}\dot{p}_i^s(\theta_j^-)[s_i(\theta_j^-)-s_i(\theta_j^+)]\geq 0, \quad j=1,\ldots,J, \tag{10.37a}$$

$$H(\theta_j^-)-H(\theta_j^+)-\sum_{i\in\Theta_j^1}\dot{p}_i^b(\theta_j^+)[s_i(\theta_j^+)-s_i(\theta_j^-)] + \sum_{i\in\Theta_j^2}\dot{p}_i^s(\theta_j^+)[s_i(\theta_j^-)-s_i(\theta_j^+)]\leq 0, \quad j=1,\ldots,J. \tag{10.37b}$$

Theorem 10.2.4 and Corollary 10.2.2 can be specialized for problems with a concave Hamiltonian – the bulk of problems encountered in the economics literature. It turns out that conditions (10.37) (or (10.34) and (10.36)) are superseded by other conditions, as we now show.

Corollary 10.2.3. Consider the problem described in Theorem 10.2.4 and assume that $H^0(\mathbf{s},\boldsymbol{\pi},t)=\max H(\mathbf{s},\mathbf{c},\boldsymbol{\pi},t)$ is a concave function of s. Then if an upward jump is optimal at date θ_1, conditions (10.34) are replaced by

$$\dot{p}^b(\theta_1^-)\leq\dot{\pi}(\theta_1^-) \quad \text{and} \quad \dot{p}^b(\theta_1^+)\geq\dot{\pi}(\theta_1^+). \tag{10.38}$$

If a downward jump takes place at time θ_2, conditions (10.36) are replaced by

$$\dot{p}^s(\theta_2^-)\geq\dot{\pi}(\theta_2^-) \quad \text{and} \quad \dot{p}^s(\theta_2^+)\leq\dot{\pi}(\theta_2^+). \tag{10.39}$$

More generally if there are several state variables and several jumps as in the problem of Corollary 10.2.2 and $H^0(\mathbf{s}, \boldsymbol{\pi}, t)$ is jointly concave in all s_i, $i=1,\dots,I$, the following condition replaces (10.37):

$$\dot{p}_i^b(\theta_j^-) \le \dot{\pi}_i(\theta_j^-) \text{ and } \dot{p}_i^b(\theta_j^+) \ge \dot{\pi}_i(\theta_j^+), \quad i \in \Theta_j^1,\ j=1,\dots,J, \tag{10.40a}$$

$$\dot{p}_i^s(\theta_j^-) \ge \dot{\pi}_i(\theta_j^-) \text{ and } \dot{p}_i^s(\theta_j^+) \le \dot{\pi}_i(\theta_j^+), \quad i \in \Theta_j^2,\ j=1,\dots,J. \tag{10.40b}$$

Proof. It is clear that the necessary conditions (10.25) and (10.26) themselves imply (10.40), as argued in the proof of Theorem 10.2.3; hence, (10.40) is a necessary condition. Furthermore, we now show that, under concavity of H^0, (10.40) implies (10.37), thereby superseding it.

When $H^0(\mathbf{s})$ is a concave function, we know that

$$(\mathbf{s}_2-\mathbf{s}_1)' \bullet H_{\mathbf{s}}^0(\mathbf{s}_2) \le H^0(\mathbf{s}_2) - H^0(\mathbf{s}_1) \le (\mathbf{s}_2-\mathbf{s}_1)' \bullet H_{\mathbf{s}}^0(\mathbf{s}_1), \quad \forall \mathbf{s}_1 \ne \mathbf{s}_2.$$

Hence, letting $\mathbf{s}_2$ be $\mathbf{s}(\theta_j^+)$ and $\mathbf{s}_1$ be $\mathbf{s}(\theta_j^-)$, where θ_j is any interior jump date, and noting that $H_{\mathbf{s}} = -\dot{\boldsymbol{\pi}}$, we have

$$\begin{aligned}[\mathbf{s}(\theta_j^+)-\mathbf{s}(\theta_j^-)]' \bullet \dot{\boldsymbol{\pi}}(\theta_j^+) &\ge H^0(\theta_j^-) - H^0(\theta_j^+) \\ &\ge [\mathbf{s}(\theta_j^+)-\mathbf{s}(\theta_j^-)]' \bullet \dot{\boldsymbol{\pi}}(\theta_j^-).\end{aligned} \tag{10.41}$$

Condition (10.40) implies that

$$[s_i(\theta_j^+)-s_i(\theta_j^-)]\dot{p}_i(\theta_j^-) \le [s_i(\theta_j^+)-s_i(\theta_j^-)]\dot{\pi}_i(\theta_j^-)$$

and

$$[s_i(\theta_j^+)-s_i(\theta_j^-)]\dot{p}_i(\theta_j^+) \ge [s_i(\theta_j^+)-s_i(\theta_j^-)]\dot{\pi}_i(\theta_j^+),$$

where $p_i = p_i^b$ when $i \in \Theta_j^1$ and $\dot{p}_i = \dot{p}_i^s$ when $i \in \Theta_j^2$. Therefore, with the same summary notation,

$$\begin{aligned}[\mathbf{s}(\theta_j^+)-\mathbf{s}(\theta_j^-)]' \bullet \dot{\mathbf{p}}(\theta_j^-) &\le [\mathbf{s}(\theta_j^+)-\mathbf{s}(\theta_j^-)]' \bullet \dot{\boldsymbol{\pi}}(\theta_j^-), \\ [\mathbf{s}(\theta_j^+)-\mathbf{s}(\theta_j^-)]' \bullet \dot{\mathbf{p}}(\theta_j^+) &\ge [\mathbf{s}(\theta_j^+)-\mathbf{s}(\theta_j^-)]' \bullet \dot{\boldsymbol{\pi}}(\theta_j^+).\end{aligned} \tag{10.42}$$

Together (10.41) and (10.42) imply (10.37). We can in turn specialize this to (10.38) and (10.39) by selecting $I=1$, $J=2$, $\Theta_1^1=1$, $\Theta_2^1=\emptyset$, $\Theta_1^2=\emptyset$, $\Theta_2^2=1$; this means that the single state variable has an upward jump at θ_1 and a downward jump at θ_2.

Example 10.2.4: interior jump with a strictly concave Hamiltonian. We use again the model of Example 10.2.3. The general solution (10.29) still applies and, without jumps, the costate ranges from $\frac{6}{7}$ to $\frac{3}{7}$. Let us construct an example that will make a downward jump optimal at $t=0.5$, say. Since $\pi^*(0.5) \simeq 0.70756$, let us choose a larger price, say $p^s(0.5) = 0.8$. In order to solve for the new solution consistent with a jump at $t = 0.5$, we must distinguish two branches: the first is valid on $[0, 0.5)$, while the second holds on $(0.5, 1]$. $\hat{s}(t)$ will be discontinuous at $t=0.5$, but $\hat{\pi}(t)$ will not be. It is essential to understand that the lack of discontinuity of

$\hat{\pi}(t)$ at $t=0.5$ is consistent with the existence of two distinct branches of $\hat{\pi}(t)$, which will meet with a kink at $t=0.5$. Both branches must obey $\pi(0.5)=0.8$; this imposes the following restriction on α and β:

$$3\alpha^{-1}(\beta-0.5\alpha)^{1/3}=0.8.$$

In addition, the first branch must satisfy $s(0)=4$, or

$$\beta^{2/3}=4, \quad \beta=8,$$

while the second branch must satisfy $s(1)=1$, or

$$(\beta-\alpha)^{2/3}=1, \quad \beta=\alpha+1.$$

Substitution yields, for the first branch,

$$3(8-0.5\alpha)^{1/3}=0.8\alpha, \quad \text{or} \quad \alpha\simeq 6.339, \ \beta=8,$$

and for the second branch,

$$3(1+0.5\alpha)^{1/3}=0.8\alpha, \quad \text{or} \quad \alpha\simeq 5.937, \ \beta\simeq 6.937.$$

Using the pairs of values for α and β, we can calculate the values of $\hat{s}$ near the jump point; we have $\hat{s}(0.5^-)=2.8575$ and $\hat{s}(0.5^+)=2.5066$ – thus a downward jump of 0.3509. We can also verify that the costate values, $\hat{\pi}(0.5^-)=0.800015$ and $\hat{\pi}(0.5^+)=0.800012$, are approximately equal at the jump time, using these α and β values. The control in the first half is $c\simeq 2.113$ and that in the second half is $c\simeq 1.979$. The maximum value is $V=2(\ln 2.113+\ln 1.979)/2+0.8(0.3509)\simeq 1.7114$, an improvement over the lack of jumps. In order to choose an appropriate price path we need to know the slope of π before and after the jump. We have $\dot{\pi}(0.5^-)=-0.3499$ and $\dot{\pi}(0.5^+)=-0.3989$. We must choose a price path with a slope less steep than that of π before the jump and steeper than that of π after the jump. Furthermore, we must have $p^s(t)<\pi(t)$, $0\le t<0.5$, and $p^s(t)<\pi(t)$, $0.5<t\le 1$. For instance,

$$p^s(t)=\begin{cases} 0.90-0.2t, & 0\le t<0.5,\\ 1.1-0.6t, & 0.5<t\le 1.\end{cases}$$

Note that $p^s(t)$ is continuous and that $p^s(0.5)=0.8$. The relative shapes of $p^s(t)$ and $\hat{\pi}(t)$ before and after the jump are plotted approximately in Figure 10.3a, and the jump area is scaled up in Figure 10.3b.

To illustrate Corollary 10.2.3 we now verify that (10.36) applies. We have

$$H^0(\theta^-)-H^0(\theta^+)=\ln s(\theta^-)-\ln s(\theta^+)$$

here; hence,

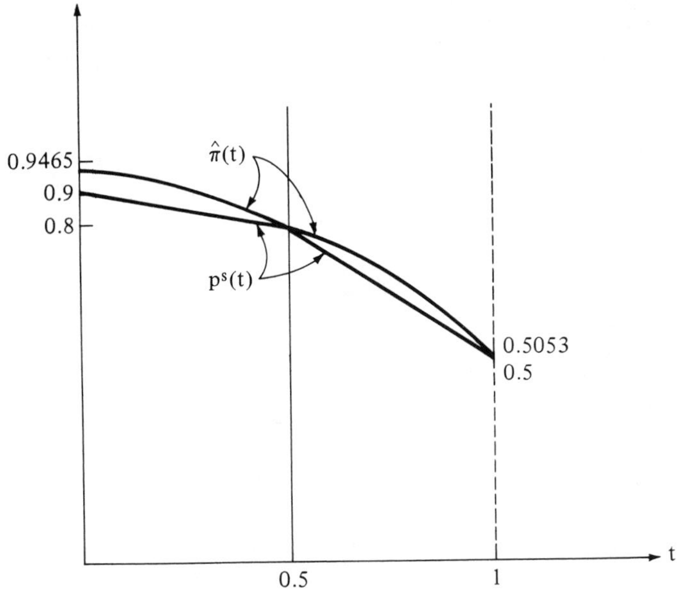

(a)

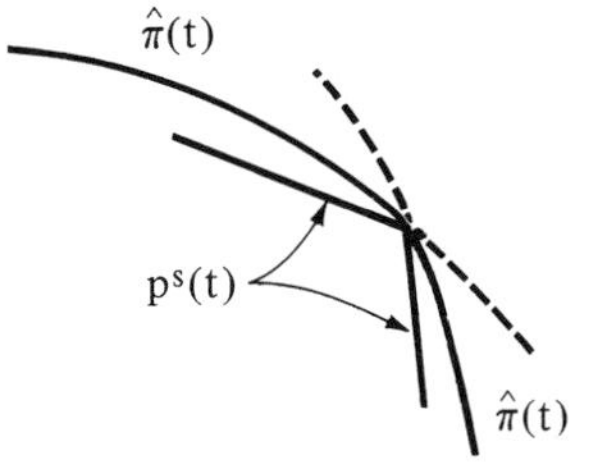

(b)

Figure 10.3

$$H^0(0.5^-) - H^0(0.5^+) + \dot{p}(0.5^-)[s(0.5^-) - s(0.5^+)]$$
$$= \ln(2.8575/2.5066) + (-0.2)(2.8575 - 2.5066)$$
$$= 0.13102 - 0.07018 = 0.06084 > 0;$$

$$H^0(0.5^-) - H^0(0.5^+) + \dot{p}(0.5^+)[s(0.5^-) - s(0.5^+)]$$
$$= 0.13102 - 0.6(2.8575 - 2.5066)$$
$$= -0.07952 < 0.$$

Smooth trading in capital goods. We wish to offer an alternative to jumps in the state variables, which still allows trading on the capital market. We argue that this possibility, not the discontinuities themselves, is the important feature. After all, a continuous-time model is at best an idealization of reality, and similarly an instantaneous transaction can be represented by a very fast, but smooth change in the value of the state variable.

Consider the (strictly concave) control problem of finding c_t that maximizes

$$\int_0^T u(c_t, s_t)e^{-\delta t}\,dt + e^{-\delta T}\phi(s_T)$$

subject to

$$\dot{s}_t = f(c_t, s_t), \quad s_0 \text{ fixed}, \ s_T \text{ free}.$$

Suppose we wish to consider altering the capital stock s by outside purchases or sales. We propose to introduce a new state variable X_t that indicates the net stock of capital purchased from outside, to date. This trade is not instantaneous in the sense that there is an upper bound on the number of units of stock that can be traded in an instant of time. We maximize

$$\int_0^T [u(c_t, s_t + X_t) - \mathfrak{p}_t^b x_t + \mathfrak{p}_t^s z_t]e^{-\delta t}\,dt + e^{-\delta T}\phi(s_T + X_T) \tag{10.43}$$

subject to

$$\dot{s}_t = f(c_t, (s_t + X_t)), \quad s_0 \text{ fixed}, \ s_T \text{ free}, \tag{10.44}$$

$$\dot{X}_t = x_t - z_t, \quad X_0 = 0, \ X_T \text{ free}, \tag{10.45}$$

$$\bar{x} \geq x_t \geq 0, \quad \bar{z} \geq z_t \geq 0.$$

The current-value Hamiltonian is

$$H = u(c, s+X) - \mathfrak{p}^b x + \mathfrak{p}^s z + \pi f(c, s+X) + \varphi(x - z). \tag{10.46}$$

The optimal solution is characterized by (10.44), (10.45), and

$$(c, x, z) \text{ maximize } H \text{ in (10.46) subject to } 0 \leq x \leq \bar{x},\ 0 \leq z \leq \bar{z}; \tag{10.47}$$

$$\dot{\pi} = -u_2' + (\delta - f_2')\pi, \quad \pi_T = \phi_T' \tag{10.48}$$

and

$$\dot{\varphi} = -u_2' + \delta\varphi - \pi f_2', \quad \varphi_T = \phi_T', \tag{10.49}$$

where

$$u_2' = \frac{\partial u}{\partial(s+X)}, \quad f_2' = \frac{\partial f}{\partial(s+X)}, \quad \text{and} \quad \phi_T' = \frac{\partial \phi(s_T + X_T)}{\partial(s_T + X_T)}.$$

Quite obviously $\pi = \varphi$ along the optimal path, so that the Hamiltonian can be expressed as

$$H = u(c, s+X) - \mathfrak{p}^b x + \mathfrak{p}^s z + \pi[f(c, s+X) + x - z].$$

We can reduce this problem to one with a single state variable, say $y \equiv s+X$. We maximize

$$\int_0^T [u(c_t, y_t) - \mathfrak{p}_t^b x_t + \mathfrak{p}_t^s z_t] e^{-\delta t}\, dt + e^{-\delta T}\phi(y_T)$$

subject to

$$\dot{y}_t = f(c_t, y_t) + x_t - z_t, \quad y_0 = s_0, \quad y_T \text{ free}, \qquad (10.50)$$
$$0 \le x_t \le \bar{x}, \quad 0 \le z_t \le \bar{z}.$$

The current-value Hamiltonian is

$$H = u(c, y) - \mathfrak{p}^b x + \mathfrak{p}^s z + \psi[f(c, y) + x - z], \qquad (10.51)$$

and the optimality conditions are (10.50) and

$$(c, x, z) \text{ maximize } H \text{ (in 10.51) subject to } 0 \le x \le \bar{x},\ 0 \le z \le \bar{z}, \qquad (10.52)$$
$$\dot{\psi} = -u_2' + (\delta - f_2')\psi, \quad \psi_T = \phi_T'. \qquad (10.53)$$

These conditions are identical to those of the preceding problem with the Hamiltonian (10.46), $y = s + X$ and $\pi = \varphi = \psi$. The values of the bounds $\bar{x}$ and $\bar{z}$ are arbitrary; thus, the transactions can be made as abrupt as desired, but not instantaneous. We must assume $\mathfrak{p}^b > \mathfrak{p}^s$; otherwise, the firm would simply buy and sell at the maximum rate throughout the horizon. Expanding (10.52) yields

$$u_1' + \psi f_1' = 0; \qquad (10.54)$$

$$-\mathfrak{p}^b + \psi < 0 \Rightarrow x = 0,$$
$$-\mathfrak{p}^b + \psi > 0 \Rightarrow x = \bar{x}, \qquad (10.55)$$
$$-\mathfrak{p}^b + \psi = 0 \Leftarrow 0 < x < \bar{x};$$

$$\mathfrak{p}^s - \psi < 0 \Rightarrow z = 0,$$
$$\mathfrak{p}^s - \psi > 0 \Rightarrow z = \bar{z}, \qquad (10.56)$$
$$\mathfrak{p}^s - \psi = 0 \Leftarrow 0 < z < \bar{z}.$$

It is impossible for x and z to be positive at the same time, since this implies $\psi \ge \mathfrak{p}^b$ from (10.55) and $\psi \le \mathfrak{p}^s$ from (10.56), which together contradict $\mathfrak{p}^b > \mathfrak{p}^s$.

We now specialize the model for illustrative purposes by taking $u(c, y) = \ln c$ and $f(c, y) = f(y) - c$; $\mathfrak{p}^s$ and $\mathfrak{p}^b$ are constant. Equations (10.53)–(10.56) become

$$\dot{\psi} = (\delta - f'(y))\psi, \qquad \psi_T = \phi_T'; \qquad (10.57)$$

$$c^{-1} = \psi, \qquad c_T = (\phi_T')^{-1}; \qquad (10.58)$$

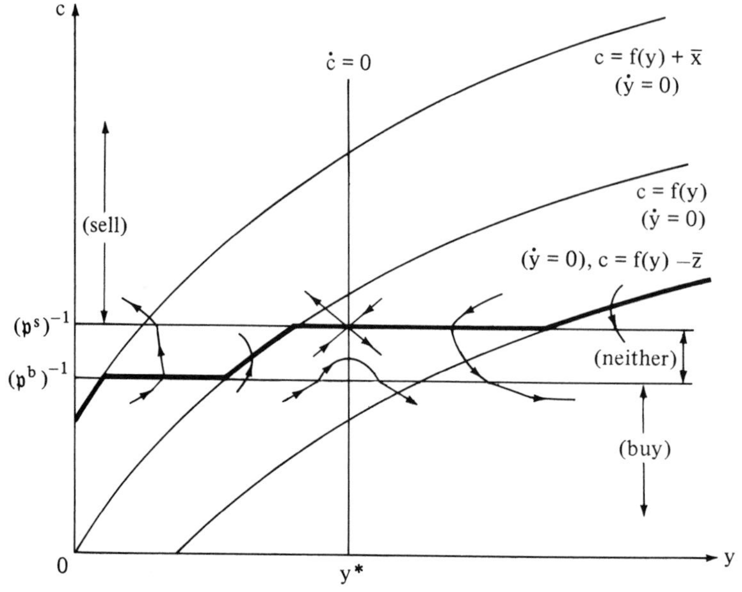

Figure 10.4

$$
\begin{aligned}
&-\mathfrak{p}^b+\psi<0 \text{ or } c>(\mathfrak{p}^b)^{-1} \Rightarrow x=0,\\
&-\mathfrak{p}^b+\psi>0 \text{ or } c<(\mathfrak{p}^b)^{-1} \Rightarrow x=\bar{x},\\
&-\mathfrak{p}^b+\psi=0 \text{ or } c=(\mathfrak{p}^b)^{-1} \Leftarrow 0<x<\bar{x};
\end{aligned}
\tag{10.59}
$$

$$
\begin{aligned}
&\mathfrak{p}^s-\psi<0 \text{ or } c<(\mathfrak{p}^s)^{-1} \Rightarrow z=0,\\
&\mathfrak{p}^s-\psi>0 \text{ or } c>(\mathfrak{p}^s)^{-1} \Rightarrow z=\bar{z},\\
&\mathfrak{p}^s-\psi=0 \text{ or } c=(\mathfrak{p}^s)^{-1} \Leftarrow 0<z<\bar{z}.
\end{aligned}
\tag{10.60}
$$

From these a phase diagram can easily be constructed; this is done in Figure 10.4. When c is above $(\mathfrak{p}^s)^{-1}$, capital is sold at the maximal rate $\bar{z}$; when c lies between $(\mathfrak{p}^s)^{-1}$ and $(\mathfrak{p}^b)^{-1}$, no trade in capital takes place; when c is below $(\mathfrak{p}^b)^{-1}$, capital is purchased at the maximal rate $\bar{x}$. We can have $0<x<\bar{x}$ only along the $c=(\mathfrak{p}^b)^{-1}$ line and $0<z<\bar{z}$ along the $c=(\mathfrak{p}^s)^{-1}$ line. There are potentially three $\dot{y}=0$ loci, but only the thick portions are relevant because of the above restrictions on capital trading. Similarly, only along the thick portions of the $c=(\mathfrak{p}^s)^{-1}$ and $c=(\mathfrak{p}^b)^{-1}$ lines does $\dot{y}$ change sign (it does not become zero, however); along other portions of these lines, trajectories have a kink but no change in the direction of y. The values of $\mathfrak{p}^s$ and $\mathfrak{p}^b$ affect the topography of the diagram; only one instance is represented in Figure 10.4. There is never a sudden shift from buying to selling (or vice versa), but an intermediate no-trade phase occurs between the two, although not all three phases need be present. The buying or selling of capital need not be bunched at the beginning of the horizon. If $\bar{z}$ and $\bar{x}$ are taken to be large (so as to better approximate a jump) the curves $c=f(y)+\bar{x}$ and $c=f(y)-\bar{z}$ become irrelevant.

One final remark concerns the scrap value function. We had set a current value $\phi(y_T)$ arbitrarily for a terminal stock y_T, but perhaps some argument can be provided to tie the scrap value to that of other exogenous prices. Let us call $\phi(y)$ the current scrap value of y units of capital; $\mathfrak{p}^s$ and $\mathfrak{p}^b$ are the prices of one unit of capital. Presumably whoever buys y can generate an economic rent of $\mathfrak{p}^s f(y)$ forever without altering the capital stock. (We have implicitly assumed that depreciation, if any, is accounted for by f; see exercise 11.) Thus, it seems reasonable to define the capitalized value of y as

$$\begin{aligned}\phi(y) &= \int_0^\infty e^{-\delta t}[\mathfrak{p}^s f(y)]\,dt \\ &= \frac{\mathfrak{p}^s f(y)}{\delta}. \qquad (10.61)\end{aligned}$$

When this scrap value function is used, the transversality condition is

$$\psi_T = \partial\phi/\partial y_T = \mathfrak{p}^s \delta^{-1} f'(y_T),$$

or

$$c_T = (\mathfrak{p}^s)^{-1}\delta/f'(y_T). \qquad (10.62)$$

Recall that y^* is defined by $\delta = f'(y^*)$. Therefore, if $y_T = y^*$, then $c_T = (\mathfrak{p}^s)^{-1}$, and if $y_T > y^*$ $(< y^*)$, then $c_T > (\mathfrak{p}^s)^{-1}$ $(< (\mathfrak{p}^s)^{-1})$ by concavity of f. This is illustrated in Figure 10.5; the crossed line represents the transversality condition (10.62). It is now possible to restrict trajectories further. Inspection shows that all optimal trajectories must lie in the shaded area. A different configuration is drawn in Figure 10.6. We observe that in both these figures, buying or selling occurs first, if at all, followed (but not always) by a no-trade period. Therefore, in this model, and for fixed trading prices and this scrap value function, trading does take place toward the beginning of the horizon. Our method of analysis, if not the results, can be applied to other models in which trade in capital goods is possible.

10.2.3 *A final remark*

In this section we have attempted to model additions (or subtractions) to the stock of capital from outside sources. In some cases it was presumed that lumps of capital could simply be added onto existing capital. This is not always reasonable; for instance, if the existing capital stock is the

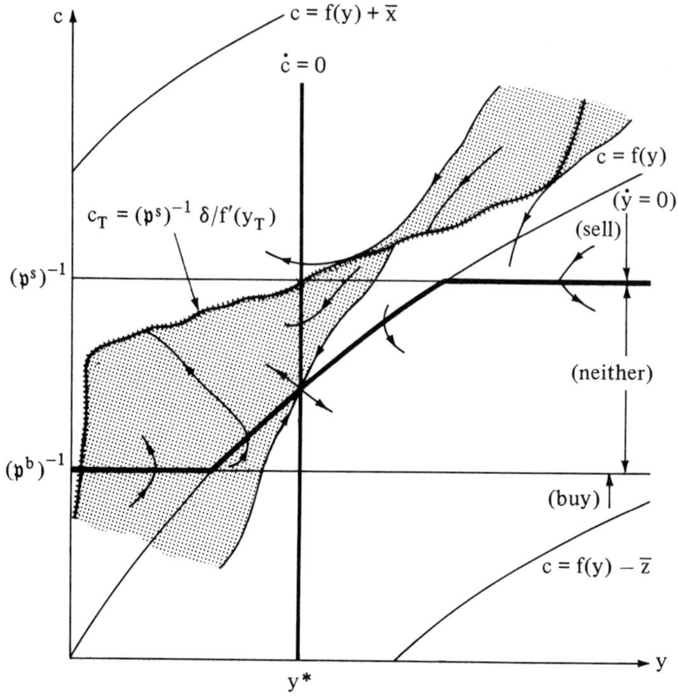

Figure 10.5

total reserves of ore in a mine, the purchase of another mine would not normally just add to the stock (unless, perhaps, it were an adjacent strip mine). In some such cases the formulations of this section would be inappropriate.

10.3 Constraints on the state variables

In all preceding chapters and sections we assumed that there were no constraints of the form

$$\phi^k(\mathbf{s}, t) \geq 0, \quad k = 1, 2, \ldots, K. \tag{10.63}$$

This is a constraint on the state variables. It differs from the constraints introduced in Chapter 6, which were of the form

$$g^j(\mathbf{s}, \mathbf{c}, t) \geq 0, \quad j = 1, 2, \ldots, m, \tag{10.64}$$

in that it does *not* involve control variables. As explained in Section 6.1 constraints such as (10.64) restrict the controls, for given values of s and t,

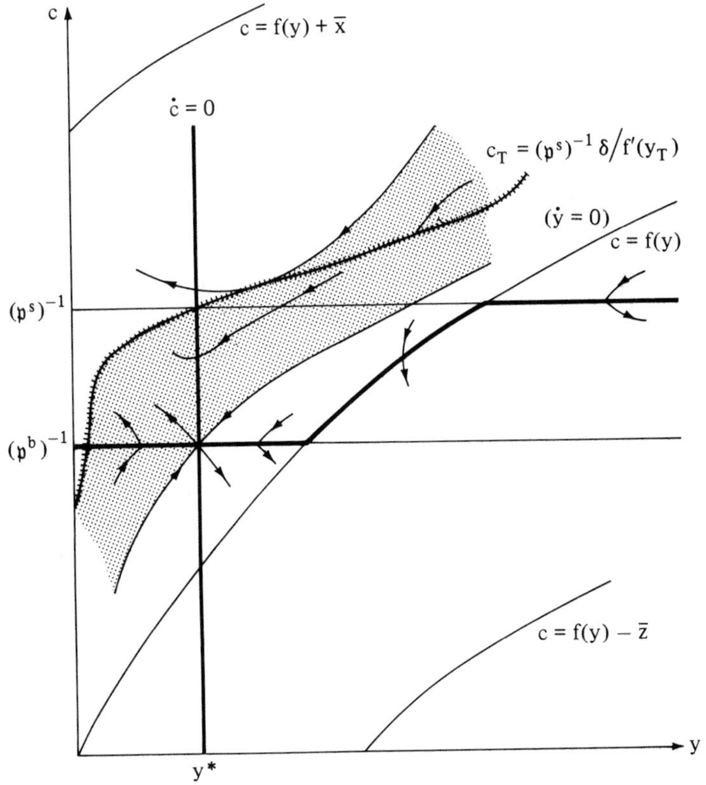

Figure 10.6

whereas (10.63) simply restricts **s**, given t. Because of this, such constraints do not satisfy the constraint qualifications of Chapter 6, which required, among other things, that the vector of partial derivatives $\partial g^j/\partial \mathbf{c}$ be nonzero. As a consequence we are unable to use the control variables to satisfy these constraints, and the path of $\mathbf{s}(t)$ may lead to a "collision" with some constraints. When this occurs, the costate variables may exhibit *jump discontinuities*.

There are several methods of dealing with problems involving state variable inequality constraints. The method presented in this chapter consists of attaching a multiplier to each constraint; it yields relatively simple necessary conditions that are valid if the optimal path is "well-behaved," in a sense to be made clear later. This method can be used to identify candidates for the optimal path; if one of these also satisfies Theorem 10.3.2 (sufficiency), it is an optimal path. References to other, more complicated methods are given at the end of the section in case the above fails.

We now state the problem. Find a vector of controls $\mathbf{c}(t)$ that maximizes

$$V=\int_0^T v(\mathbf{s},\mathbf{c},t)\,dt \tag{10.65}$$

subject to

$$\dot{s}_i=f^i(\mathbf{s},\mathbf{c},t),\quad i=1,2,\ldots,n, \tag{10.66a}$$

$$g^j(\mathbf{s},\mathbf{c},t)\geq 0,\quad j=1,2,\ldots,m, \tag{10.66b}$$

$$\phi^k(\mathbf{s},t)\geq 0,\quad k=1,2,\ldots,K, \tag{10.66c}$$

$$s_i(0)=s_{i0},\quad i=1,2,\ldots,n, \tag{10.66d}$$

$$s_i(T)=s_{iT},\quad i=1,2,\ldots,n', \tag{10.66e}$$

$$s_i(T)\geq s_{iT},\quad i=n'+1,\ldots,n'', \tag{10.66f}$$

$$s_i(T)\text{ free},\quad i=n''+1,\ldots,n. \tag{10.66g}$$

As usual we assume that the functions v, $\mathbf{f}$, $\mathbf{g}$, ϕ are twice differentiable and that the constraint qualifications on $\mathbf{g}$ are satisfied (see Section 6.1).

The dates t_J, $J=1,\ldots,M$, at which some state variable constraints become binding or cease to be so are called *junction points;* the initial and terminal times (0 and T) are also classified as junction points. The costate variables corresponding to the state variables involved at a junction point may exhibit a jump discontinuity – except that for fixed initial $s_i(0)$, there can be no jump in π_i at $t=0$. In what follows we also assume that the controls are continuous while $\phi^k(\mathbf{s},t)=0$.

10.3.1 *Necessary conditions*

The method expounded here is due to Jacobson, Lele, and Speyer (1971, p. 267). First form the Lagrangean

$$\mathcal{L}=H+\sum_{j=1}^{m}\lambda_j g^j+\sum_{k=1}^{K}\mu_k\phi^k\equiv H+\boldsymbol{\lambda}'\cdot\mathbf{g}+\boldsymbol{\mu}'\cdot\boldsymbol{\phi}, \tag{10.67}$$

where

$$H=v+\sum_{i=1}^{n}\pi_i f^i\equiv v+\boldsymbol{\pi}'\cdot\mathbf{f}.$$

Theorem 10.3.1: necessity. Let $(\mathbf{s}^*, \mathbf{c}^*)$ be a solution to the problem (10.65)–(10.66); then there exist costate variables $\boldsymbol{\pi}$ and multipliers $\boldsymbol{\lambda}$ and $\boldsymbol{\mu}$ such that

(i) $\mathbf{c}^*$ maximizes H subject to (10.66b); (10.68a)

(ii) $\lambda_j\geq 0$, $g^j(\mathbf{s}^*,\mathbf{c}^*,t)\geq 0$, $\lambda_j g^j(\mathbf{s}^*,\mathbf{c}^*,t)=0$, $j=1,2,\ldots,m$; (10.68b)

(iii) $\mu_k\geq 0$, $\phi^k(\mathbf{s}^*,t)\geq 0$, $\mu_k\phi^k(\mathbf{s}^*,t)=0$, $k=1,\ldots,K$; (10.68c)

(iv) π_i is both continuous and piecewise-differentiable except possibly at junction points; whenever $\dot{\pi}_i(t)$ exists, it satisfies

$$\dot{\pi}_i = -\partial\mathcal{L}/\partial s_i$$
$$= -v_{s_i} - \boldsymbol{\pi}'\bullet\mathbf{f}_{s_i} - \boldsymbol{\lambda}'\bullet\mathbf{g}_{s_i} - \boldsymbol{\mu}'\bullet\boldsymbol{\phi}_{s_i}, \quad i=1,\ldots,n, \tag{10.68d}$$

and at junction points t_J, $J=1,\ldots,M$, the jumps in the costate variables are given by

$$\pi_i(t_J^-) - \pi_i(t_J^+) = \boldsymbol{\beta}'(t_J)\bullet\boldsymbol{\phi}_{s_i}, \quad i=1,\ldots,n, \tag{10.68e}$$

where

$$\boldsymbol{\beta}'(t_J)\bullet\boldsymbol{\phi}_{s_i} = \sum_{k=1}^{K} \beta_k(t_J)\frac{\partial\phi^k}{\partial s_i}$$

and $\beta_k(t_J)$ satisfies

$$\beta_k(t_J) \geq 0, \quad \beta_k(t_J)\phi^k(\mathbf{s}^*(t_J), t_J) = 0, \quad J=1,\ldots,M; \tag{10.68f}$$

(v) $$\pi_i(T) \geq 0, \quad \pi_i(T)[s_i^*(T) - s_{iT}] = 0, \quad i=n'+1,\ldots,n'', \tag{10.68g}$$
$$\pi_i(T) = 0, \quad i=n''+1,\ldots,n. \tag{10.68h}$$

Conditions (10.68e) and (10.68f) relating to the jumps in costate variables at junction points are the only new elements. Because of the form taken by these conditions, one must often first make a guess about the optimal path and then check it against the necessary conditions, as the following numerical example illustrates.

Example 10.3.1: a law-abiding speed enthusiast. A driver wishes to maximize his enjoyment of a ride, and this depends on speed $s(t)$ and acceleration $c(t)$. There is a speed limit, and acceleration is also restricted. Find $c(t)$ to maximize

$$V = \int_0^2 [2(s(t))^{1/2} + 0.005(c(t) - 0.1)]\,dt \tag{10.69}$$

subject to

$$\dot{s}(t) = c(t) - 0.1, \tag{10.70}$$
$$1.1 - c(t) \geq 0, \tag{10.71a}$$
$$2 - s(t) \geq 0, \tag{10.71b}$$
$$s(0) = 1 \quad \text{and} \quad s(2) \text{ free.} \tag{10.71c}$$

Condition (10.71b) is the constraint on the state variable (speed), which is required to hold at all times. Since the maximand is an increasing function of speed s, it is natural to guess that the speed limit should be reached in minimum time and thereafter maintained. Specifically,

$$c(t) = 1.1 \text{ until time } t_1, \text{ at which } s(t_1) = 2;$$
$$c(t) = 0.1 \text{ when } t_1 < t \leq 2. \tag{10.72}$$

To find t_1, set

$$\int_0^{t_1} \dot{s}(t)\,dt = 2 - s(0) = \int_0^{t_1} (1.1 - 0.1)\,dt = 2 - 1; \quad \text{hence, } t_1 = 1.$$

If we follow policy (10.72) we have

$$s^*(t) = t + 1, \quad 0 \leq t \leq 1, \quad \text{and} \quad s^*(t) = 2, \quad 1 \leq t \leq 2.$$

We now proceed to verify that this solution satisfies Theorem 10.3.1. Let $\mathcal{L} = 2s^{1/2} + 0.005(c - 0.1) + \pi(c - 0.1) + \lambda(1.1 - c) + \mu(2 - s)$; we require

$$\frac{\partial \mathcal{L}}{\partial c} = 0.005 + \pi - \lambda = 0, \tag{10.73a}$$

$$\dot{s} = c - 0.1, \tag{10.73b}$$

$$\lambda \geq 0, \quad 1.1 - c \geq 0, \quad \lambda(1.1 - c) = 0, \tag{10.73c}$$

$$\mu \geq 0, \quad 2 - s \geq 0, \quad \mu(2 - s) = 0, \tag{10.73d}$$

$$\pi(2) = 0 \text{ and } \dot{\pi} = -s^{-1/2} + \mu \text{ whenever } \dot{\pi} \text{ exists}, \tag{10.73e}$$

and at each junction point t_J

$$\pi(t_J^-) - \pi(t_J^+) = \beta(t_J)(-1), \quad \beta(t_J) \geq 0, \quad \beta(t_J)(2 - s(t_J)) = 0. \tag{10.73f}$$

According to our solution, there are two junction points at $t = 1$ and $t = 2$. In the second half of the horizon, the constraint on c is not binding; hence, by (10.73c) $\lambda = 0$ and by (10.73a) $\pi = -0.005$. This contrasts with (10.73e), and there is a jump in π at $t = 2$. Specifically, $\pi(2^-) - \pi(2) = \beta(2)(-1) = -0.005$; hence, $\beta(2) = 0.005 > 0$ and $s(2) = 2$, satisfying (10.73f). Since $\pi = -0.005$ when $1 \leq t < 2$, (10.73e) yields $\mu = 1/\sqrt{2}$ on that interval. In the first half $(0 \leq t < 1)$ $\mu = 0$ and $\dot{\pi} = -(t+1)^{-1/2}$; thus, $\pi = -2(t+1)^{1/2} + A$. If there is no jump in π at the junction point $t = 1$, this must satisfy $\pi(1) = -0.005$; hence,

$$\pi(t) = -2(t+1)^{1/2} + 2\sqrt{2} - 0.005.$$

Substituting this in (10.73a) yields $\lambda = 2\sqrt{2} - 2(t+1)^{1/2} \geq 0$ when $0 \leq t \leq 1$ and (10.73c) is satisfied. Therefore, there is no jump in π at $t = 1$ and $\beta(1) = 0$. The only jump occurs at $t = 2$. In the next section we shall see that the sufficient conditions are also met and that (10.72) is indeed the unique optimal policy. Before we leave this example, note that if we had required that $s(2) = 2$ in (10.71c), the transversality condition $\pi(2) = 0$ would have disappeared and no jump in π would have been required, although the outcome would have been unchanged.

10.3.2 *Sufficiency results*

Theorem 10.3.2. If a path $(\mathbf{s}^*, \mathbf{c}^*)$ and associated costates and multipliers satisfy the conditions of Theorem 10.3.1 and also

(i) $\pi_i^*(T)[s_i(T) - s_i^*(T)] \geq 0$ for all feasible $s_i(T)$, $i = 1, \ldots, n$;
(ii) the Lagrangean is concave in $(\mathbf{s}, \mathbf{c})$; and
(iii) $\phi^k(\mathbf{s}, t)$ has the property $(k = 1, \ldots, K)$

$$\phi^k(\mathbf{s}_2, t) \geq \phi^k(\mathbf{s}_1, t) \Rightarrow \phi_{\mathbf{s}}^k(\mathbf{s}_1, t) \cdot (\mathbf{s}_2 - \mathbf{s}_1) \geq 0, \quad \text{all } \mathbf{s}_1 \text{ and } \mathbf{s}_2$$

(which is automatically satisfied if ϕ^k is concave in $\mathbf{s}$),

then this path represents an optimal solution to the problem (10.65)–(10.66).

Proof. For simplicity we assume that there are only two junction points t_1 and t_2, in addition to T. We have the by now familiar arguments,

$$\begin{aligned}
\int_0^T (v^* - v)\, dt &= \int_0^T [(H^* - \boldsymbol{\pi}^* \cdot \dot{\mathbf{s}}^*) - (H - \boldsymbol{\pi}^* \cdot \dot{\mathbf{s}})]\, dt \\
&\geq \int_0^T (\mathcal{L}^* - \mathcal{L})\, dt + \int_0^T \boldsymbol{\pi}^* \cdot (\dot{\mathbf{s}} - \dot{\mathbf{s}}^*)\, dt \\
&\qquad \text{(by (10.66a) and (10.66b))} \\
&\geq \int_0^T \left[\left(\frac{\partial \mathcal{L}^*}{\partial \mathbf{s}} \right)' \cdot (\mathbf{s}^* - \mathbf{s}) + \left(\frac{\partial \mathcal{L}^*}{\partial \mathbf{c}} \right)' \cdot (\mathbf{c}^* - \mathbf{c}) \right] dt \\
&\quad + \int_0^T \boldsymbol{\pi}^* \cdot (\dot{\mathbf{s}} - \dot{\mathbf{s}}^*)\, dt, \quad \text{(by concavity of } \mathcal{L}) \\
&\geq \int_0^T [\dot{\boldsymbol{\pi}}^* \cdot (\mathbf{s} - \mathbf{s}^*) + \boldsymbol{\pi}^* \cdot (\dot{\mathbf{s}} - \dot{\mathbf{s}}^*)]\, dt \quad \text{(by (10.68d))} \\
&= \int_0^T \frac{d}{dt} [\boldsymbol{\pi}^* \cdot (\mathbf{s} - \mathbf{s}^*)]\, dt \\
&= -\boldsymbol{\pi}^*(0) \cdot [\mathbf{s}(0) - \mathbf{s}^*(0)] + [\boldsymbol{\pi}^*(t_1^-) - \boldsymbol{\pi}^*(t_1^+)] \cdot [\mathbf{s}(t_1) - \mathbf{s}^*(t_1)] \\
&\quad + [\boldsymbol{\pi}^*(t_2^-) - \boldsymbol{\pi}^*(t_2^+)] \cdot [\mathbf{s}(t_2) - \mathbf{s}^*(t_2)] + \boldsymbol{\pi}^*(T^-) \cdot [\mathbf{s}(T) - \mathbf{s}^*(T)] \\
&\geq 0. \qquad\qquad (10.74)
\end{aligned}$$

The right-hand side of (10.74) is the sum of nonnegative terms. To see this recall the following:

(i) $s(0) = s^*(0)$.
(ii) For each t_J ($J = 1, 2, T$ in our case),

$$\begin{aligned}
&[\boldsymbol{\pi}^*(t_J^-) - \boldsymbol{\pi}^*(t_J^+)] \cdot [\mathbf{s}(t_J) - \mathbf{s}^*(t_J)] \\
&\qquad = \sum_{k=1}^{K} \beta_k(t_J)[\phi_{\mathbf{s}}^k(\mathbf{s}^*(t_J), t_J)] \cdot [\mathbf{s}(t_J) - \mathbf{s}^*(t_J)] \quad \text{by (10.68e).}
\end{aligned}$$

In order to sign this expression, first note that $\beta_k(t_J) \geq 0$ by (10.68f); second, recall that at a junction point $\phi^k(\mathbf{s}^*(t_J), t_J) = 0$, whereas $\phi^k(\mathbf{s}(t_J), t_J) \geq 0$ since $\mathbf{s}$ is feasible; then $\phi^k(\mathbf{s}(t_J), t_J) - \phi^k(\mathbf{s}^*(t_J), t_J) \geq 0$ and assumption (iii) of Theorem 10.3.2 implies that the expression is nonnegative.

(iii) $$\boldsymbol{\pi}^*(T^-)\cdot[\mathbf{s}(T)-\mathbf{s}^*(T)] = [\boldsymbol{\pi}^*(T^-)-\boldsymbol{\pi}^*(T)]\cdot[\mathbf{s}(T)-\mathbf{s}^*(T)] + \boldsymbol{\pi}^*(T)\cdot[\mathbf{s}(T)-\mathbf{s}^*(T)].$$

The first term is nonnegative by the same argument as used in (ii). (T is a junction point and by convention $\pi(T) = \pi(T^+)$; the second term is nonnegative by assumption (i) of Theorem 10.3.2.) □

Corollary 10.3.1. For infinite-horizon problems, replace assumption (i) of Theorem 10.3.2 with

$$\lim_{t\to\infty} \pi_i^*(T)[s_i(t)-s_i^*(t)] \geq 0 \text{ for all feasible } s_i(t), \quad i=1,\ldots,n, \tag{10.75a}$$

or, more generally, in case a limit does not exist for (10.75a),

$$\liminf_{t\to\infty} \pi_i^*(t)[s_i(t)-s_i^*(t)] \geq 0 \text{ for all feasible } s_i(t), \quad i=1,\ldots,n. \tag{10.75b}$$

For more general sufficiency results the reader is referred to Seierstad and Sydsaeter (1977). We now turn to a diagrammatic analysis of a control problem involving a simple constraint on the state variable and show how the sufficiency results can be applied.

Example 10.3.2. Consider the fishery model of Section 9.5 and assume that the firm exploiting this renewable resource is required by law to maintain the stock of fish above a predetermined level, say $\underline{s}$ – we must have $\underline{s} \leq 1$ since a stock $s > 1$ cannot be maintained indefinitely (see Section 9.5 for notation and assumptions). The problem facing the firm is to find $n(t)$ that maximizes

$$V = \int_0^\infty e^{-rt}(2s^{1/2}n^{1/2} - wn)\,dt \tag{10.76}$$

subject to

$$n \geq 0, \tag{10.76a}$$

$$\dot{s} = s(1-s) - 2s^{1/2}n^{1/2}, \tag{10.76b}$$

$$s(t) \geq \underline{s}, \tag{10.76c}$$

$$s(0) = s_0 \; (> \underline{s}). \tag{10.76d}$$

Applying Theorem 10.3.1 we obtain a set of necessary conditions with the proviso that we work with the current-value Lagrangean; thus, (10.68d)

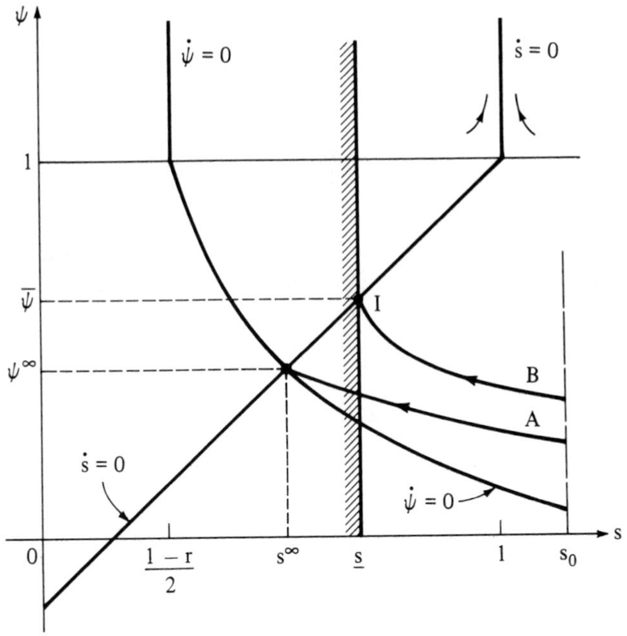

Figure 10.7

is modified to $\dot{\psi} = r\psi - \partial\mathcal{L}/\partial s$. The current-value Lagrangean is $\mathcal{L} = 2s^{1/2}n^{1/2} - wn + \psi[s(1-s) - 2s^{1/2}n^{1/2}] + \mu(s - \underline{s})$ and we obtain

$$\partial\mathcal{L}/\partial n = (1-\psi)(s^{1/2}n^{-1/2}) - w \leq 0 \;\; (=0 \text{ if } n > 0), \tag{10.77a}$$

$$\mu \geq 0, \;\; \mu(s - \underline{s}) = 0, \quad s - \underline{s} \geq 0, \tag{10.77b}$$

$$\dot{\psi} = -(1-\psi)s^{-1/2}n^{1/2} + \psi(r - 1 + 2s) - \mu, \tag{10.77c}$$

and at junction points

$$\psi(t_J^-) - \psi(t_J^+) = \beta(t_J), \tag{10.77d}$$

where

$$\beta(t_J) \geq 0, \;\; \beta(t_J)(s - \underline{s}) = 0, \;\; s - \underline{s} \geq 0. \tag{10.77e}$$

It is often advisable to build on the simpler analysis of the case without state constraints. Here we begin by ignoring (10.77b) and the multiplier μ. Then the phase diagram of Figure 9.1 is applicable; it is reproduced here in Figure 10.7, except that only the region to the right of the vertical line $s = \underline{s}$ is relevant. For concreteness assume that $\underline{s} > s^{\infty}$, where the latter is the steady-state stock of the unconstrained problem of Section 9.5 (a lower bound $\underline{s} < s^\infty$ would have no effect). In Figure 10.7, path A leads to the steady state, but it is now blocked by the state variable constraint $s \geq \underline{s}$. We claim that the optimal trajectory, for any initial stock size $s(0) \geq \underline{s}$, follows path B, which is above path A and leads to the intersection of the $\dot{s} = 0$ locus with $s = \underline{s}$, at point I. The intuition behind this guess is that the steady state is now out of bounds and a substitute must be found. At I, $\dot{s} = 0$ but it is off the $\dot{\psi} = 0$ locus. However, equation (10.77c) reveals that $\dot{\psi}$ can be adjusted by a change in μ when I is reached, which is in finite time. Note that ψ does *not* jump at that point.

We now proceed with explicit calculations. For coordinates, the intersection point I has $(\bar{\psi}, \underline{s}) = (1-(1-\underline{s})w/2, \underline{s})$; this follows from (9.31). At this point, if there were no μ, $\dot{\psi}$ would take the value given by (9.29), with $\psi = \bar{\psi}$,

$$\dot{\psi} = \bar{\psi}(r-1+2\underline{s}) - (1-\bar{\psi})^2/w, \tag{10.78}$$

which is positive because I is above the $\dot{\psi} = 0$ locus where $\dot{\psi} > 0$. The jump in μ, to the value equal to the right-hand side of (10.78), just makes $\dot{\psi} = 0$. The value of ψ does not jump and $\beta = 0$. Note that Theorem 10.3.2 with Corollary 10.3.1 applies to this example because $\mathcal{L}$ is concave in (s, n), the state constraint (10.76c) is linear, and the transversality condition (10.75a) is satisfied because all s values are bounded by $\underline{s}$ and $s(0)$ and

$$\lim_{t \to \infty} \pi^*(t) = \lim_{t \to \infty} e^{-rt}\bar{\psi} = 0.$$

The economic implications of imposing the state constraint $s \geq \underline{s}$ are interesting. Denote the catch by $c = 2s^{1/2}n^{1/2}$; then along either path A or path B we have, by (10.77a), $c = 2s(1-\psi)w^{-1}$ and since s decreases and ψ increases, the catch must decrease through time on the way to equilibrium. Along $\dot{s} = 0$, the catch must necessarily be $c = s(1-s)$. Hence, the long-run value of the catch depends on s alone and is largest at $s = 0.5$. We know that $(1-r)/2 < s^\infty < \underline{s} < 1$, but we cannot ascertain whether s^∞ or $\underline{s}$ yields the higher catch. Since ψ is higher along path B, we know that for the same s values, the catch is smaller than on path A, but it may converge to a larger equilibrium value.

10.3.3 *Concluding comments*

We attempt to provide an intuitive explanation of condition (10.68e) that indicates possible discontinuities of the costate variables at junction points. For simplicity we consider a problem with one inequality constraint on the state variables. Assume further that the time horizon is $[0, T]$ and that over $[0, t_1)$ the state constraint is slack while it binds over $[t_1, T]$. Let $(\mathbf{s}^*, \mathbf{c}^*)$ be an optimal path. Then over $[0, t_1)$, we must find $\mathbf{c}$ that maximizes

$$V_0(\mathbf{s}_0, \mathbf{s}^*(t_1)) = \int_0^{t_1} v(\mathbf{s}, \mathbf{c}, t)\, dt \tag{10.79}$$

subject to

$$\dot{\mathbf{s}} = \mathbf{f}(\mathbf{s}, \mathbf{c}, t), \tag{10.80a}$$

$$\mathbf{s}(0) = \mathbf{s}_0, \quad \mathbf{s}(t_1) = \mathbf{s}^*(t_1) \text{ fixed.} \tag{10.80b}$$

Note that the state constraint $\phi(\mathbf{s}, t) \geq 0$ can be ignored in this problem as $\phi(\mathbf{s}^*, t) > 0$ by assumption. Denoting the costate variables by $\boldsymbol{\pi}(t)$, we know that

$$\partial V_0 / \partial s_i^*(t_1) = -\pi_i(t_1^-). \tag{10.81}$$

The second subproblem is to find **c** that maximizes

$$V_1(\mathbf{s}^*(t_1), \mathbf{s}_T) = \int_{t_1}^{T} v(\mathbf{s}, \mathbf{c}, t)\, dt \tag{10.82}$$

subject to

$$\dot{\mathbf{s}} = \mathbf{f}(\mathbf{s}, \mathbf{c}, t), \tag{10.83a}$$

$$\phi(\mathbf{s}, t) \geq 0, \tag{10.83b}$$

$$\mathbf{s}(t_1) = \mathbf{s}^*(t_1), \quad s(T) = \mathbf{s}_T. \tag{10.83c}$$

Again, we know that if the costate variables for this problem are $\mathbf{p}(t)$, we have

$$\partial V_1 / \partial s_i^*(t_1) = p_i(t_1^+). \tag{10.84}$$

If, however, $\mathbf{s}^*(t_1)$ is optimal, it must be the solution of the "static" problem to find $\mathbf{s}^*(t_1)$ that maximizes

$$V_0(\mathbf{s}_0, \mathbf{s}^*(t_1)) + V_1(\mathbf{s}^*(t_1), \mathbf{s}_T) \tag{10.85}$$

subject to

$$\phi(\mathbf{s}^*(t_1), t_1) \geq 0. \tag{10.86}$$

Let $\beta(t_1)$ be the multiplier associated with (10.86). Then problem (10.85) yields

$$\partial V_0 / \partial s_i^*(t_1) + \partial V_1 / \partial s_i^*(t_1) + \beta_i(t_1) \partial \phi / \partial s_i^*(t_1) = 0. \tag{10.87}$$

From (10.81), (10.84), and (10.87),

$$-\pi_i(t_1^-) + p_i(t_1^+) + \beta_i(t_1) \phi_{s_i}(\mathbf{s}^*(t_1), t_1) = 0, \tag{10.88}$$

which we recognize as a special case of (10.68e).

It is interesting to observe the parallel between jumps in state variables when costates hit the price boundary $p(t)$ (the selling or buying price of the capital stocks), as in Section 10.2, and jumps in the costate variables when the state variables hit the admissible state boundary $\phi(\mathbf{s}, t) = 0$. In both cases we were able to explain the jumps by using the value function. It should be borne in mind that this approach is not meant to be rigorous and that in general value functions may not be differentiable.

Finally, we note that there exist more general methods for dealing with state constraints. One such method uses the observation that while a state constraint $\phi^k(s,t)\geq 0$ is binding, its time derivative is identically zero; hence,

$$\phi^k_{\mathbf{s}'}\bullet\dot{\mathbf{s}}+\phi^k_t=0$$

or

$$\phi^k_{\mathbf{s}'}\bullet\mathbf{f}(\mathbf{s},\mathbf{c},t)+\phi^k_t=0. \tag{10.89}$$

Condition (10.89) is a constraint involving the control variables and can replace the state constraint on the relevant interval. For details see Neustadt (1976) and Russak (1970).

Exercises

1. Consider the two-state-variable problem of maximizing $\int_0^T U(s_1,s_2,c_1,c_2)e^{-\delta t}\,dt$ subject to $\dot{s}_1=F^1(s_1,s_2,c_1)$, $\dot{s}_2=F^2(s_1,s_2,c_2)$ with conditions on $s_1(0)$, $s_2(0)$, $s_1(T)/s_2(T)$. Assume that U is homogeneous of degree 0 in (s_1,s_2) and that both F^1 and F^2 are homogeneous of degree 1 in (s_1,s_2). Reduce this problem to a one-state-variable problem. (*Hint:* Let $\theta=s_1/s_2$; find $\dot{\theta}$ and use the homogeneity assumptions.)
2. Reconsider the problem of Example 10.2.1 with a buying price of capital $p(t)=0.8+(t-1)^2$. Show that an upward jump takes place at $t=1$; find the size of the jump and the total revenue during the planning horizon.
3. Reconsider the problem of Example 10.2.1 when discounting is introduced. The Hamiltonian is now $H=2c^{1/2}e^{-0.25t}-\pi c$. The buying price remains $p(t)=1+(t-1)^2$. Show that an upward jump still takes place at $t=1$. Find the size of the jump; how does it compare with the jump in Example 10.2.1? Verify that all conditions of Theorem 10.2.1 apply.
4. Reconsider the problem of Example 10.2.1 with depreciation. The state equation now reads $\dot{s}=-c-0.3s$, and the buying price is $p(t)=e^{0.3t}(1+(t-1)^2)$. Find the timing and size of an upward jump.
5. Consider the problem of maximizing $\int_0^2 2c^{1/2}e^{-\delta t}\,dt+p(\theta)(s(\theta^+)-s(\theta^-))$ subject to $\dot{s}=-c-ms$, $s(0)=2$, and $s(2)=1$, where m and δ are specified positive constants and $p(t)$ is a specified function. Suppose that a downward jump takes place at time θ and derive the values of $s(\theta^+)$ and $s(\theta^-)$ in terms of m, δ, θ, and $p(\theta)$. Indicate the restrictions to be placed on $p(t)$, $\dot{p}(\theta)$ and the values of δ, m, θ, and $p(\theta)$ which guarantee that the jump is optimal and that $s(\theta^-)\geq s(\theta^+)\geq 0$. Choose a functional form for $p(t)$ and values for δ and m that induce a downward jump at some point θ, $0<\theta<2$, and calculate the size of that jump.
6. Obtain the general solution (i.e., with two arbitrary constants) to the following problem: Maximize $\int_0^2 3(2)^{-2/3}(s(t))^{1/3}(c(t))^{1/3}\,dt$ subject to $\dot{s}(t)=-c(t)$ with

$s(0)$ and $s(2)$ to be specified. Show that the path of the state variable is of the form $s(t)=(At+B)^{1/2}$, where A and B are arbitrary constants with $A<0$. Check that the maximized Hamiltonian $H^0(s(t),\pi(t))$ is strictly concave in $s(t)$. In the remainder of the exercise use $s(0)=4$ and $s(2)=0$. Derive the exact solution for $s(t)$, $\pi(t)$, and $c(t)$ when no jumps are permitted. Calculate $\pi(1)$.

Suppose now that the capital good s can be bought at price $p(t)=0.5e^{r(1-t)}$, where

$$r=\begin{cases}1 & 0\le t\le 1,\\ 0.4 & 1<t\le 2.\end{cases}$$

Show that an upward jump at time $\theta=1$ is optimal. Calculate the size of the jump. Compare the slopes of $p(t)$ and $\pi(t)$ before and after the jump.

Alternatively, suppose that the capital good can be sold at price $p(t)=e^{r(1-t)}$, where

$$r=\begin{cases}0.2 & 0\le t\le 1,\\ 0.6 & 1<t\le 2.\end{cases}$$

Show that a downward jump at time $\theta=1$ is optimal. Calculate the size of the jump. Compare the slopes of $p(t)$ and $\pi(t)$ before and after the jump.

7. Obtain the general solution (i.e., with two arbitrary constants) to the following problem: Maximize $\int_0^2 \sqrt{3}(s(t))^{1/6}(c(t))^{1/2}\,dt$ subject to $\dot{s}(t)=-c(t)$ with $s(0)$ and $s(2)$ to be specified. Show that the path of the state variable is of the form $s(t)=(At+B)^{3/4}$, where A and B are arbitrary constants, with $A<0$. Check that the maximized Hamiltonian $H^0(s(t),\pi(t))$ is strictly concave in $s(t)$. In the remainder of the exercise use $s(0)=8$ and $s(2)=0$. Derive the exact solution for $s(t)$, $\pi(t)$, and $c(t)$ when no jumps are permitted. Calculate $\pi(1)$.

   Suppose now that the capital good can be bought at price $p(t)=(0.06)^{1/4}\times e^{r(1-t)}$, where

$$r=\begin{cases}1.1 & 0\le t\le 1,\\ 0.2 & 1<t\le 2.\end{cases}$$

   Show that an upward jump is optimal at time $\theta=1$. Calculate the size of the jump. Compare the slopes of $\pi(t)$ and $p(t)$ before and after the jump. Choose another price path that would elicit the same jump (e.g., a piecewise-linear path).

   Alternatively, suppose that the capital good can be sold at price $p(t)=(0.44)^{1/4}e^{r(1-t)}$, where

$$r=\begin{cases}0.1 & 0\le t\le 1,\\ 0.3 & 1<t\le 2.\end{cases}$$

   Show that a downward jump at time $\theta=1$ is optimal. Calculate the size of the jump. Compare the slopes of $p(t)$ and $\pi(t)$ before and after the jump. Choose another price path that would elicit the same jump (e.g., a piecewise-linear path).

8. Consider the problem of maximizing $\int_0^9(12)^{1/3}(s(t))^{1/6}(c(t))^{1/3}\,dt$ subject to $\dot{s}(t)=-c(t)$, $s(0)$ and $s(9)$ to be specified. Show that the general form of the

solution for the state is $s(t) = (At+B)^{2/3}$. Solve the problem with $s(0)=16$, $s(9)$ free and calculate $s(9)$. Now suppose that $s(t)$ must never go below 1; i.e., we have a pure state constraint $s(t) \geq 1$. Show that it is optimal for $s(t)$ to remain above 1, for all $t<9$. What happens to the value of $\pi(t)$ when t reaches 9?

9. Consider the problem of maximizing $\int_0^T 2(s(t))^{1/2}(c(t))^{1/2}\,dt$ subject to $\dot{s}(t) = 1-c(t)$, with boundary conditions to be specified. Derive the solution to this problem. Show that the general solution for the state variable is $s(t) = 2t - A + B\sqrt{A-2t}$, where A and B are arbitrary constants. In the first instance let $T=2$, $s(0)=4$, and $s(2)$ free. Derive the specific solution. Now impose the pure state constraint $s(t) \geq 3$, $\forall t$. Obtain the necessary conditions. Examine the following two possible solutions: (a) Follow the previous solution until $s=3$; switch to $c=1$ henceforth; (b) reach $s=3$ at time 2, which is a junction point. Determine which proposal is optimal. Calculate the paths of all variables and point out all discontinuities.
10. Reconsider the general solution to the problem of exercise 9, but now use the specification $T=1.5$, $s(0)=2$, $s(T)=2$. In the first instance calculate the solution without constraints on the state variable. Now impose the capacity constraint $s(t) \leq 2.16$, $\forall t$. Obtain the necessary conditions. Show that the optimal solution involves two junction points at which both the control and the costate are discontinuous – a phase diagram in (π, s) might give a hint. Calculate the paths of all variables.
11. The model where smooth trading in capital goods was analyzed did not mention depreciation explicitly; this is the object of this exercise. We replace equations (10.44) and (10.45) by $\dot{s} = F(c, s+X) - ms$ and $\dot{X} = x - z - mX$, respectively. Discuss how these pairs of equations differ. Derive the optimality conditions in the case presented here; can you reduce it to a one-state-variable problem?

Bibliography

Arrow, K. J., and M. Kurz. *Public Investment, the Rate of Return, and Optimal Fiscal Policy.* Baltimore, Johns Hopkins University Press, 1970.

Athans, M., and P. L. Falb. *Optimal Control.* New York, McGraw-Hill, 1966.

Bellman, R. *Dynamic Programming.* Princeton, N.J., Princeton University Press, 1957.

Bellman, R., and S. Dreyfus. *Applied Dynamic Programming.* Princeton, N.J., Princeton University Press, 1962.

Benhabib, J., and K. Nishimura. "The Hopf Bifurcation and the Existence and Stability of Closed Onbits in Multisector Models of Optimal Economic Growth." *Journal of Economic Theory* 20 (1979), 421–44.

Bensoussan, A., E. G. Hurst, and B. Naslund. *Management Applications of Modern Control Theory.* Amsterdam, North Holland, 1974.

Benveniste, L. M., and J. A. Scheinkman. "Duality Theory for Dynamic Optimization Models of Economics: The Continuous Time Case." *Journal of Economic Theory* 27 (1982), 1–19.

Bliss, G. A. *Lectures on the Calculus of Variations.* Chicago, University of Chicago Press, 1946.

Brauer, F., and J. A. Nohel. *Qualitative Theory of Ordinary Differential Equations.* New York, Benjamin, 1969.

Calvo, G. A., and M. Obstfeld. "Optimal Time-Consistent Fiscal Policy with Finite Lifetimes." *Econometrica* 56 (1988), 411–32.

Clark, Colin W. *Mathematical Bioeconomics: The Optimal Management of Renewable Resources.* New York, Wiley, 1976.

Clark, C. W., H. C. Frank, and G. R. Munro. "The Optimal Exploitation of Renewable Resource Stocks: Problems of Irreversible Investment," *Econometrica* 47 (1979), 25–47.

Coddington, Earl A., and Norman Levinson. *Theory of Ordinary Differential Equations.* New York, McGraw-Hill, 1955.

Cohen, D., and P. Michel. "How Should Control Theory Be Used to Calculate a Time Consistent Government Policy?" *Review of Economic Studies* 54 (1988), 263–74.

Das, S. P., and Y. Niko. "A Dynamic Analysis of Protection, Market Structure and Welfare." *International Economic Review* 27 (1986), 513–23.

Dasgupta, P., and G. Heal. "The Optimal Depletion of Exhaustible Resources." *Review of Economic Studies,* 1974 Symposium, 3–29.

Diewert, W. E. "Duality Approaches to Microeconomic Theory." In K. J. Arrow and M. D. Intriligator (eds.), *Handbook of Mathematical Economics,* vol. 2. Amsterdam, North Holland, 1982.

Dixit, A. *Optimization in Economic Theory.* Oxford, Oxford University Press, 1976.

Dorfman, Robert. "An Economic Interpretation of Optimal Control Theory." *Americal Economic Review* 59 (1969), 817–31.

Dorfman, R., P. A. Samuelson, and R. M. Solow. *Linear Programming and Economic Analysis.* New York, McGraw-Hill, 1958.

Feichtinger, G. (ed.). *Optimal Control Theory and Economic Analysis.* Amsterdam, North Holland, 1982.

Feichtinger, G. (ed.). *Optimal Control Theory and Economic Analysis,* vol. 2. Amsterdam, North Holland, 1985.

Feichtinger, G. (ed.). *Optimal Control Theory and Economic Analysis,* vol. 3. Amsterdam, North Holland, 1988.

Fourgeaud, C., B. Lenclud, and P. Michel. "Technological Renewal of Natural Resource Stocks." *Journal of Economic Dynamics and Control* 4 (1982), 1–36.

Forster, Bruce A. "On a One State Variable Optimal Control Problem: Consumption–Pollution Trade-Offs." In J. D. Pitchford and S. J. Turnovsky (eds.), *Applications of Control Theory to Economic Analysis.* Amsterdam, North Holland, 1977.

Goldberg, Samuel. *Introduction to Difference Equations.* New York, Wiley, 1958.

Guesnerie, R., and J.-J. Laffont. "A Complete Solution to a Class of Principal-Agent Problems with an Application to the Control of the Self-managed Firm." *Journal of Public Economics* 25 (1984), 329–69.

Hadley, George. *Linear Programming.* Reading, Mass.: Addison-Wesley, 1962.

Hadley, G., and M. C. Kemp. *Variational Methods in Economics.* Amsterdam, North Holland, 1971.

Halkin, Hubert. "Necessary Conditions for Optimal Control Problems with Infinite Horizons." *Econometrica* 42 (1974), 267–72.

Hamada, K. "On the Optimal Transfer and Income Distribution in a Growing Economy." *Review of Economic Studies* 34, no. 3 (1967), 295–9.

Harris, Milton. "Optimal Planning under Transaction Costs: The Demand for Money and Other Assets." *Journal of Economic Theory* 12 (1976), 298–314.

Hartl, R. "A Simple Proof of the Monotonicity of the State Trajectories in Autonomous Control Problems." *Journal of Economic Theory* 41 (1987), 211–15.

Hestenes, Magnus R. *Calculus of Variations and Optimal Control Theory.* New York, Wiley, 1966.

Hirsch, M. W., and S. Smale. *Differential Equations, Dynamical Systems, and Linear Algebra.* New York, Academic Press, 1974.

Hotelling, H. "The Economics of Exhaustible Resources." *Journal of Political Economy* 39 (1931), 137–75.

Jacobson, D. H., M. M. Lele, and J. L. Speyer. "New Necessary Conditions of Optimality for Control Problems with State Variable Inequality Constraints." *Journal of Mathematical Analysis and Applications* 35 (1971), 255–84.

Jensen, R., and M. Thursby. "A Strategic Approach to the Product Life Cycle." *Journal of International Economics* 21 (1986), 269–84.

Jovanovic, B., and S. Lach. "Entry, Exit and Diffusion with Learning by Doing." *American Economic Review* 79 (1989), 690–9.

Kamien, M. I., and N. L. Schwartz. "Optimal Exhaustible Resource Depletion with Endogenous Technical Change." *Review of Economic Studies* 45 (1978), 179–96.

Kamien, M. I., and N. L. Schwartz. *Dynamic Optimization: The Calculus of Variations and Optimal Control in Economics and Management.* Amsterdam, North Holland, 1981.

Kemp, M. C., and N. V. Long. "Optimal Control Problems with Integrands Discontinuous with Respect to Time." *Economic Record* 53 (1977), 405–20.

Kemp, Murray C., and N. V. Long. *Exhaustible Resources, Optimality and Trade.* Amsterdam, North Holland, 1980.

Kemp, M. C., and N. V. Long. "On the Evaluation of National Income in a Dynamic Economy." In G. Feiwell (ed.), *Samuelson and Neoclassical Economics.* Boston, Nijhoff, 1982.

Kemp, Murray C., and N. V. Long (eds.). *Essays in the Theory of Exhaustible Resources.* Amsterdam, North Holland, 1984.

Kemp, M. C., and N. V. Long. "Union Power in the Long Run." *Scandinavian Journal of Economics* 89 (1987), 103–13.

Koopmans, T. C. "Proof of a Case When Discounting Advances Doomsday." *Review of Economic Studies* 41 (1974), 117–20.

Kurz, M. "The General Instability of a Class of Competitive Growth Processes." *Review of Economic Studies* 35 (1968), 155–74.

Kydland, F. E., and E. C. Prescott. "Dynamic Optimal Taxation, Rational Expectations and Optimal Control." *Journal of Economic Dynamics and Control* 2 (1980), 79–91.

Lefschetz, S. *Stability of Nonlinear Control Systems.* New York, Academic Press, 1965.

Leland, H. E. "The Dynamics of a Revenue Maximizing Firm." *International Economic Review* 13 (1972), 376–85.

Léonard, Daniel. "The Signs of the Costate Variables and Sufficiency Conditions in a Class of Optimal Control Problems." *Economic Letters* 8 (1981), 321–5.

Léonard, Daniel. "Costate Variables Correctly Value Stocks at Each Instant: A Proof." *Journal of Economic Dynamics and Control* 11 (1987), 117–22.

Léonard, Daniel. "Market Behaviour of Rational Addicts." *Journal of Economic Psychology* 10, no. 2 (1989), 117–44.

Long, N. V. "International Borrowing for Resource Extraction." *International Economic Review* 15 (1974), 168–83.

Long, N. V. "Resource Extraction Under the Uncertainty About Possible Nationalization." *Journal of Economic Theory* 10 (1975), 42–53.

Long, N. V. "Optimal Exploitation and Replenishment of a Natural Resource." In J. D. Pitchford and S. J. Turnovsky (eds.), *Applications of Control Theory to Economic Analysis.* Amsterdam, North Holland, 1977.

Long, N. V., and H. Siebert. "Optimal Foreign Borrowing: The Impact of the Planning Horizon on the Half and Full Debt Cycle." *Zeitschrift für Nationalökonomie* 49 (1989), 279–97.

Long, N. V., and N. Vousden. "Optimal Control Theorems." In J. D. Pitchford and S. J. Turnovsky (eds.), *Applications of Control Theory to Economic Analysis.* Amsterdam, North Holland, 1977.

Mangasarian, O. L. "Sufficient Conditions for the Optimal Control of Nonlinear Systems." *SIAM Journal of Control* 4 (1966), 139–52.

Mangasarian, Olvi. *Nonlinear Programming.* New York, McGraw-Hill, 1969.

Manning, R. "Optimal Aggregative Development of a Skilled Workforce." *Quarterly Journal of Economics* 89 (1975), 504–11.

Michel, P. "On the Transversality Condition in Infinite Horizon Optimal Control Problems." *Econometrica* 50 (1982), 975–85.

Milne, Frank. "The Adjustment Cost Problem with Jumps in the State Variable." In J. D. Pitchford and S. J. Turnovsky (eds.), *Applications of Control Theory to Economic Analysis.* Amsterdam, North Holland, 1977.

Mirrlees, J. A. "Optimum Growth When Technology Is Changing." *Review of Economic Studies* 34 (1967), 95–124.

Mirrlees, J. A. "The Optimum Town." *Swedish Journal of Economics* 74 (1972), 114–35.

Neustadt, L. W. *Optimization: A Theory of Necessary Conditions.* Princeton, N.J., Princeton University Press, 1976.

Nordhaus, W. "The Political Business Cycle." *Review of Economic Studies* 42 (1975), 165–90.

Pitchford, J. D. *Population in Economic Growth.* Amsterdam, North Holland, 1974.

Pitchford, John D. "Two State Variable Problems." In J. D. Pitchford and S. J. Turnovsky (eds.), *Applications of Control Theory to Economic Analysis.* Amsterdam, North Holland, 1977.

Pitchford, J., and S. Turnovsky (eds.). *Applications of Control Theory to Economic Analysis.* Amsterdam, North Holland, 1977.

Pollak, R. A. "Consistent Planning." *Review of Economic Studies* 35 (1968), 201–8.

Pontryagin, L. S. *Ordinary Differential Equations.* Reading, Mass.: Addison-Wesley, 1962.

Pontryagin, L. S., V. G. Boltyanskii, R. V. Gamkrelidze, and E. F. Mishchenko. *The Mathematical Theory of Optimal Processes.* New York, Wiley, 1962.

Quyen, N. V. "The Optimal Depletion and Exploration of a Nonrenewable Resource." *Econometrica* 56 (1988), 1467–71.

Ramsey, F. "A Mathematical Theory of Savings." *Economic Journal* (1928); reprinted in K. J. Arrow and T. Scitovsky (eds.), *Readings on Welfare Economics.* Homewood, Ill., Irwin, 1969.

Russak, I. B. "On General Problems with Bounded State Variables." *Journal of Optimization Theory and Applications* 6 (1970), 424–52.

Ryder, Harl E., Jr., and Geoffrey M. Heal. "Optimal Growth with Intertemporally Dependent Preferences." *Review of Economic Studies* 40 (1973), 1–31.

Sampson, A. A. "A Model of Optimal Depletion of Renewable Resources." *Journal of Economic Theory* 12 (1976), 315–24.

Seierstad, A. "A Sufficient Condition for Control Problems with Infinite Horizons." Memorandum from Institute of Economics, University of Oslo, Jan. 26, 1977a.

Seierstad, A. "Transversality Conditions for Control Problems with Infinite Horizons." Memorandum from Institute of Economics, University of Oslo, Jan. 27, 1977b.

Seierstad, A. "Sufficient Conditions in Free Terminal Time Optimal Control Problems." *Journal of Economic Theory* 32 (1984), 367–71.

Seierstad, Atle, and Knut Sydsaeter. "Sufficient Conditions in Optimal Control Theory." *International Economic Review* 18 (1977), 367–91.

Seierstad, A., and K. Sydsaeter. *Optimal Control Theory with Economic Applications.* Amsterdam, North Holland, 1987.

Sethi, S. P., and G. L. Thompson. *Optimal Control Theory Applications to Management Science.* Boston, Nijhoff, 1981.

Shell, Karl (ed.). *Essays on the Theory of Optimal Economic Growth.* Cambridge, Mass., MIT Press, 1967.

Smith, Vernon L. "An Optimistic Theory of Exhaustible Resources." *Journal of Economic Theory* 9 (1974), 384–96.

Stokey, N. L. "Learning by Doing and the Introduction of New Goods." *Journal of Political Economy* 96 (1988), 701–17.

Strotz, R. H. "Myopia and Inconsistency in Dynamic Utility Maximization." *Review of Economic Studies* 23 (1955–6), 165–80.

Takayama, Akira. *Mathematical Economics.* Cambridge University Press, 1985.

Tu, Pierre N. V. *Introduction to Optimization Dynamics: Optimal Control with Economics and Management Science Applications.* Berlin, Springer, 1984.

Uzawa, H. "Optimal Growth in a Two-Sector Model of Capital Accumulation." *Review of Economic Studies* 31 (1964), 1–24.

Vind, Karl. "Control Systems with Jumps in the State Variables." *Econometrica* 35 (1967), 273–7.

Vousden, Neil. "Basic Theoretical Issues of Resource Depletion." *Journal of Economic Theory* 6 (1973), 126–43.

Vousden, Neil. "Resource Depletion with Possible Nonconvexities in Production." In J. D. Pitchford and S. J. Turnovsky (eds.), *Applications of Control Theory to Economic Analysis.* Amsterdam, North Holland, 1977.

Weitzman, M. "Welfare Significance of National Product in a Dynamic Economy." *Quarterly Journal of Economics* 90 (1976), 156–62.

Index

BERT MOVSESIAN
1922 HUNTINGTON DR.
CHERRY HILL, N.J. 08003